PENGUIN COOKERY LIBRARY

Jane Grigson's Fruit Book

Jane Grigson was brought up in the north-east of England, where there is a strong tradition of good eating, but it was not until many years later, when she began to spend three months of each year working in France, that she became really interested in food. *Charcuterie and French Pork Cookery* (also published by Penguin) was the result, exploring the wonderful range of cooked meat products on sale in even the smallest market towns. This book has also been translated into French, a singular honour for an English cookery writer.

After taking an English degree at Cambridge in 1949, Jane Grigson worked in art galleries and publishers' offices, and then as a translator. In 1966 she shared the John Florio prize (with Father Kenelm Foster) for her translation of Beccaria's *Of Crime and Punishment*. Since 1968 she has been writing cookery articles for the *Observer Colour Magazine*; *Good Things* and *Food with the Famous* are based on these highly successful series. In 1973, *Fish Cookery* was published by the Wine and Food Society, followed by *English Food*, for which she was voted Cookery Writer of the year in 1977, and *The Mushroom Feast* (1975), a collection of recipes for cultivated, woodland, field and dried mushrooms. Jane Grigson's *Fruit Book* and highly acclaimed *Vegetable Book* (1978) have both won the André Simon Memorial Fund Book Award and the Glenfiddich Writer of the Year Award. All of these are published in Penguins.

Jane Grigson's reputation rests on elegance, erudition and her infectious enthusiasm for quality and simplicity in food and cooking. The *Sunday Telegraph* has said of her, 'Jane Grigson is one of the best of the British cookery writers', *The Times* has called her 'one of the three most influential and popular cookery writers in Britain', and the *Sunday Times* has commented, 'Like Elizabeth David, she writes like an angel'.

Jane Grigson was married to the poet and critic the late Geoffrey Grigson.

JANE GRIGSON'S

FRUIT BOOK

ILLUSTRATED BY YVONNE SCARGON

PENGUIN BOOKS

PENGUIN BOOKS

Published by the Penguin Group
27 Wrights Lane, London W8 5TZ, England
Viking Penguin Inc., 40 West 23rd Street, New York, New York 10010, USA
Penguin Books Australia Ltd, Ringwood, Victoria, Australia
Penguin Books Canada Ltd, 2801 John Street, Markham, Ontario, Canada L3R 1B4
Penguin Books (NZ) Ltd, 182–190 Wairau Road, Auckland 10, New Zealand

Penguin Books Ltd, Registered Offices: Harmondsworth, Middlesex, England

First published by Michael Joseph 1982
Published in Penguin Books 1983
5 7 9 10 8 6

Made and printed in Great Britain by
Richard Clay Ltd, Bungay, Suffolk
Photoset in Linotron 202 Times

For

Geoffrey
Sophie
Laure
& Jean-Sébastien

'A handsomely contrived, and well furnished Fruit Garden is an Epitome of Paradise, which was a most glorious place without a palace.' JOHN EVELYN

'Boats have entered harbour heavy with the ripe fruits of lands unknown to us. Quick, set them ashore, for us to enjoy at last.' ANDRÉ GIDE

CONTENTS

ACKNOWLEDGEMENTS

Friends who will find themselves in this book include a number of generous colleagues. First Elizabeth David, whose taste in the matter of fruit is unequalled, then Claudia Roden, Anne Willan, Josceline Dimbleby, Elisabeth Lambert Ortiz, Alan Davidson and Anton Mosimann.

Other friends who have provided information and help are Betty Bolgar, Elisabeth Bond, Anna and Luca Benedetto, Antoinette Daridan, the Hatchwell family, Adey Horton, Tao Tao Liu, Annette Macarthur-Onslow, Ariane Nisberg, Victoria Renard, Edward and Phyllis Schafer, Jean-Marie and Geneviève Stehli, Baillie Tolkien, Marjorie White, Margarette Worsfold, Paul Bailey and David Healey, Stephanie Hayman and Phyllis Hanes.

Special help was given by Pat Bellman who cooked many of the dishes, by Hilary and Toby Paine who provided much fruit, by Jenny Dereham who edited the text with tact and enthusiasm, and by Caroline Schuck who supplied the American measurements.

Acknowledgement also to the estate of Robert Frost, to Edward Connery Lathem, editor of *The Poetry of Robert Frost*, and to Jonathan Cape Ltd, the publishers, for permission to include lines from 'Blueberries'.

Other debts I acknowledged through the book, so that readers could more easily follow up points that catch their interest. Six books provided a basis of information and ideas:

Edward Bunyard, *The Anatomy of Dessert* (Chatto & Windus, 2nd revised edition, 1933)

André Gide, *Les Nourritures Terrestres* (Mercure de France, 1897, and Gallimard, 1935: trans. by Dorothy Bussey as *Fruits of the Earth*, Secker & Warburg, 1949)

Geoffrey Grigson, *A Dictionary of English Plant Names*, (Allan Lane, 1974)

Hortus Third, by the staff of the L. H. Bailey Hortorium, Cornell University (Macmillan, New York, 1976)

Leslie Johns and Violet Stevenson, *The Complete Book of Fruit* (Angus & Robertson, 1979)

G. B. Masefield, M. Wallis, S. G. Harrison, B. E. Nicholson, *The Oxford Book of Food Plants* (Oxford University Press, 1969)

INTRODUCTION

I grew up in a town devoid of fruit. There were of course apples, oranges and bananas in the shops, and one or two friends had kitchen gardens, but fruit trees were not part of our lives. There was nothing to raid when summer came along. The few blackberries were dry and sooty.

This, I imagine, is why certain experiences of fruit in my childhood remain bright, an orchard in Gloucester where old trees bent into tunnels and tresses of plums, a huge basket of strawberries that an uncle produced one day when we were visiting him in Worcestershire, raspberry canes blobbed with red and yellow fruit that met over our heads in a Westmorland cottage garden, unending peaches and water-melons of a student summer in Florence.

This special feeling towards fruit, its glory and abundance, is I would say universal. We have to bear the burden of it being good for us – though I would not think many people in man's history agreed with one sixteenth-century doctor who said that fruit should not be eaten for pleasure, but because it did us good. An apple a day, an orange a day have not spoilt our feelings. We respond to strawberry fields or cherry orchards with a delight that a cabbage patch or even an elegant vegetable garden cannot provoke.

Such feelings towards fruit have made this book more fun to write than any of the others. My only regret is the fruit I have not tasted, the fruit I have had to leave out. Take the durian – people leaving Bangladesh or Burma by air are not allowed to pack a durian or two, so pervasive and appalling is the smell. I should like to have smelled that smell, and then tasted the delicious flesh inside. I am sad at having to abandon the *Monstera deliciosa* that an Australian friend drew for me: in this country, we only know the leafing plant, not the long cone of a fruit. She told me, also, about the sweet quandong that they make into jellies: pigeons eat it too, and grow fat and well-flavoured, extra good for the pot. The one fruit in the book I have not tasted is the feijoa. I saw the tree once in the Cambridge Botanic Gardens – it had a single deep red orchid-like flower – but so common is it in the South of France that many people will have tasted the fruit there as I hope to myself one day.

Fruit that is eaten principally as a vegetable – the tomato, for instance – came into an earlier book. Olives seemed out of the scheme of this book, and there was no room to deal properly with nuts. Preserving fruit is so large a subject that it needs a book on its own (and there are one or two good ones in print). I have only given particular favourites when it came to jams, fruit pastes, drinks and so on. Dried fruit is another matter, recipes for that are included.

Sections in the book vary in length. There are the great cooking fruits: there are fruits that have been much loved for hundreds, sometimes thousands of years: there are the new tropical fruits that are now being added to our experience (travellers, Indian civil servants and soldiers apart, how many people would have known what to do with a mango five years ago?). These all have long sections, though not necessarily many recipes. The raspberry and strawberry are rarely improved by cooking or complication. Neither is the peach – conscientiously I tried peaches with fish, peaches with pork, chicken and duck, and found the combinations regrettable (although a spiced peach goes quite well with ham and smoked poultry). At one smart restaurant in France, I tried a salad of lobster and white peaches: it looked pretty – yellowish-white and pink on a bed of deep bright green summer spinach – but tasted pointless. It was evident that lobster and peaches have little to say to each other.

As with vegetables, what moved me about fruit was the centuries of patient work that have built up the repertoire of apple, pear and strawberry varieties, that have developed cherries, peaches, plums and citrus fruit for different tastes and places. Before farming began, people cleared space round certain trees so that their fruit could grow in better light with less competition. Later soldiers, travellers and explorers brought new fruit, or better versions of familiar fruit home with them. The excitement of the Renaissance extended to gardening too, which is something that historians leave out of their accounts. Northern gardeners took from Italy the idea of planting apricots, peaches and so on against walls so that they might benefit from the stored warmth of the bricks and stones, and the espalier method of growing fruit trees was developed particularly in France. Such methods extended the range of fruit-growing. 'No longer do we have to travel down to Touraine for a Bon Chrétien pear,' wrote one enthusiastic Parisian – pears from Burgundy and Anjou as well were being grown on the outskirts of Paris, and sold in abundance in the markets.

It was in Paris that I discovered what had moved the skilful gardeners of that period. Not just the pleasure of the fruit or the triumph of intelligence, but a grand design. I was in the Bibliothèque Nationale reading room at the end of a long day. Knowing it would be several months before we'd be back in France, I had crashed through a pile of French books without moving from the chair. Twenty minutes was left for one English book, John Evelyn's *Compleat Gard'ner*, of 1693, his translation of a book by Jean de la Quintinie, one of Europe's great gardeners, expert on melon cultivation and the conduct of orangeries. He had made the gardens at Versailles from mounds of rubble, and was in charge of all Louis XIV's other gardens. No sinecure, as Louis had a passion for fruit and vegetables and inspected their progress regularly.

I opened the book in a tired blur, started at the beginning with resignation – and suddenly woke up. Here were these two voices speaking about what had moved them to their labours, de la Quintinie producing fruit with the least faults for 'the man who has most merits' – i.e. the Sun King – and Evelyn trying to turn Britain into an Elysian parkland, dotted with fine gardens. To explain what they were about, they chose fruit – 'Fruit, as it was our primitive, and most excellent as well as most innocent food, whilst it grew in Paradise; a climate so benign, and a soil so richly impregnated with all that the influence of Heaven could communicate to it; so it has still preserved, and retained no small tincture of its original and celestial virtue.' Even in its fallen state, fruit is still the most 'agreeable closure' to a meal, however grand and princely. And so it is the gardener's labour 'to repair what the choicest and most delicious fruit has been despoiled of, since it grew in Paradise'. To aim, in other words, at recovering the original flavour of Eden, even if such transcendent perfection can never quite be achieved.

I suppose we no longer believe that God Almighty first planted a garden. Heavy cropping seems to be the aim of fruit breeders today, rather than fine quality of flavour. For this reason I am grateful to have been able to live and work part of my life in France during the last twenty years, in one of the great fruit-growing areas. In that genial climate, in private gardens at least, it is possible to sense what Evelyn and de la Quintinie and their tribe were after. People in Kent and Herefordshire, in Gloucestershire, Evesham and the Carse of Gowrie, may sometimes have a chance to sense it, too – so will people who stop by the occasional box of perfect fruit, nectarines perhaps or muscat grapes, on a market stall or in a supermarket, if they are not in too much of a hurry.

Jane Grigson

October 1981 Broad Town and Trôo

QUANTITIES

The recipes in this book are intended for six people as a general rule, though you must allow for the rest of the meal, or the position of a dish in the meal, when making your calculations. When a dish is, exceptionally, enough for two or for twenty, warning is given.

When measuring ingredients, follow either metric *or* imperial *or* American cup systems, not a mixture. The metric measurements are given first, followed by the imperial and American cup measurements in brackets: where there are only two measurements given, the one in brackets is the American measurement.

The cup, of 8 oz, is half an American pint of 16 oz. The Imperial pint is 20 oz and the litre is approximately 36 oz. I have worked on the basis of 1 oz = 30 g.

Our spoons are slightly larger than American spoons. But at these small quantities, the difference does not much matter. Spoon measurements are level.

Note When preparing fruit, always use stainless steel knives.

APPLE

The apple was the first fruit of the world, according to Genesis, but it was no Cox's orange pippin. God gave the crab apple and left the rest to man. This is perhaps what has made some suspect that the Tree of the Knowledge of Good and of Evil was really a banana. After all, it was the earliest fruit taken into cultivation.

It is perhaps ridiculous to attempt to apply science or archaeology to myths and legends. Let us stock Paradise with a decent apple. Let us imagine it was as good as a Cox's or a Blenheim orange, or the Orléans reinette. Many people would agree with Edward Bunyard, the fruit grower and writer, that they are the best apples of all. No wonder France sends us only her Golden Delicious. She has the sense to keep the Orléans and other reinettes for herself, only taking to 'le Golden' when the reinettes are finished.

The row over English versus French apples has had much publicity during the last few years. Letters and articles in *The Times*, paragraphs in the *Sun*. One *Times* correspondent observed that if people buy Golden Delicious, it is because Cox's orange pippins have been so down-graded by our growers. Everyone sees Cox's for sale that are in no way up to standard: presumably heavy cropping strains have been promoted. They may technically be Cox's orange pippins, but to anyone who grows his own fruit or can remember back beyond twenty or so years, they are travesties. The Golden Delicious may be boring, but it is reliable. French growers have turned its defect into a virtue. Moreover, if you keep it until it is yellow, it is not so bad, and even better for cooking. Our growers have turned the richness of the Cox into a boring crunch: it no longer comes up to expectation. That puts people off. All the smart packaging in the world will not disguise its descent.

The food trade makes the egalitarian mistake, which is also a convenience for itself, of thinking that every food has to be as cheap and inoffensive as every other similar food. This mistake has ruined chicken and potatoes and bread. No wine merchant sells only plonk, no flower shop sticks to daisies. In the matter of vegetables and fruit, we seem often to be reduced to a steady bottom of horticultural plonk.

Given this attitude, it is not surprising that, with all the researches and resources of fruit stations, nobody has come up with an apple to touch the Cox's orange pippin of *circa* 1825 and the Orléans reinette that was around before 1776. Humiliating, if you think about it. Every time I come across an apple worth eating that is new to us, it turns out be an older variety. Charles Ross apples were in abundance the autumn I was testing apple recipes. It was raised by Mr Ross and introduced in 1899. He originally named it after Thomas Payne Knight, the great horticulturist who first worked on hybridization, and was a founder of the Horticultural Society. Another sharp eater, or sweet cooker, the Belle de Boskoop, a splendid apple, was a favourite in Holland before Edward Bunyard noted it growing in England in 1920. The French Calville is an extremely old variety and was about in Normandy in the sixteenth century: from the Calville blanche and Cox's orange pippin came Ellison's orange introduced in 1911. The Blenheim orange goes back to before 1818, the Devonshire Quarrenden to before 1650.

All these magnificent apples – and we are offered Golden Delicious which is not much of a gustatory experience, and those thick red-skinned apples of various kinds with collapsing flesh. When we do get a good one such as Discovery, it has too short a season to make the impact it should. Perhaps the new Mutsu alias Crispin will dazzle us. It was raised in Japan and one parent was Golden Delicious. I hope the other parent, a Japanese apple, has more go to it. 'It is considered by some authorities that, in order to compete with overseas producers, it will be necessary to grow much heavier cropping apples than the present commercial range. Crispin shows every promise of fulfilling this need as it yields regularly and heavily, and produces fruit that grades out very well on size.' Ah well.

It does not have the air of a future star. I doubt anyone will be thrown out of Paradise on its account, or will risk the dangerous way to some garden of the Hesperides to taste it.

I came across a new apple myth and custom recently in the folk museum at Paris. A dazzling, irritating place where nothing is properly labelled, and bright objects are set against black, and half-lit in a chic artistic gloom. There in one case was a polished wooden 'tree', sticks bristling upwards all round a central stick, and a glossy red and yellow apple stuck on the end of each one. *Arbre des morts*, said the label, adding that it had been given by the church at Plougastel-Daoulas in Brittany when they replaced it with a new tree of the dead in 1952.

Up in the office, no one knew about it. No catalogue. Nothing in the library. In the end we set off for Plougastel, which had once been a main centre for early strawberries (*qv*). No new *arbre des morts* in the church. Not even the cleaner knew about it – but in the Maison de la Presse there was a helpful book.

In years gone by, it seems (and now once again with the new passion for *le folklo*) people brought apples and medlars to their church on All Saints Day, a bringing-in of the harvest to offer to the amiable crowd of dead souls – the Anaon, as it is called – who are as thick around us 'as the grass in the fields and the drops of rain in a shower'. After mass, there was a simple meal outside, under a lean-to. Apples and medlars were sold to pay for masses for the dead. Suddenly, a man would arrive carrying the apple 'tree' – the *arbre des morts* – stuck with new apples, the best from his orchard. A reserve price was set and the bidding began. Families would get together in groups to bid as the price would go quite high. To conclude the auction, an assistant snatched off the top apple. Then there was a procession, a procession for the whole community, the living and the dead close around them.

When an orchard in Brittany was picked, one last and best apple was left at the end of the highest branch. If it clung to the branch until all the leaves fell in the autumn winds, there would be a good crop next year, not just on that tree but in the whole orchard, pear trees and plum trees as well. This mother of apples 'is the sign of fertility, the apple of good luck. An old man once said to me that Adam and Eve were driven out of the Earthly Paradise because they had eaten this miraculous apple.'

HOW TO CHOOSE APPLES

In his section on temperate fruit in *The Oxford Book of Food Plants* – a book of a reasonable price that everyone should possess – Michael Wallis gives the best summary of apples that I have ever come across. In five pages of text and five of illustration, he clearly shows what the possibilities can be, and covers many varieties in text and clear coloured drawings.

He points out that there are over 3000 named varieties of apple, and then quotes the simple scheme of dividing apples into groups proposed by Edward Bunyard for the National Fruit Trials at Brogdale:

i. The Lord Derby group. The smooth sour green apples, a few of which turn white or yellow when ripe. None of them are sweet and none show any striping, though some may have a brown flush in the sun.

ii. The Granny Smith group. The green sweetly flavoured apples, some of which have brown or orange flush.

iii. The Lane's Prince Albert group. The striped sour green apples – like the last group but striped red and flushed, and sometimes crimson all over like Crimson Bramley.
iv. The Peasgood group. Smooth, striped and sweet apples.
v. The Golden Noble group. Smooth golden apples, without stripes, sometimes brown or pink flushed.
vi. Worcester Pearmain group. A group of red-skinned fruit, some shiny and some rough.
vii. The Blenheim and Cox's or Reinette group. Apples with a dry or rough skin with some russet, with orange red or crimson flush or stripes.
viii. The Russet group. More or less russetted on a golden or green ground, without red.

With the help of this grouping, and with a careful reading of Mr Wallis's pages, you will be quite well armed for the puzzling battles with the commercial apple trade. And if you want to grow apples in your garden, you will have a better idea of what to look for, the varieties that coincide with your opinions in the matter, and with the groups you prefer.

As a general rule, I avoid bright red shiny apples (tough skin and collapsing woolly flesh) and bright green shiny apples (too hard, undeveloped flavour, though a green Golden Delicious can be kept until it improves to its correct honeyed colour and taste). Russets and apples streaked with red and orange are usually the best for eating.

These days there is much emphasis on uniformity of size and shape in the production of fruit and vegetables. It makes for easier packing, but not necessarily for better flavour. And sad to say that, as with vegetables, the big-yielding varieties are the poorest in flavour. For this reason you may often do better with a small greengrocer, who knows people with apples to sell privately, than with the big supermarkets. A very slight wrinkling, the odd blemish or crack, do not indicate a poor-tasting apple – though it is of course common sense to avoid fruit with bruises and the kind of damage that comes from rough handling.

Always try to buy apples a little ahead, so that they can ripen at home for a day or two. If you go out picking apples on any scale, or have your own fruit, wrap them loosely in newspaper and stand them on slatted wooden trays, one above the other, in a dry airy place. One winter, I tried the plastic bag method sometimes recommended, but the apples soon acquired a brown rash under the skin and tasted unpleasant. Newspaper seems to work better in the less than perfect atmosphere of the average back kitchen or garage. Fruit storage is tricky in our climate, no wonder we developed a passion for dried fruit to fill that third of a year between the last of the apples (except after good seasons) and the first gooseberries. We have a habit of cursing modern fruit storage for spoiling its taste: really we should be grateful.

TO PREPARE APPLES

The main point to remember is that apples discolour fast as you peel them, so before you begin, prepare a bowl of water lightly-salted, or acidulated with lemon or vinegar, that you can slip the pieces into as they are prepared.

When recipes specify the use of lemon juice, you can omit the bowl of water so long as you work quickly and turn the pieces over in the lemon juice. If you are slow, or are preparing a large quantity, it is best to stick to the bowl of water. When all the apples are prepared, you can drain them and then sprinkle them with lemon juice which in this sort of recipe is intended as a flavouring rather than a way of keeping the fruit white.

When you need to core an apple whole, do this before you peel it – the apple is less likely to break. Most people will then proceed to peel the apple in a spiral, I usually do and try to keep the peel in one piece – why? The other day I saw a chef peeling his cored apples down in stripes from top to bottom, much quicker and much more even.

CONSOMMÉ À L'INDIENNE

The other day, a friend said to me how much she liked Boulestin's cookery books, because he did not waste time in the long explanation that we do today. I felt the reproach. There is an elegance about his writing, a lack of fuss that one also finds in other writers of the Thirties – such as Ruth Lowinsky, whose recipe this is (other recipes of hers appear elsewhere in the book). It all seems quick, easy, smart.

This recipe is given, ingredients and all, in seven lines. Fast food. Yet, like fast food today, which is decidedly not of Mrs Lowinsky's standard, it depends on labour or machinery or both, even if they are provided by somebody else. Mrs Lowinsky could go into her kitchen and expect that the cook would always keep stock in readiness. My friend who admires Boulestin has been providing her family with good food for so many years that she fits stock-making into her life without trouble or much thought. She can interpret seven lines. She has the taste and judgment to add seasonings not specified; the experience, the confidence to alter and amplify the brief words. She will not use this recipe until the day after the family has had roast chicken or game, when there is a litre of stock in the freezer from last week's boiled beef or cockie-leekie, and a couple of egg whites surviving from an hollandaise sauce.

And I will cheat. in a way, by pointing you towards your basic cookery book for a stock recipe.

1¼ litres (generous 2 pt/5 cups) beef or veal stock
2 large onions, sliced
1 large cooking apple, sliced
1 tablespoon desiccated coconut
1 dessertspoon (2 tsp) curry powder
bones from chicken or game carcase
2 large egg whites
salt, pepper, Cayenne, lemon juice
bits of chicken or game picked from the carcase
3–4 tablespoons (¾ cup) boiled rice

Simmer stock, onion, apple, coconut, curry powder and bones in a covered pan for an hour. Strain and remove any grease with kitchen paper. Beat in the egg whites, put back on the stove and continue whisking over a moderate heat until a dirty white cloud forms on top. Leave to simmer for 5 minutes, then pour off the clear liquid below the egg-white cloud through a cloth-lined strainer. Taste and adjust seasoning with salt and pepper, a pinch or two of Cayenne if you like the liquid hotter, and some lemon juice to bring out the flavour.

While all this was going on, cut the chicken or game into neat little shreds. Add them with the rice to the clarified soup and reheat without boiling.

CAUCHOISE SALAD WITH AVOCADO CREAM
SALADE CAUCHOISE À LA CRÈME D'AVOCAT

Being fond of *salade cauchoise*, which is a mixture of celery, potatoes and smoked country ham with a cream dressing, I was nervous of trying this version at the Hotel Marine at Tancarville on the Seine. With the avocado pear dressing, it seemed, as it were, a long way from home (which is the Pays de Caux, chalk country, north of Rouen).

This version has what could be described as a Waldorf note – apple and walnut. There are people who consider that, by introducing fruit, i.e. sweetness, into salads, the *maître d'hôtel* of the Waldorf-Astoria in New York was responsible for the corruption of American food towards the over-sweet. This needs to be watched, I agree. Few things are more disgusting than peaches canned in heavy syrup as a partner to duck coated with honey. On the other hand, meat and fish have a sweetness of their own, and some vegetables are as sweet as fruit (nobody grumbles about duck or lamb with young green peas). It's a case of keeping the balance, nothing too much.

If you do not like avocados, or cannot buy them, use a dressing made of half soured, half double cream, with oil and wine vinegar, plus seasoning. The following ingredients are enough for six to eight, and make a salad to be served as a first course:

1 head of celery
375 g (12 oz/2 cups) cooked new potatoes
3 reinettes, Cox's or Blenheim orange, or other aromatic crisp apple
lemon juice
100 g (3½ oz/⅔ cup) walnuts, coarsely chopped
60 g (2 oz) smoked country ham, Danish smoked pork loin, Bayonne or Westphalian ham
2 small avocados, ripe
oil and wine vinegar
salt, pepper
a crisp lettuce

String or strip away coarse outer stalks of the celery. You may be able to use all of one of the small summer heads of celery.

Mix with the diced potatoes. Peel, core and dice the apples, sprinkling the pieces with lemon juice before they have a chance to discolour. Put them into the basin with the celery and potatoes and mix them well. Add the walnuts, and the ham cut into short matchstick strips.

Process or liquidize the stoned and peeled avocados, adding 1½ tablespoons (2 tbs) of white wine vinegar, and enough oil to make the consistency thick but pourable. Season, and season the salad as well.

Just before serving, line six to eight bowls, or one large salad bowl, with lettuce leaves. Mix the avocado cream with the salad, turning the whole thing gently, and put into the bowl(s).

APPLE AND HORSERADISH SAUCE

We keep apple sauce for pork, duck, goose, the acidity balancing the fatness of the meat. In Eastern Europe, cold apple and horseradish sauces set off hot boiled beef, and other hot boiled meat, such as smoked tongue and pig's head. In the last of the four versions given below, it can also be served with game and boiled fish, and makes a good sauce for a Christmas buffet table.

Two things to be careful about, first the horseradish. It should be freshly grated from the root (the outside is milder than the central core), and if possible the root should be newly dug from the ground. Decidedly a gardener's sauce. The second point is to make sure that horseradish and apple have no chance to discolour: lemon juice should be mixed in rapidly and thoroughly.

1) Cook ½ kg (1 lb) apples – sharp apples such as Grenadiers or Bramley's – peeled, cored and quartered, with a tablespoon of sugar and lemon to sharpen slightly. Mash to a slightly lumpy purée. When cold, add 2 or 3 tablespoons finely grated horseradish shreds. Add more sugar and lemon to taste, perhaps a dash of wine vinegar. Some people like the sauce quite sweet, but the general effect should be sweet-sour in balance.

2) To the recipe above, add peeled, cored and finely-grated apple at the end.

3) Squeeze the juice of a lemon into a basin. Grate in about 2 or 3 tablespoons of horseradish, then 3 peeled and cored apples. Stir from time to time, so that nothing discolours. Mix in 4 tablespoons (⅓ cup) beef stock, a pinch of salt, sugar to taste.

4) Follow recipe 3, but mix in soured cream instead of beef stock, and stir everything well together to make a firm sauce. If you want the sauce runnier, add a little beef stock, a tablespoon at a time.

STUFFED APPLES

DOLMEH SIB

Americans have two categories of cookery book, one the result of the other, that we lack. First the recollections of immigrants. They are Americans now, and proud: recreating the cookery of their origins is an acceptable outlet for the pain of homesickness. Like St John Perse, the French poet, they cannot return home. They sail round it by way of cooking, as he sailed round his Guadeloupe but could not bring himself to land. There are many famous examples in this category. One I particularly like for its fruit and meat recipes is *In a Persian Kitchen*, by Maideh Mazda.

The second category consists of books which try to smooth disparate elements to an 'American' style. Bedding down Koenigsberger Klops with sweetcorn and elk, gremolata with racoon, Sally Lunns with Johnny cake, is a mammoth enterprise. Admirable perhaps, understandable certainly, but in its results not very sympathetic. It comes out as cooking-by-committee.

I have made a few slight alterations in Mrs Mazda's recipe, as in the similar recipe for stuffed quinces on p. 387. Usually I stuff the apples with a good pork sausagemeat rather than beef, because it seems to me to blend more happily with apple. Good Persians would not approve as Muslims do not eat pork. The most attractive aspect of the dish is the use of yellow split peas, to give a slight crunch to the sweet-sour blend of fruit, vinegar, meat and sugar.

Dolmeh is Persian for 'stuffed', like 'dolma' in Turkish and Greek cookery.

12 medium apples of a firm acid sweetness
90 g (3 oz/⅓ cup) split peas
1 large onion, chopped
3 tablespoons butter
½ kg (1 lb) lean beef, minced
1 teaspoon cinnamon
salt
4 tablespoons (⅓ cup) wine vinegar
3 tablespoons sugar

Slice a lid from the stalk end of the apples. Scoop out the core and pulp from the apples, leaving ½ cm (¼ inch) walls. Cook peas in unsalted water until tender.

Fry onion gently in 2 tablespoons butter until soft and golden. Push to one side. Raise the heat, add the beef in a layer. Allow it to brown, then turn it over and brown the other side. Mix meat, onion, drained peas and cinnamon, and season to taste. Fill the apples and replace their lids.

Put into a buttered ovenproof serving dish with a thin layer of water in the base. Bake for 15 minutes at gas 5, 190°C (375°F). Meanwhile, bring remaining butter, vinegar, sugar and 4 tablespoons (⅓ cup) of water to boiling point. Remove the apple lids and pour the sweet-sour liquid into each apple. Replace the lids and return to the oven for 10–20 more minutes, turning round the apples occasionally so that they cook evenly. Try to catch them before they crack too much: this does not spoil the taste, but makes the final dish look a little messy.

RILLONS WITH APPLE

A thing that surprises me about English cooking is that it has never developed a pork dish that in any way resembles the delectable French *rillons*. For simplicity and cheapness, we have nothing to beat it.

Buy a large piece of belly of pork as fat as you can find, and ask the butcher to remove the skin. (Take the skin home, too, and use it in small rolls to add a good texture to beef stews.) Cut it into rectangles and squares measuring roughly 8 cm (3 inches) and put them in a heavy pan with a thin layer of water in the base to prevent sticking in the early stages. Add a level teaspoon of *quatre-épices*, or half a teaspoon of pepper and a generous pinch of ground nutmeg, ground cloves and either cinnamon or ginger (these are the constituent spices of *quatre-épices*).

Cover and cook either over a very low heat for 3 hours, or in a slow oven, gas 1–2, 140–150°C (275–300°F). By this time the cubes should be tender.

Raise the heat, remove the lid and brown the cubes all over to an appetising rich colour. In the oven, raise the heat to gas 7, 220°C (425°F) and turn the cubes from time to time.

Have ready either baked firm eating apples (they can be cooked in the hot oven, if it's on), or lightly fried slices of apple, sprinkled with cinnamon or nutmeg, and buttery mashed potato.

Make a large potato ring on a hot serving dish, put the *rillons* in the middle, and arrange the apple slices round the *rillons* or at intervals in the potato if baked.

One of the best midday dishes I know, but you do need fat interlarded belly of pork, rather than the thin lean belly from some modern breeds such as Q pigs.

SMOKED PORK LOIN WITH APPLES

Although *hamburgerryg* – smoked pork loin, sold principally by the Danish food centres in London and Manchester – is usually served uncooked in thin slices (like the similar German *lachschinken*), it makes an excellent hot dish for special occasions. The Danes cook a variety of vegetables to go with it, but this apple garnish, which may be augmented with some new potatoes, has the advantage of simplicity and piquant flavour.

6–8 firm sharp apples
250 g (8 oz/1¼ cups) sugar

150 ml ($\frac{1}{4}$ pt/$\frac{2}{3}$ cup) water
chopped candied peel
$\frac{3}{4}$–1 kg (1$\frac{1}{2}$–2 lb) smoked pork loin

Peel the apples. Cut into halves down their middle and scoop out the pips and core plus some of the surrounding flesh, so that you end up with 12–16 bowls. Meanwhile, bring sugar and water to the boil in a wide pan and simmer for two minutes, before putting in the apples. Poach them for about 5 minutes each side until they are cooked, but not disintegrating. Drain them and fill the hollows with peel.

Bring a long pan half-full of water to boiling point. Put in the pork (minus plastic wrappings). Bring the water back to simmering point, cover and leave to cook for 20 minutes, turning the pork over after 10 minutes. Switch off the heat and leave the pan undisturbed for another 30 minutes.

Drain the pork and carve it into slices. Arrange them down the centre of a hot serving dish, with the reheated apples down each side, interspersed if you like with potatoes. Serve a salad afterwards.

Note The apple boats may be also filled with a mixture of peel and chopped cranberries that have been lightly cooked in water and sugared. Or with peel and a dab of redcurrant jelly.

NORMANDY PORK WITH APPLES AND CREAM
FILET DE PORC À LA NORMANDE

There are numerous versions of this famous combination of pork, pheasant or guineafowl with apples, Calvados and cream (*see also* p. 12–13). The apples should be a reinette variety, and the cream a very thick form of *crème fraîche*. If you can get these two important items, use them, otherwise follow the suggestions below:

2 large tenderloins, or 6–8 thick slices of boned, rolled loin of pork
salt, pepper
1 large chopped onion
clarified butter
4 tablespoons ($\frac{1}{3}$ cup) Calvados
1 round tablespoon flour
$\frac{1}{2}$ litre ($\frac{3}{4}$ pt/2 cups) dry cider
250 ml (8 fl oz/1 cup) veal, poultry or light beef stock
150 ml (5 fl oz/$\frac{2}{3}$ cup) mixed sour and double cream

3 large Cox's orange pippins
1 rounded tablespoon sugar
lemon juice

Trim and slice the tenderloins thickly, on the diagonal. Season the pork slices with salt and pepper, whichever kind you are using, and leave for several hours.

Soften the onion in some butter. Raise the heat and push the onion to one side while you brown the pork. You may have to use two sauté or frying pans.

Warm the Calvados, set it alight and pour it over the pork. Stir everything about until the flames die down. Sprinkle on the flour, cook a minute or two, then deglaze the pan(s) with the cider and stock. Lower the heat, cover the pans and simmer until the pork is tender. This will not take long for the tenderloin, but allow 10 to 15 minutes for the loin. Turn the slices over occasionally. Remove them to a serving dish and keep them warm. If you have been using two pans, amalgamate the juices in one of them, and reduce it to concentrate the flavour. Stir in the cream, season with salt, pepper and lemon juice and pour over the pork.

While the pork cooks, prepare the apple garnish. Peel, cut up and slice the apples neatly. Fry them in butter, not too fast, so that they become lightly browned and tender at the same time. As they begin to look soft and golden, pour on the sugar and turn the slices. Continue to cook, and as the slices brown, raise the heat to increase the caramelized effect. Arrange them round and among the pork slices. Serve very hot. If you manage to make this dish when summer reinettes or the first Cox's come in, you can cook new potatoes to go with it (apple and potato make a splendid partnership). In the winter, if you cannot get good potatoes, serve egg noodles or rice, or plenty of good home-made bread.

Note You can adapt this recipe to a joint of pork. Bone the loin and remove the skin – if you keep a brine crock, put it in overnight. Then tie it up. After browning the onion and meat, and deglazing the pan with cider and stock, transfer to a moderate oven to finish cooking. Cover with foil or use a lidded casserole.

When it comes to making the sauce, remove the fat from the cooking juices before reducing them and adding the cream.

NORMANDY PHEASANT
FAISAN À LA NORMANDE

This is the best way to cook pheasant, which can be a dry bird. If Calvados is not available, use whisky – preferably a single malt. When *crème fraîche* is available, use it instead of the two creams below.

1 large oven-ready pheasant
125 g (4 oz/½ cup) butter
salt, pepper
1 kg (2 lb) reinette or Cox apples, peeled, cored, sliced
cinnamon
125 ml (4 fl oz/½ cup) soured cream
125 ml (4 fl oz/½ cup) double cream
4 tablespoons (⅓ cup) Calvados

Melt half the butter in a frying pan. Brown the pheasant all over and season well. Melt the remaining butter in another pan. Add the apple and fry until golden. Sprinkle with a pinch of cinnamon.

Put a layer of apples in the base of a close-fitting casserole. Lay the pheasant on top, breast down, and pack the remaining apples down the sides. Mix the creams and pour in half. Cover the casserole and leave in a moderate oven, gas 4, 180°C (350°F), for an hour, or until pheasant is cooked. It is prudent to check after 50 minutes.

Remove the casserole from the oven and raise the heat to gas 8, 230°C (450°F). Pour the remaining cream over the pheasant, then the Calvados. Taste the apple mash for seasoning and make necessary adjustments.

Cover the casserole and return to the oven for 5 minutes. Serve from the casserole.

Should the pheasant have to wait while you eat a first course, turn off the oven and leave the door slightly ajar. The pheasant will be all the easier to carve and won't have lost much heat. Make sure the plates are good and hot.

BUTTERED APPLES
POMMES AU BEURRE

Only after the great public holiday of 15 August, the Assumption of the Virgin, do you begin to notice the apple trees showing their fruit. At least in the Val du Loir on the northern edge of Touraine. At market the following week, the first red and yellow reinette apples from neighbouring Anjou are on sale. Like our Cox's or Blenheims, they are ideal for this family pudding that I have never come across outside France. To me, it is perfect cooking, what food at home should always be like. Good ingredients treated simply and with affection.

My one concession to culinary art is to clarify the butter before I start cooking, by boiling it up in a small pan and then tipping it through a lined sieve into the sauté pan I use for the apples. If the butter goes in straight from the packet, something happens to make me turn my back on the stove – cats, children, telephone – and the apples burn. With clarified butter, you have some leeway, as its catching point is higher.

Peel, core and cut into eight wedges one apple for each person. Sprinkle with lemon or orange juice, or drop straight into lightly salted water. Drain and dry before frying.

Cook the slices gently in one layer – two batches, or two pans may be needed – in a wide sauté or frying pan. When the undersides cook to an autumn brown, turn the slices over and sprinkle the pan with 2 generous tablespoons of caster sugar. It soon begins to caramelize. Watch the pan carefully. Sprinkle the apples with a little cinnamon if you like.

Put the slices on a serving dish. Deglaze the pan with water or cider or apple juice; if a glass of Calvados can go in as well, so much the better. Keep stirring. Pour in some cream – single, whipping or double – to make a little sauce, and add a knob of butter to freshen the buttery flavour. Spoon over the apples and serve.

A similar version for pears is on p. 316.

BAKED APPLES ROYALE

POMMES AU BEURRE À LA ROYALE

Peel and core large reinette or Cox's orange or other firm eating apples of the pippin type, one per person.

Butter a gratin dish generously and sprinkle it with sugar. Mash roughly equal quantities of butter and sugar, and push into the empty core of the apples. Stand them in the baking dish and sprinkle with crushed macaroons. Bake in a moderate oven until tender, basting occasionally. Prepare an apricot sauce (p. 58), unsieved with nice pieces of fruit in it, and bring to just under boiling point.

When the apples are cooked, put the pieces of apricot on top of the now empty holes, or into them if they are small enough, and serve with the sauce.

BLACK CAPS

I do not understand why this delightful way of baking apples has slipped from our repertoire – it must have had something to do with the growing nineteenth-century use of Bramleys for cooking, as they turn to a foam and need the skin to contain them. In earlier times, other firmer varieties were preferred. The seventeenth-century cook used John apples which go back to Shakespeare's time. In the eighteenth century, pippins or reinettes were the thing, especially the golden reinette, 'the farmer's greatest favourite apple, because, when all others miss bearing, this generally stands his friend and bears him large quantities on one tree'. Eliza Acton, writing in 1845, suggested the Norfolk biffin – she grew up in East Anglia – or the winter greening, which had come to take the place of the John apple. Her recipe for black caps *par excellence* is a winner; you will find it in her *Modern Cookery* (reprinted by Elek in 1966) or in my *Food with the Famous*.

Miss Acton's method produces a candied apple that can be kept for a while in the refrigerator. I often turn back nowadays to the simpler hot dish of eighteenth-century cooks and use Cox's orange pippin or Reine de Reinettes, any dessert apple with good acidity and firm flavour.

8 large apples
thinly-cut peel of a large lemon or orange
juice of the lemon or orange
2 tablespoons orange-flower water, or 1 tablespoon triple-strength rose-water
175 g (6 oz/¾ cup) caster sugar

Core and cut the apples in half, horizontally across. Range them closely together, cut side down, in an ovenproof serving dish. Chop the peel and put it into the core cavities. Mix the juice – there should be at least 2 tablespoons – with the orange-flower water, or with the rose-water plus a tablespoon of water. Pour some over each apple dome, to moisten the skin. Sprinkle over the sugar, making sure that plenty falls on the apple tops. Bake at the top of a hot oven, gas 6–7, 200–220°C (400–425°F) for 25 minutes. Check after 10–15 minutes, and add a little extra fruit juice if there seems any danger of the apples drying up. Taste the syrup before you add any extra flower water – they are strong and should not be overdone. If the tops are not brown by the time the apples are tender, finish them under the grill.

Serve hot with egg custard, or with cream. Enough for 8 people.

BRAMLEY'S SEEDLING IMPROVED

Southwell with its famous leafy Chapter House has another less known distinction. It's the home of Bramley's Seedling, which was raised there in the early 1870s I imagine, as it was noted and introduced in 1876.

In *The Apples of England*, revised edition of 1948, Mr H. V. Taylor had this to say in spite of his earlier praise of certain dessert apples for cooking: 'During the twentieth century, Bramley's Seedling has become the finest of all culinary apples in England, the most planted, the heaviest cropper, the best cooker, and the easiest to sell.' To the second, third and fifth propositions, I feel like saying 'Yes, alas!'

Mr Taylor's first and fourth propositions are matters of opinion, and his opinion is a grand and knowledgeable opinion. I wish I could agree with him. To disagree seems presumptuous. But at least one reader shares our family's opinion and makes a good, practical suggestion:

'I feel I must tell you that the grated rind, flesh and juice of one orange added to each pound of stewed Bramley apples much improves that uninteresting fruit, especially when tons of it land in the freezer.'

In fairness to Mr Taylor and myself, I suspect that the Bramleys we buy these days are often as debased a version as the wretched little apples that we are supposed to accept as Cox's orange pippins, Class I. Poor Mr Cox, it's as well he is dead. Perhaps we should say the same about Mr Bramley? His apple in its first half century at any rate kept firm and well, and cooked 'to perfection – with plenty of juice and no hard pieces . . . the best for making dumplings and canned apples'.

APPLE STRUDEL
APFELSTRUDEL

It is not as tricky to make a passable apple strudel as you might think. A really fine one is, of course, another matter, needing skill and regular practice, a developed *tour de main*. I have an Austrian friend whose mother is a dab hand at strudel pastry but she herself is not, and I can understand that such skills may skip a generation. If mother can do it well, why bother, why take her on? This friend was visiting a bread factory in preparation for a broadcast, and was surprised to find that the strudel pastry they sold was hand-made. Women in white coats threw the dough over their shoulders to get it thinner and thinner and thinner. She found it puzzling and admirable, this skill raised to art, those people in their white coats throwing pastry over their shoulders, and again, and again, day after day.

The practice of layering thin sheets of pastry with melted butter belongs very much to the Middle East and North Africa, and to Eastern Europe as well. Their fila dough is plainer and thinner than strudel, but it can be used instead and is to be found here in some delicatessens. Frozen puff pastry can also be rolled out and then stretched. I saw fila pastry being made one day in a small shop at the back of the Chania market in Crete. The boy making it did not throw the paste over his shoulder. He pulled it out over a great tump of pastry layers, covered every so often with canvas cloths, presumably to prevent them drying out. The baker would throw a lump of dough on top of the pile, then the floury boy would walk round and round, pulling the dough out rapidly until it covered the whole thing. He was completely relaxed, he knew where exactly to pull, how fast to pull, how to coax the delicate cloth of dough to muslin transparency. Obediently the dough stretched, stretched without any obstinacy, without breaking into holes. Customers came and went, made jokes, the boy went on pulling and walking as he answered them. The baker wrapped things up, took money, without the rhythm of that shop being in the least disturbed. I stood and watched. Both men enjoyed their small drama, showmen conniving in an exhibition of skill, enjoying the sunny day, the chat and, all the time, this steady walk round the tump that grew higher by millimetres, yet what speedy millimetres.

To make a beginning with strudel dough, choose a day when you have time, peace and a companion. Clear the kitchen table, or one end of it – enough to accommodate eventually and with ease an oblong of roughly 50 × 60 cm (20 × 24 ins). Make the dough first, and prepare the filling as it rests. My Austrian friend suggests substituting roasted chopped walnuts and other nuts instead of breadcrumbs – I use both and always leave out the sultanas as I cannot bear them with apples, or cooked in that kind of way.

DOUGH
125 g (4 oz/1 cup) plain flour
½ large beaten egg
30 g (1 oz/2 tbs) melted butter or appropriate nut oil
generous pinch salt

FILLING
100 g (3½ oz/2 cups) fresh breadcrumbs
butter
250 g (8 oz) sharp cooking apples
100 g (3½ oz/½ cup) caster sugar
60 g (2 oz/½ cup) sultanas (optional)
100 g (3½ oz/¾–1 cup) chopped walnuts, almonds, hazelnuts etc, toasted if you like
caster sugar to finish
melted butter for basting

To make the dough, use an electric beater or the processor for quick and even mixing. Use warm water to mix the ingredients to a soft and supple dough, with a waxy look. Put it on to a board in a warm place. Rinse out the mixing bowl with hot water, dry it and turn it upside down over the dough. Leave for half an hour.

For the filling, fry the breadcrumbs lightly in 100 g (3½ oz/½ cup) butter until golden rather than brown. Peel, core and slice the apples thinly. Mix sugar, sultanas if used, and nuts. The nuts should be blanched to get rid of their skins, and all or some can be toasted in a low oven or under the grill. There is no law about strudel fillings: Austrians make their own variations – so can we.

Spread a cloth over the table where you intend to work. Sprinkle it with plain flour to stop the dough sticking. Place the dough in the centre and roll it out into a fairly thin oblong. If there are two of you, oil your hands and stand either side of the table. Slip your hands underneath the dough, raising it and stretching it partly by its own weight, partly by the gentle pulling movement of your hands. I find that it is best to use the back of the hand, so that you work palms down. This prevents the dough being broken by an involuntary movement of the fingers. Gradually it becomes so transparent that you can see the pattern of the cloth through it. Eventually, you may be able to read the newspaper through it – text not headlines: this is the counsel of perfection.

You will have noticed that when making this pastry, warmth is the thing. Warm dough, warm hands, as if you were making bread.

When you first experiment, the dough is bound to develop a few holes. These can be patched at the end, if they are large, or close together, with bits of dough cut from the edge and rolled or pulled thin. When you have stretched the dough as far as you dare, trim off the thick edges (they can be dried, grated into soup as pasta). Do not worry about making a neat oblong. Only the final edge needs to be cut straight.

To fill the pastry, brush it over first with melted butter. Then spread the breadcrumbs over it, leaving a wide rim at the straightened edge and a narrower rim at the opposite edge and the sides. Put the apple on top of the breadcrumbs, then the sugar, sultanas and nuts.

Flip the long edge opposite to the straightened edge over the filling, then fold over the sides (this is why you left rims free). Start rolling carefully, using the cloth to turn the dough over into a roll. When you come to the end, brush the edge with water, so that it will stick to the dough, then flip the whole thing on to a piece of Bakewell paper and slide it on to a baking sheet. If the roll is on the long side, bend it gently to fit the tray.

Melt 60 g (2 oz/¼ cup) butter and brush it over the strudel. Bake at gas 4, 180°C (350°F) for about 35 minutes until beautifully browned and crisp. Baste from time to time with butter. Sprinkle with caster sugar before serving, and provide cream.

Although strudels can be eaten cold, they are at their very best when hot or warm straight from the oven. Any left over can be reheated. The quantity above is enough for 10 people.

Note Other fruit can be used; all the northern favourites, plums, apricots, gooseberries, rhubarb, black currants, bilberries. For a spiced filling, turn to the strudel recipe on p. 124.

APPLE DUMPLINGS

Try making them with Cox's orange pippins or reinette varieties, or with Charles Ross, rather than sharp apples like Bramley's Seedling or Calvilles which cook to a foam. With fine apples, you do not need additional spicing such as cloves.

Make a plain shortcrust pastry with 500 g (1 lb) of flour, butter and lard. You can use a sweet pastry if you prefer it, but I find the plain kind best for fruit pies and dumplings. Form the dough into a roll, and rest it in the refrigerator for at least half an hour.

Peel and core an apple for each person. Block the cavities with dabs of butter. Pour in caster sugar, and close with dabs of butter.

Cut the roll of pastry across into roughly 125 g (4 oz) pieces, as many as you require for the apples. Of course, if the apples are extra large or extra small, you can adjust the size of the pieces accordingly.

Roll out each piece, not too thinly, and enclose an apple in each one. Trim away any surplus pastry, pressing the cut edges firmly together. Put the dumplings smooth side up on to a greased baking sheet. Make a central hole and brush over with egg wash (an egg yolk or an egg beaten with a teaspoon of water) or thin cream. Decorate with a few pastry leaves, brushing them over with egg or cream in their turn.

Bake for 30–50 minutes, according to the size of the apples, at gas 6, 200°C (400°F). Alternatively, lower the heat after 20 minutes, to gas 4, 180°C (350°F) and allow extra cooking time. Test them with a skewer through the central hole. The apple should be tender. Serve with cream or with brandy butter (hard sauce), p. 66.

You can always make a tart on the dumpling principle. Line a tin with pastry, put in the stuffed apples close together, cover with pastry and bake as above, allowing extra time.

APPLE DUMPLINGS FROM ANJOU

BOURDONS, BOURDAINES

Upside-down apple dumplings are made in Anjou with the local variety, the reinette du Mans. You will not find it in this country. Use other firm aromatic eating apples instead. Or look out for Ribston pippins which were first grown, it is said, from reinette pips brought from Normandy in 1709. The original tree at Ribston Hall, Knaresborough, lasted for over a hundred years. A shoot from it grew into a tree that was blown over for good and all in 1928.

Peel and core the apples; stuff the cavities with butter and sugar mashed together, or with jam – Elizabeth David suggests plum or quince – which can be given a little extra fire by a judicious spoonful of brandy or other appropriate alcohol.

Place each apple on a square of pastry. Draw up the points and fasten them together on top. Press the edges together as well. You can put a neat circle of moistened pastry over the points, if you like. Make a central hole in the top. Brush with top of the milk or single cream.

Bake as for apple dumplings in the previous recipe.

APPLE PUFFS

RISSOLES DE POMME

Make a firm apple purée of apples of the Blenheim orange and Cox's type. Flavour it with a little cinnamon, candied orange peel (p. 198), sugar and rose-water (this is a seventeenth-century recipe, as you might guess). Put in a few dill seeds if you like that anise flavour.

Use to fill small turnovers of shortcrust or puff or fila pastry, and fry – puff and fila are best deep-fried, but shortcrust works well with a shallow depth of oil, about a centimetre or half an inch. Drain well, keep hot in the oven on kitchen paper, and serve sprinkled with sugar.

APPLE CRUMBLE TART

TARTE BON ACCUEIL

A speciality of the hotel Au Bon Accueil at Yport on the Normandy coast, not far from Fécamp. The recipe is given by Simone Morand in her book *Gastronomie Normande*. The apple filling is covered with an almond crumble, which should be popular with English cooks and their families.

In Normandy, tarts are most often made with puff pastry, but you can quite well use shortcrust. If you use bought puff pastry which is not made with butter, butter the base before you put the apples on to it.

375 g (12 oz) puff pastry
5–6 well-flavoured eating apples, reinette or Cox type rather than Golden Delicious
100 g (3 oz/6 tbs) unsalted butter
100 g (3 oz/¾ cup) flour
100 g (3 oz/¾ cup) chopped almonds
200 g (6 oz/1 cup) granulated sugar from the vanilla pod jar
¼ teaspoon cinnamon
generous pinch salt

Line a 23–25 cm (9–10 inch) tart tin with pastry. Fill it two-thirds full, or slightly more as they will sink in the cooking, with peeled, cored apple slices.

Mix together the remaining ingredients to a crumble and distribute over the top. Bake in a fairly hot oven, gas 5, 190°C (375°F), until crumble and pastry are brown and appetizing, by which time the apples will be cooked – about 35 minutes.

SOUR CREAM APPLE PIE

Jane Austen wrote about a new cook, that she was glad she had begun so well and could make apple pies – 'Good apple pies are a considerable part of our domestic happiness.' Here is an unusual one, from New Zealand, I believe.

Line a deep pie plate with shortcrust pastry made with all lard, or half butter half lard.

Fill it with thin slices of sharp apples, put nicely in rows, and mounded up in the centre. Mix 60 g (2 oz/½ cup) plain flour with 90 g (3 oz/½ cup) granulated sugar and a level teaspoon of cinnamon. Scatter this over the top fairly thickly. Pour over it 125 ml (4 fl oz/½ cup) soured cream and bake it at gas 6, 200°C (400°F) for 15 minutes, then reduce the heat to gas 4–5, 180–190°C (350–375°F) until the crust is nicely browned. It makes an unusual top, not as hefty as a crumble topping, but lighter and more agreeable altogether. Try it also with pears.

Depending on the size of your pie plate, which should be about 2½ cm (1 inch) deep, you will need from ¾–1 kg (1½–2 lb) apples.

APPLE CHEESE CAKE, OR APPLE CURD TART

The charm of this particular version of what southern Americans would call a transparent pie is the sharp freshness of the apple. The filling is a kind of custard made with melted butter rather than milk or cream. When it cooks, it does not really become transparent, but rather translucent.

Such tarts were by origin English. We still make one or two of them, but not as many as present-day Americans. A pity, as the idea is a practical one – you have the basic mixture always to hand; in summer you flavour it with grated apple, in winter with nuts and candied or dried fruit, glacé orange peel or dates.

Make a *pâte sucrée* based on 125 g (4 oz/1 cup) flour, and line either 18 small tartlet tins or one 20–23 cm (8–9 inch) flan tin with a removable base. (*See* last section of the book.)

60 g (2 oz/¼ cup) lightly salted butter
60 g (2 oz/¼ cup) caster sugar from vanilla pod jar
1 large egg
1 large cooking apple

Melt butter and sugar together over such a low heat that they become no more than tepid. Remove from the stove and beat in the egg. Quickly peel, core and grate the apple coarsely and stir thoroughly into the butter mixture.

Divide between the pastry cases or spread over the pastry-lined flan tin. Bake at gas 7, 220°C (425°F) for about 15 minutes until golden brown. At this stage, the tartlets may be ready. For the single tart, lower the heat to gas 4–5, 180–190°C (350–375°F) and cook for a further 15 to 20 minutes, until nicely coloured.

Good to eat whether hot, warm or cold.

CROUSTADE DE MOISSAC

In the south-west of France, the part that some now call Occitania, there is a habit of making both puff pastry and a strudel or fila type of dough using olive oil or goose fat instead of butter. Indeed the renowned *serpent quercynol* – the Quercy snake – is exactly an apple strudel. The sheet of stretched, transparent paste is spread with chopped apples, raisins soaked in rum and dabbed with honey and goose fat. The whole thing is rolled up and coiled into a spiral like a sleeping snake.

This thin dough can also be cut and built up in layers like East Mediterranean pies, each layer kept separate from its neighbours by a thin brushing of fat. As with puff pastry, in which many more layers are built up by a different technique, the aim is lightness and crispness. The earliest French recipe to employ this system is given by La Varenne in *Le cuisinier françois*, first published in 1651. His *tourte de franchipane* has six thin layers of pastry (made from flour and egg whites) brushed with butter, then a filling of almond and pistachio custard, then six more pastry layers. To finish, 'sugar the top and serve it with flowers'.

Croustades and similar tarts do sometimes have the look of 'a giant overblown tea-rose', as the thin pastry edges brown and buckle in the hot oven, and the curls of pastry on top twist as they colour.

We were in Moissac last summer, to see again the sculptured doorway of the church with its scalloped centre shaft, the figure of Lust eaten away by frogs, serpents and time, and the oval-limbed saints. No *croustade* at the pâtisserie nearby, but very light prune tarts. The original of this recipe, which I have altered slightly, is to be found at Miramont, to the north of Moissac, in the south-west of France.

500 g (1 lb) puff pastry
4 large reinettes or Cox's or Blenheim orange
2 heaped tablespoons (3 tbs) sugar
3 tablespoons (¼ cup) rum
2 tablespoons goose fat
2 tablespoons (3 tbs) ground-nut oil

The night before you intend to cook the *croustade*, spread the kitchen table with a cloth and flour it, as if you were going to make strudel dough. Roll out the puff pastry very thinly, then stretch it with your hands. Cut out at least 10 circles, more if you like. Spread them out and leave them to dry overnight on the cloth, or move them on floured trays to a larder. Peel, core and slice the apples very thinly. Put them in a dish with the sugar and 2 tablespoons of rum. Mix them, cover and leave until next day.

To assemble the *croustade*, take a tart tin just a little larger than your circles. Mix the last of the rum together with the goose fat and oil, beating them well together. Brush the base of the tin. Put on a pastry circle. Brush it with the goose fat mixture, etc, then put on another circle and brush that too. Put on some of the apple slices, then two more layers of dough, and so on, finishing with a double layer of pastry. From the trimmings of the dough, cut strips about 4 cm (2 in) wide, curl them lightly and put on top of the *croustade* after brushing them with goose fat mixture.

Put into a hot oven, gas 7, 220°C (425°F) for half an hour, or until nicely browned and cooked. To finish the *croustade*, you can brush it with a caramel made by boiling 125 g (4 oz/½ cup) sugar with 5 tablespoons of water until it turns a golden brown.

A beautiful and delicious tart, this *croustade*, which can as well be made with pears. If you have no goose fat, use duck or chicken fat or lard: sunflower oil can be substituted for ground-nut, which is so much more common in France than here.

APPLE GALETTE

GALETTE AUX POMMES

An apple tart made with a yeast dough instead of pastry, in the style of an old-fashioned Alsatian quiche. Other fruit can be used – greengages, mirabelle plums, apricots or peaches; the juices soak into the upper part of the dough and make a good contrast to the crisp surface underneath. It is worth putting nuts into the dough, and scattering them over the fruit, too, but they are an optional ingredient. Galettes of this type do not re-heat well – they should be eaten straight from the oven or warm. If you have to prepare them in advance, store the rolled-out dough and the lightly cooked fruit separately in the refrigerator. Finish the assembly, and put into the oven just before you start the meal.

6–8 large eating apples of the pippin or reinette type
150 g (5 oz/⅔ cup) butter
4 tablespoons (⅓ cup) sugar
2 level teaspoons cinnamon
½ quantity brioche dough made with hazel or walnuts, p. 463, well-chilled hazelnut or walnut oil or melted butter
150 ml (5 fl oz/⅔ cup) whipping cream
1 large egg
extra sugar
about 30 g (1 oz/¼ cup) chopped hazel or walnuts

Peel, core and cut the apples into thin wedges. Divide the butter between two frying or sauté pans, and when hot put in the apple wedges to fry over a moderate heat. They should turn golden rather than brown, though the odd light patch doesn't matter. Mix the sugar and cinnamon and sprinkle over the apples. Continue to cook, turning the slices, until they are lightly caramelized. Remove from the heat and cool to tepid.

Roll out the dough thinly, to fit a 30-cm (12-inch) shallow pan or pizza plate. Grease the pan or plate generously with nut oil or melted butter. Put in the dough and arrange the apple wedges and their juice over the top, leaving a little rim. Put in a warm place for 20 minutes, then bake at gas 6, 200°C (400°F) for 25 minutes. Beat cream, egg and a little sugar together and pour over the galette: the free dough edge should have risen slightly, enough to contain the cream mixture, or most of it. Scatter with nuts and put back into the oven for 10 minutes, or until the cream sets.

Note Instead of a final scatter of nuts, try coarsely grated farmhouse Cheddar cheese – same quantity. Or Wensleydale if you can get it, which is the right cheese for apple tarts and pies. It sets off the apple well, without being intrusively savoury. Decidedly a Yorkshire finish.

APPLE CHARLOTTE

The Charlotte brown, within whose crusty sides
A belly soft the pulpy apple hides.

This splendid pudding crept into our repertoire between 1761, when George III married Charlotte Sophia of Mecklenburg Strelitz (the name was uncommon before that) and 1796, when the lines above were written. Mrs Rundell seems to have been the first writer to publish a recipe, in the first decade of the nineteenth century, in *A New System of Domestic Cookery*. The recipe is still the same.

My early experience of apple charlotte had been so bad – sour apples over-sweetened, margarine, factory bread – that I had never made it myself until a short time ago. By lucky chance, we had a basket of huge Charles Ross apples, weighing three to a kilo, good to eat, even better for cooking, and a bloomer loaf made from unbleached flour. This gave me the impulse to make a charlotte. And it was a success, a great success. If you have no Charles Ross, try a mixture of cookers and Cox's orange pippins, or similar eating apples with a good rich flavour. Quince pulp can be added, too. And good white bread can be used.

175 g (6 oz/¾ cup) butter
¾ kg (1½ lb) apples
caster sugar
*2 egg yolks (*see *recipe)*
half a bloomer loaf

Look out a metal tin – charlotte mould, cake tin, oval bread tin – of a generous litre, (2-pt or 4-cup) capacity.

The apple can be cooked in advance. Put a knob of butter into a heavy pan with just enough water to cover the base thinly. Peel core and cut up the fruit into the pan. Stir in 125 g (4 oz/½ cup) sugar. Cover and cook gently until the juices run, then more vigorously, without a lid, so that the apple cooks to as unwatery a pulp as possible. Taste and add more sugar if you like. Off the heat, stir in the egg yolks which help to cohere the pulp slightly, though they are not absolutely necessary.

Cut the remaining butter into cubes. Bring to the boil in a small pan, stirring. When the white salty crust separates from the yellow oil, pour through a damp cloth-lined strainer into an oval dish. Brush some of this clarified butter over the tin. Put in a heaped tablespoon of sugar and turn the tin about, so that it is completely coated inside.

Cut bread into slices a centimetre (generous ¼ inch) thick, and remove the crusts. Cut pieces to the height of the tin, and 3–4 cm (1½ inches) wide. Dip them on both sides into the butter, and press gently into place against the side of the tin. Leave no gaps. Cut a large piece for the base, dip it in the butter and put on the bottom. Brush all joints with lightly beaten egg white, to strengthen the palisade. Cut another piece of bread for the lid, dip it in butter, too, and set aside.

Fill the cavity with apple pulp, tepid or cold rather than hot. Put the lid on top. It doesn't matter if the edges of bread and the lid come slightly above the top of the tin.

Bake for about an hour, first at gas 6, 200°C (400°F) for about 20 minutes, then at gas 5, 190C (375°F) for the remaining time, or until it is nicely and richly browned. Should it be ready before you get to the pudding stage, apple charlotte will keep warm happily at a lower heat.

To serve, run a knife round between tin and pudding. Invert a warm plate on top, then turn the whole thing over rapidly. Caster sugar and a jug of the best cream you can find set apple charlotte off perfectly.

SORBET AU CALVADOS

Make a smooth apple sorbet (p. 474), flavouring it with a little Calvados. Pour Calvados over each serving, just as you take it to table.

KELMSCOTT CRAB APPLES

Fair is the world, now autumn's wearing,
And the sluggard sun lies long abed;
Sweet are the days, now winter's nearing,
And all winds feign that the wind is dead.

Dumb is the hedge where the crabs hang yellow,
Bright as the blossoms of the spring;
Dumb is the close where the pears grow mellow,
And none but the dauntless redbreasts sing.

Fair was the spring, but amidst his greening
Grey were the days of the hidden sun;
Fair was the summer, but overweening,
So soon his o'er-sweet days were done.

Come then, love, for peace is upon us,
Far off is failing, and far is fear,
Here where the rest in the end hath won us,
In the garnering tide of the happy year.

Come from the grey old house by the water,
Where, far from the lips of the hungry sea,
Green groweth the grass o'er the field of the slaughter,
And all is a tale for thee and me.

William Morris

APRICOT

One thing I should like to do is to eat a ripe apricot straight from the tree. And I suppose that, for perfection, that tree would have to be growing in its original homeland near Peking, or in Armenia. Or I would happily settle for a journey along the Karakoram Highway into the Hunza valley to try one of those tiny little apricots that we can only sample dried. All manner of virtues are ascribed to a diet in which these wild apricots predominate; they are classed with yoghurt in myths and reports of people living to 150 years, fathering children at 110 and bearing them at sixty. But who wants that kind of talisman? The reason for eating Hunza apricots is because they are so good.

'At its best the Apricot has a certain Eastern lusciousness,' said Edward Bunyard in his *Anatomy of Dessert*, 'a touch of the exotic which comes strangely into our homely country. In some Persian Palace whose quiet garden hears only the tinkle of a fountain, it would seem to find its right setting, fitly waiting on a golden dish for some languid Sharazade. But even in the sun-warmed Midi, it is not always grown to perfection, and much of the fruit which enters this country from abroad is quite unworthy.' Yes, that is the problem. This lovely fruit comes early in the summer (which is what its name means, deriving, like 'precocious', from the Latin *praecox*). It blossoms early, after the almond, before the peach, which means that the flowers are easily caught by frost. And then, such fruit as manages to grow is picked early, which means it will at worst be greenish and hard, at best woolly. We can rarely buy it as it should be, which is why apricots are held less in popular esteem than peaches.

Good apricots have been grown in England since Henry VIII's gardener, a French priest and plant expert, Jean Le Loup – anglicized to John Wolf – brought the apricot from Italy to England in 1542. Eighty years later in 1621 John Tradescant the Elder came home from Algiers, where he had gone with an expedition against the Barbary corsairs, with a collection of plants including the apricot that later became famous as the 'Argier Apricocke'. By the time that William Pinkerton, the orchard nurseryman of Wigan, put out his catalogue in 1782, he could offer no less than fourteen different varieties. But then he kept one of the best nurseries ever. Who nowadays could offer a choice of 121 apples, 68 pears and 48 peaches plus 18 nectarines and 20 cherries? Between EEC regulations and the mass intention of Garden Centres, we are lucky if we ever eat or grow more than half-a-dozen varieties of any fruit. Such splendour of choice, as John Harvey remarks in his book *Early Nurserymen*, is rare at any time and in any country.

I suppose the best English apricot is the Moor Park. In about 1760, Lord Anson brought home an apricot stone to his Hertfordshire garden which William Temple had described – and emulated at his Moor Park in Farnham – with such affection. The splendid results from this stone became so renowned in northern Europe – its particular advantage was its late flowering – that it was even paid the compliment of French mis-spelling and turns up as Morpak (like rosbif from our roast beef). Fame indeed.

My regret about the fruit sold here is that it does not – at least in my experience – include the wonderful peach-apricot, a luscious combination of the two fruits. The ones on sale in Paris are, I imagine, an apotheosis of the *alberge* of Touraine and Anjou which must have been a smaller fruit, if a prized one. In *Eugénie Grandet*, set in Saumur, Balzac placed alberge jam at the height of the things one could spread on bread, the most recherché and delicate, at the opposite end of the scale from butter which was the commonest, most everyday. A paste made from alberges was once used for stuffing the *pruneaux de Tours*. Nowadays, ordinary dried apricots are used. Every time I go to Tours in the summer, I ask around for alberges but I am always just too late or too early. It makes me wonder if they are really gone for ever, except from a garden or two where fine fruit is the high aim.

HOW TO CHOOSE AND PREPARE APRICOTS

Choice between varieties is usually impossible, one kind only being offered for sale at a time. If you can pick out apricots yourself, choose ones that look ripe and glowing and are firm. The stronger the colour, the sweeter the apricot as a rule. Less ripe fruit with a pale or greenish look should be kept for cooking.

As things are in this country, you will perhaps be wise to regard apricots as a fruit for the kitchen, then you will not be disappointed. They are the favourite of pastrycooks and confectioners who have any pretensions to skill. Apricot tarts, apricot paste, apricot fillings for almond meringues, apricot jam, apricot glaze as a barrier between icing and cake or as a sweetness in bitter chocolate cake, apricot sorbet or ice cream, apricot dumplings, apricot bavaroise, apricot soufflé, amardine cream, apricot sauce are some of the commonest apricot dishes that you are likely to come across in the season, once you set foot on mainland Europe.

Apricot with lamb is an idea from the Middle East that we have much taken to in this country, the sweet sharpness of the dried fruit goes well with the sweetness of the meat. Apricots do not go so perfectly with other meats, apart from duck, ham and cured pork dishes which are well set off by pickled apricots or an apricot salad dressed with yoghurt and tarragon, or soured cream with tarragon or lemon thyme, salt and pepper.

From some of the best dried apricots, you gain an idea of what the fresh original fruit must have been. And you will discover that there is a modest variety to choose from. The small dark wizened apricots are sharper than the large fleshy 'ears', and for that reason they go well with meat and, as far as flavouring is concerned, they can be used in smaller quantity. You will notice that certain packages are printed with the words 'sulfur dioxide'. In some countries, Australia for instance, treatment of dried fruit in this way, to prevent discoloration and spoilage, is compulsory. If you want to avoid this, stick to the whole Hunza apricots but remember to pick them over before you put them to soak and discard any with holes in them.

Apricots, like other dried fruit, can be soaked overnight in water, or water flavoured with wine or orange juice. Cover the fruit by 2½ cm (about 1 inch) and use the remaining liquid when cooking the fruit. To speed up the process, pour boiling liquid over the apricots in the pan you intend to cook them in. After an hour, you can re-heat the liquor to boiling point. Or you can keep the pan on a very low heat that keeps the water hot, not boiling.

If you are on holiday in the Middle East, you may see strips or sheets of apricot leather or *amardine*. This can be soaked and used like dried apricots. I give a recipe for making it at the end of the mango section in case you should ever be faced with a glut of apricots and in despair about keeping them. You can make it from dried apricots too – or dried peaches, or a mixture – but I cannot see the point: the result is too strong to eat pleasurably as a sweet, since the extra sugar added, as you knead it, emphasizes the sharpness of the fruit.

Almonds are the natural partner, replacing the almond-flavoured kernel. If you strike a variety of apricots with sweet kernels, they should be used even it it is a nuisance cracking the stones. If the kernels are bitter, too many are harmful – best to discard them.

Kirsch, brandy and orange liqueurs all go with apricots. You can make an apricot liqueur yourself, to pour over ices and creams, *see* the end of the section.

CRACKED WHEAT AND APRICOT STUFFING

Kibbled or cracked wheat, or *burghul*, or *bulgur*, or *pourgouri*, cooks to a moist rich graininess which is ideal for duck in particular, but also for lamb. In addition it has the lovely nut flavour that many cereals have before they are polished and refined. Buy it from health food shops, or from a Cypriot or Greek delicatessen.

Soften a medium chopped onion in 60 g (2 oz/$\frac{1}{4}$ cup) butter. Stir in 250 g (8 oz/1$\frac{3}{4}$ cups) cracked wheat, and keep stirring over a low heat for about 10 minutes.

Pour in a generous 300 ml ($\frac{1}{2}$ pt/1$\frac{1}{4}$ cups) duck or appropriate stock, season lightly, add a tablespoon of pine kernels or blanched almonds and 6 whole dried apricots cut into strips. Cover and simmer for 10–15 minutes, until liquid is absorbed.

Stir in another 60 g (2 oz/$\frac{1}{4}$ cup) butter. Put a cloth over the top of the pan, jam on the lid tightly and leave over a very low heat indeed for a further 20–30 minutes.

The duck or lamb should be stuffed with some of the mixture. The rest should be reheated later and served round the meat.

APRICOT STUFFING

For chicken, duck and, above all, lamb. Quantities are generous, to allow for stuffing with some left over to be served round the meat. If you cannot buy brown rice, use basmati.

90 g (3 oz/$\frac{3}{4}$ cup) chopped onion
125 g (4 oz/$\frac{1}{2}$ cup) butter
250 g (8 oz/1$\frac{1}{4}$ cups) brown rice
60 g (2 oz/$\frac{1}{2}$ cup) seedless raisins or blanched split almonds
125 g (4 oz/$\frac{3}{4}$ cup) dried apricots, soaked
2 tablespoons chopped parsley
salt, pepper

Soften the onion in half the butter, stir in the rice and then add 600 ml (1 pt/2$\frac{1}{2}$ cups) water. Bring to the boil, then turn down heat and leave to simmer gently, uncovered, until cooked. Top up the liquid if necessary.

Brown rice takes about 40 minutes, longer than the basmati. Aim to end up with the water absorbed, and the top of the rice very smooth with occasional holes in it.

While the rice cooks, simmer the raisins gently in the rest of the butter until they are puffy. Or brown the almonds lightly in the butter. Add to the rice about 5 minutes before it is cooked. Cut the drained apricots into strips and add them too. Remove the pan from the heat, put in parsley and season to taste.

Cool before stuffing the poultry or meat. Reheat the left-over stuffing either in a buttered, covered dish in the oven, or in a double boiler or pudding basin set over a pan of simmering water.

LAMB POLO

'The apricot has a special affinity for lamb,' writes Claudia Roden in her *Book of Middle Eastern Food*. 'Its sweet acidity blends with, and seems to bring out, the mild sweetness of the meat. The early Arab Abbasid dynasty, centred in Persia, greatly favoured the combination and created a series of dishes on this theme called *mishmishaya* (*mishmish* is Arabic for apricot). It is still a great favourite as a partner to lamb in modern Persia.' As in this lamb and apricot polo – i.e. pilau – that she gives.

125 g (4 oz/½ cup) butter
1 onion, finely chopped
½ kg (1 lb) lean lamb, cubed
salt and pepper
½ teaspoon ground cinnamon
½ teaspoon crushed coriander seeds
2 tablespoons (3 tbs) seedless raisins
125 g (4 oz/¾ cup) dried apricots
½ kg (1 lb) long grain rice
extra lump of butter

Melt the butter in a large sauté pan, add the onion and cook gently until soft and golden. Raise the heat, add the lamb cubes and brown them lightly. Season with salt and pepper; stir in the spices, then the raisins and apricots. Continue to fry for a minute or two, stirring everything about. Pour in enough water barely to cover, put a lid or a double sheet of foil on the pan and simmer for 1½ hours. The meat should be very tender, and well-flavoured with the fruit and spices. Should the sauce be on the liquid side strain it off and boil it down. Return it to the meat pan and correct the seasoning.

Meanwhile, boil the rice in salted water with the lump of butter. After 10 or 15 minutes, the rice should be almost but not quite cooked. Drain off any surplus water, then arrange the rice in alternate layers with the meat in a heavy pan, beginning and ending with rice. Put a doubled cloth or muslin over the top, then jam on the lid to make a tight fit. Set over a very low heat, to steam for 20 minutes. The rice will be tender and succulent, having absorbed part of the meat sauce. The cloth soaks up some of the steam which helps to keep the dish light rather than on the soggy side.

CHICKEN VARIATION Joint and cook a chicken according to the first paragraph of the recipe above, reducing the time so that the chicken is not over-cooked. Bone the pieces and return them to the sauce to reheat. Meanwhile, cook the rice. Omit the layering process and serve the chicken in a ring of rice.

Use Hunza apricots and brown rice for extra good flavour. If the apricots are too hard to chop, soak them until they soften a little.

POACHED APRICOTS AND CARDAMOM

Fresh whole apricots can be poached in syrup flavoured with cardamom seeds (remove the pod first, or cut it open). Put the seeds into a muslin bag and suspend it in the syrup – 100 g (3½ oz/½ cup) sugar dissolved in 600 ml (1 pt/2½ cups) water, plus the peel thinly cut from a lemon. After several minutes' simmering, taste the syrup and remove the cardamom bag if the flavour is strong enough. If not, leave it and check later.

With a quantity of apricots, use the deep-fryer method on p. 437. Otherwise use a thermometer.

With a small quantity, use a blanching basket in a deep saucepan, and put in the apricots in batches when the syrup is simmering. Lower the heat so that the syrup remains at 90°C (200°F). Apricots need about 8 minutes if whole.

Remove them to a dish. Boil down some or all of the syrup by half, or until very thick; add orange juice to taste, and the finely grated zest of 2 oranges, and pour over the fruit. Cool and chill.

Dried apricots can be soaked, and then cooked in the same way. Use the soaking water that remains to make up the syrup. Keep a check on the sweetness. Add pine kernels or blanched, split almonds before serving with clotted cream, or chopped walnuts or pecans.

THE CHARTER

On 10 August 1786, Parson Woodforde the diarist and his niece Nancy had a disaster. They were on holiday from their Norfolk parish, Weston Longville, where Nancy kept house for her uncle, and were staying with the family in Somerset. A dinner party was planned for that evening. The two of them had volunteered to make a special pudding that they had sampled at a friend's house near Norwich. 'Nancy and self very busy this morning in making the Charter . . . but unfortunately the cellar door being left open while it was put in there to cool, one of the greyhounds (by name Jigg) got in and eat the whole with a cold tongue etc. Sister Pounsett and Nancy mortally vexed at it.' The only other sweet dish they had for the party was a barberry tart.

For years I wondered about the Charter, and whether it was indeed a pudding or a meat dish. At one time, I came across a recipe for Cornish charter pie, chickens in cream and parsley, and wondered if it were something of that kind. The answer came at last in the 1980 autumn bulletin of the Parson Woodforde Society. A founder member, Mrs Baker, had after much searching and effort acquired a Norwich manuscript cookery book of 1821 with a recipe in it for the Charter. It turns out to be a custard of the kind used for burnt cream (p. 133), but with apricots on top rather than caramelized sugar. In my version below, ingredients and relative quantities are identical. I have shortened the method slightly, and use a homely style of preserved apricots, rather than the glacé apricots that I assume were intended.

1 litre (1¾ pt/4½ cups) single cream
thinly-cut peel of a lemon
4 large eggs plus 4 large egg yolks
sugar
24 dried apricot halves

Bring cream and peel slowly to boiling point. Beat eggs and yolks together, then pour on the cream through a strainer. Whisk often at first, then occasionally, to make a smooth cream. Add sugar to sweeten the mixture slightly (bear in mind that the apricots will be very sweet). Pour into a soufflé dish of at least 1½ litres, 2½ pints or 6-cup capacity, or into eight custard cups. Stand in a roasting pan half full of boiling water. Bake at gas 2, 150°C (300°F) until just firm – allow at least 45 minutes, but be guided by appearance. The custard should not be too wobbly and a knife stuck into the centre should come out clean. This can take longer than 45 minutes depending on the depth and size of the dish: small pots should be checked after 30 minutes. Take the dish(es) from the pan and cool.

Meanwhile pour boiling water on the apricots to cover them generously. Leave them for an hour at least, then simmer until the apricots are tender. Keep them covered with water, topping it up if necessary. Stir in 3 rounded tablespoons of sugar, raise the heat and boil hard until the liquor is reduced to a little syrup and the apricots are glossy and candied. Drain them on a rack. Arrange them on top of the cream when everything is cold.

BAVAROISE POMPADOUR

A favourite book of mine is the *Mémorial historique et géographique de la pâtisserie*, by Pierre Lacam, published in 1890. It is a blend of commonplace book, entertainment and register of the cakes and puddings that were around at the end of the century. I discovered from it that the angelica sculpture of Niort, the angelica punts with angelica cows, the angelica lobsters and so on, is at least a hundred years old, that Pithiviers cake became popular in the time of Louis XIV, that the *talmouses* (little pies) of Saint-Denis were celebrated by the poet Villon, that to the French – or some of them – our Christmas pudding is known as 'pudding tête de mort'. Apparently Rossini had a special syringe made for garnishing pasta with mashed foie gras (there was emotional high bidding for it at the sale after his death); and the perfume makers of Paris who were alarmed at the growing popularity of vanilla in cake-making, tried to keep the monopoly for their own products by spreading the rumour that the pods contained a poison.

The one thing Monsieur Lacam does not tell us, is why some apricot dishes are called after the Pompadour. I have always imagined that it is because of the colour of her silk clothes, the roses, flowers, delicate porcelain, the grace that still delights us in her portraits.

The actual bavaroise recipe is simplicity. Monsieur Lacam set it in pistachio jelly. If you like making elaborate puddings, try it out, using syrup in which you have poached the apricots for the jelly, and stirring in chopped pistachios when it is half-set. From the flavour point of view, this has nothing to recommend it, it adds little but prettiness. To me, the plain apricot colour of the bavaroise, scattered with pistachio, is fresher and more appealing.

WITH RIPE APRICOTS Stone 375 g (12 oz) apricots and liquidize them.

You should end up with 250 ml (8 fl oz/1 cup) purée. Sweeten to taste. Put a packet of gelatine (½ oz/2 envelopes) into the liquidizer – do not bother to rinse it out – and add 6 tablespoons (½ cup) very hot water. Whizz until dissolved, add the purée and whizz again. Beat 300 ml (10 fl oz/1¼ cups) whipping or double cream and a tablespoon of icing or caster sugar. When it is stiff, taste again and add extra sugar if you like. Taste the apricot mixture, and add extra sugar to that if you think it needs it. If the apricots are on the tasteless side lemon juice can be dripped in.

Fold the apricot purée into the cream. Pour into a lightly-oiled decorative mould to set. Turn out when chilled and firm, and scatter with a couple of tablespoons of chopped pistachio nuts.

WITH UNRIPE APRICOTS Stone ½ kg (1 lb) apricots and cook briefly with 2 tablespoons of water and a little sugar. Drain when soft, and liquidize as above, adding extra cooking liquor if needed to bring the purée up to 250 ml (8 fl oz/1 cup). Complete the recipe as above.

Note The gelatine quantity above sets the bavaroises firmly enough for turning out. If you prefer a softer consistency, put in a further 60 ml (2 oz/¼ cup) apricot purée, and turn the bavaroise into a glass bowl from which it can be served. Float a layer of cream on top and scatter with pistachio nuts. Individual glasses could also be used.

THE POMPADOUR'S RIBBONS
BAVAROISE POMPADOUR RUBANÉE

This is one of the prettiest puddings I know, only to be rivalled by Queen Claude's ribbons on p. 360. Layers of apricot-coloured and creamy-white bavaroise shaped in a mould of modest decoration are exactly right for dinner out of doors. The tricky part is turning the pudding out – one careless movement and the layers separate and blur the clear lines of the shape. That is why you should not choose too fancy a mould. With something simple, you stand a better chance of correcting trouble of that kind.

If you cannot bear the thought of making a mistake, alternate the layers in glasses or a glass serving bowl. Nineteenth-century chefs liked to keep their layers accurate, as if the final assembly had been made by machine. It's more to modern taste to have them uneven, as if the Pompadour's ribbons were blowing and twirling in a rococo breeze.

Make a *bavaroise Pompadour* as in the preceding recipe. Also make the firm *coeur à la crème* mixture on p. 471. (A simpler alternative is to beat a ½ litre (15 fl oz/2 cups) yoghurt or *fromage frais* with sugar to taste and then to mix in a packet (½ oz/2 envelopes) of gelatine dissolved in 6 tablespoons (½ cup) very hot water. Finally fold in 150 ml (5 fl oz/⅔ cup) whipping or double cream, beaten stiff.)

Choose a mould of 1¼ litres (generous 2 pt or 5 cup) capacity. Brush it out with almond oil, or sunflower oil. Turn it upside down to drain.

Pour a centimetre or a little less (¼ inch) of the apricot mixture into the mould. Put in the freezer or coldest part of the fridge and take it out when almost firm. This is the boring part – you should try to catch the layers before they are so firm that they stand no chance of sticking together. Repeat with a cream layer, and so on, until both mixtures are used up. If they harden as they wait, stand them in warm water and stir.

Turn out when thoroughly chilled. Scatter with chopped pistachios. Enough for 10 or 12 people.

APRICOT PUDDING

Two summers ago, a friend in France lent me a rarity, *Le Cuisinier anglais universel ou le Nec plus ultra de la gourmandise*, published in Paris in 1810. It was a translation of *The Universal Cook* of 1792, written by Mr Francis Collingwood and Mr James Woollams. They are pictured in wigs and profile in frontispiece medallions. Good credentials are listed – they had worked at the popular London Tavern and at the Crown and Anchor in the Strand, a great meeting place for Whig supporters and politicians.

Considering the date and war with Napoleon, the title seems pushy and tactless, but the reputation of London food at least stood high with foreigners. When the first smart restaurant in Paris was opened – in the year that *The Universal Cook* was published – it was called La Grande Taverne de Londres. After the London Tavern perhaps, where John Farley was the famous chef who, in 1783, had brought out his own most successful book, *The London Art of Cookery*.

In spite of all the fame and glory, the French publisher had qualms. In his introduction, he wrote: 'The English must eat well, look at their embonpoint! If occasional recipes seem odd, they will at least, cher lecteur, broaden your experience, acquainting you with *le catchup* and *le browning*, which are unknown even to our best chefs.' Yes, indeed. What the French are likely to have appreciated more – judging by our experience – is the pudding and tart sections. From them, one can pick up recipes like this, as well as the idea of boiling fruit peelings and debris in water to make the best use of all the flavour of apples or pears when making pies.

250 g (8 oz) puff pastry
250 g (8 oz) stoned apricots
caster sugar
3 large egg yolks
250 ml (8 fl oz/1 cup) whipping or double cream
2 large egg whites, whipped stiff

Line a 25-cm (10-inch) tart tin with puff pastry. Roll it on the thin side. Prick the base lightly with a fork. Preheat the oven to gas 8, 230°C (450°F).

Choose a saucepan that will take the apricots in a single layer, cut side down. Put them in, and set the pan over a very low heat until the juices begin to run. Avoid adding water if you can. Sprinkle on 2 tablespoons sugar, and cook uncovered until the apricots are tender. Raise the heat to evaporate any wateriness. Liquidize or process with yolks and cream. Or sieve and add yolks and cream. Sweeten further to taste, bearing in mind the dulling effect of the egg whites, which should now be folded in with a metal spoon. Slip a baking sheet into the oven.

Now pour the filling into the pastry but do not overfill it. Any surplus can be used to make little custards or tarts. Slide the tin on to the baking sheet, which will have become very hot in this brief time. The heat helps to set the pastry fast, reducing any tendency to sogginess.

After 15 minutes, lower the temperature to gas 4, 180°C (350°F). If the pastry shows signs of ballooning up through the filling, prick it with a fork. Leave another 20 minutes, or until only the centre is slightly wobbly under the crust. Like a soufflé, apricot pudding benefits from the creaminess of the centre, which acts as a sauce.

OLD-FASHIONED APRICOT TART
TARTE D'ABRICOTS À L'ANCIENNE

Every year at the garlic and basil fair at Tours in July, we meet friends and have lunch out of doors at the old inn on a corner of the square where the market is held, the *place du grand marché.* Everyone brings something for the general picnic – *on peut apporter son manger* in that comfortable French style. Drink is supplied by the inn. The standard menu is hot dogs made with Arab *merguez* sausages from a stall nearby, and this apricot tart from the pastrycook opposite. What marks the tart is the butter beaten into the filling. It bakes to a rich grainy top. In France, unsalted butter is always used, but I find that the light saltiness of Danish Lurpak works better, especially with our bland English cream (soured cream is not always to be found).

Line a 25-cm (10-inch) tart tin with a rich shortcrust or sweet pastry. Bake it blind until just firm. Do not let it brown.

FILLING
16–20 apricots (approximately 1¼ lb)
light syrup 300 g (10 oz/1⅓ cups) sugar and 600 ml (1 pt/2½ cups) water
100 ml (3½ fl oz/½ cup) soured cream
200 ml (6½ fl oz/¾ cup) double cream
2 large eggs
sugar
60 g (2 oz/¼ cup) butter
icing sugar

Halve apricots, discarding the stones. Poach them lightly in the simmering syrup until they are tender but still in good shape. Drain well, allowing the liquid to fall back into the pan, and leave to cool.

Beat creams with the eggs, sweetening it to taste. Melt butter, and pour it in warm.

Arrange a ring of about half the apricots, cut side up and overlapping slightly, round the edge of the tart. Then make an inner ring, and finally a central group of halves. Pour on the egg/cream mixture gently – you may not need it all, it depends on the depth of the crust. Bake at gas 7, 220°C (425°F) for about 20 minutes, check and if the top begins to look set and slightly brown, lower the heat to gas 4, 180°C (350°F) until the filling is just set firm.

If the filling rises alarmingly, do not worry. It will fall when you take it out of the oven.

While the tart cooks, boil down the apricot cooking liquid to make a very thick syrupy glaze. When the tart comes out of the oven, let it cool a minute or two. Pieces of apricot will appear through the custard. Brush the whole thing over carefully with the apricot glaze to make it look shiny and even more delightful. Dust the pastry rim with icing sugar.

Note If you cannot buy soured cream, use all whipping cream, and add lemon juice to the apricot poaching syrup to sharpen it a little.

Apples, pears, peaches, mirabelleo plums, greengages, Chinese gooseberries (or kiwis), persimmons and Sharon persimmons, golden berries (physalis), can all be used instead of apricots.

APRICOT AND ALMOND CRUMBLE

An elegant version of the homely crumble. It is always a great success with our French and Italian friends, who ask for an English pudding but whose pioneering spirit would fail if faced with Spotted Dick or Dead Man's Leg.

24 large apricots
sugar
60 g (2 oz/½ cup) blanched, slivered almonds

CRUMBLE
125 g (4 oz/1 cup) flour
125 g (4 oz/½ cup) caster sugar
125 g (4 oz/1 cup) ground almonds
175 g (6 oz/¾ cup) butter

Peel and quarter apricots (if the skins are in good condition and not tough, you need not bother with the peeling). Arrange them in a shallow baking dish. Sprinkle with sugar. Mix dry crumble ingredients together, then rub in butter. Spread over fruit. Top with the slivered almonds. Bake at gas 6, 200°C (400°F) until the top is nicely coloured – about 35 minutes. If you find the top colouring too rapidly, lower the heat to gas 4, 180°C (350F).

Serve hot or warm, with cream or custard sauce (p. 470).

APRICOT CARAMEL CAKE

Like most upside-down cakes and puddings, this one is made on the pound cake principle of equal quantities – in the past, a pound of each, now usually 4 oz – of eggs, flour, sugar and butter. It has an extra fineness from the use of cornflour and an extra richness from ground almonds. The caramelizing of the fruit reminds me of *tarte Tatin*, and indeed you could put a pastry layer over the apricots instead of the cake mixture. Remember to prick it with a fork.

The choice of tin is important. I use a straight-sided shallow cake tin about 5 cm (2 in) deep and 23 cm (9 in) wide. If you want to use fewer apricots, choose a deeper tin, 20 cm across (8 in). If you want an even shallower layer of cake, with more apricots, use a 25 cm (10 in) tin. These differences will affect the cooking time and temperature: if the cake depth is increased, bake at mark 4, 180°C (350°F). In all cases, test it with a skewer or narrow knife blade. It's a very good tempered mixture.

Fill your chosen tin loosely with fresh apricots. By the time they are halved and poached, you will have the proper amount to cover the base closely fitted together, ¾ kg (1½ lb) will be plenty. You will also need:

175 g (6 oz/⅔ cup) sugar
300 g (10 oz/1⅓ cups) sugar and 600 ml (1 pt/2½ cups) water, boiled 5 minutes to make a syrup
2 large eggs
100 g (3½ oz/scant ½ cup) caster sugar
125 g (4 oz/½ cup) butter
60 g (2 oz/scant ½ cup) cornflour
75 g (2½ oz/½ cup) self-raising flour
1 level teaspoon baking powder
1 rounded tablespoon ground almonds

Put the 175 g (6 oz/⅔ cup) sugar into a pan with 6 tablespoons or ½ cup of water. Stir over low heat until dissolved, then boil hard to a rich caramel brown (don't stir, but watch it). Wrap your hand in a tea towel, and stir in – off the heat – 4 tablespoons (⅓ cup) water, using a wooden spoon. The caramel will sizzle, harden and look very odd. Just stir carefully, and if it does not dissolve into a clear smooth caramel, put it back over the heat until it does. Pour into the lightly greased cake tin and set aside.

Bring the syrup to the boil in a wide pan, and slip in the halved apricots, cut side down. When the syrup returns to the boil, give it a minute to simmer, then turn the apricots over. By this time, they will be cooked enough. Do not let them collapse. Remove with a slotted spoon and pack closely together, cut side down, on top of the caramel.

Put all the cake ingredients into the bowl of an electric mixer or processor and whizz until smoothly blended. If necessary, add a couple of tablespoons of the apricot syrup to make the mixture a dropping consistency. The ingredients can equally well be beaten together by hand; make sure the butter is very soft indeed.

Spread carefully and evenly over the apricots. Bake at gas 5, 190°C (375°F) for about 45 minutes. The top should brown nicely. Cool for a moment, then run a knife round the edge, put a serving plate on top and quickly turn upside down. Serve warm as a pudding, which is best, I think, or cold as a cake.

If you like, you can split and toast as many almonds as there are apricots, and put a nice brown piece in the cavity of each half of apricot.

This is best done at the end of cooking time. Or just scatter the top with chopped toasted almonds.

Note Apples or pears can be substituted for apricots. Other fruit should not be caramelized: turn to pineapple-upside-down pudding (p. 347) for a good recipe.

TRAUNKIRCHEN SHORTCAKE

TRAUNKIRCHNER TORTE

The speciality of Traunkirchen. After an Austrian friend sent me this recipe which is one of her favourites, and easy to make, I wondered why anyone ever makes an American-style shortcake and scrapped the recipe for it from this book. The filling is a witch's foam – *Hexenschaum* – made of egg white and jam, but it can just as well be strawberries and cream, or a fruit pureé mixed with the foam instead of jam. The keeping qualities of this cake are in its favour, too. Altogether a winner.

Mix together, either by hand or in a processor or with an electric beater:

240 g (8 oz/1 cup) butter
90 g (3 oz/scant ½ cup) sugar
280 g (generous 9 oz/2½ cups) plain sifted flour
3 egg yolks

Divide into four equal parts, and roll each one into a ball: they will weigh just over 150 g (5½ oz) each. Chill for half an hour.

Roll them out and shape them, on a lightly floured board, into rounds about 18 cm (generous 7 inches) each. Use flan rings if you have them. Mark one of the rounds into eight wedges for the top.

Bake at gas 3, 160°C (325°F) for about 20 minutes. The shortbread should end up golden brown and crisp. Check from time to time, and if the edges seem to be burning, turn the heat down. Remove and cool, separating the wedges along the cut marks.

For the filling, whisk the whites of the eggs to a snow with 60 g (2 oz/¼ cup) sugar and 90 g (3¼ oz/¼ cup) apricot jam – strawberry, raspberry, red currant jelly softened can also be used. Use this to sandwich the layers together, with the separated wedges on the top. Leave the cake at least overnight in a cool place, although it is reputedly best after two or three days We can never resist it that long. Enough for 8.

DRIED APRICOT JAM

A basic recipe which stands on its own, or which can be used for your own variations. Some people add a medium can of pineapple (fruit chopped, added with juice), or orange juice and zest (2 large); others blanched, slivered almonds (125 g, 4 oz or a scant cup).

Soak ¾ kg (1½ lb) dried apricots in 2 litres (3½ pt/8¾ cups) water for 24 hours, having washed them first. Put into a preserving pan with the finely-grated rind and juice of 2 large lemons. Bring to the boil and cook until the apricots are tender. Stir in 1½ kg (3 lb) sugar. Once it is melted into the liquid, raise the heat and boil vigorously until setting point is reached. Stir often to prevent sticking and burning. Stand for 10 minutes, stir up and pot as usual.

FRESH APRICOT JAM

As above, with 2 kg (4 lb) apricots, stoned weight, ½ litre (¾ pt/scant 2 cups) water, juice of a lemon, and 2 kg (4 lb) sugar. Only add the kernels if they are sweet, otherwise use almonds. The apricots do not require any soaking, of course; allow about 20 minutes for cooking with water, a further 20 minutes with sugar.

SPICED APRICOTS (OR PEACHES)

A beautiful preserve, delicious with cold ham and turkey at Christmas

1½ kg (3 lb) apricots
½ kg (1 lb/2 cups) sugar
300 ml (½ pt/1¼ cups) white wine vinegar
8 cloves
½ stick cinnamon

Wash and pick over the apricots. Bring remaining ingredients to the boil, simmer for a few minutes, then slip in the apricots and cook gently until they are tender. Do not overcook them. Pack the drained fruits into clean jam or bottling jars. Boil down cooking liquor until thick and syrupy. Break up the cinnamon stick according to the number of jars you are using, and put the pieces in with the fruit. Distribute the cloves in the same way. Pour over the boiling hot liquid. Seal and leave a month before broaching.

APRICOT LIQUEUR

There are many versions of this recipe, this one coming from *The Complete Book of Fruit* by Leslie Johns and Violet Stevenson.

16 large ripe apricots (500–625 g or 1–1¼ lb)
1 litre (1¾ pt/scant 4½ cups) dry white wine
500 g (1 lb/2 cups) sugar
½ litre (18 fl oz/2¼ cups) proof spirit – vodka, brandy
1½ level teaspoons cinnamon

Put apricots and wine in an enamel or stainless steel pan. Bring to the boil, add sugar, alcohol and cinnamon and lift straight off the fire. Pour into a warmed basin. Cover with plastic film, and infuse for four days. Strain into a jug, then filter through a cloth into a bottle. Cork tightly.

APRICOT FRITTERS Wonderful fritters can be made by soaking fresh apricot halves in this liqueur, draining them, dipping them in batter and deep-frying them. Drain on kitchen paper, put on a warm, ovenproof dish, sprinkle with sugar and glaze under the grill. Serve immediately. Another idea from the book.

DRIED WILD APRICOTS

Writing about food gives me plenty of excuses for discovery. Usually this means no more than finding a new way of cooking a familiar food, or exploring something taken too much for granted. It's not often one comes across an entirely new food, as I did one Christmas at the beginning of the seventies.

We were in the Marlborough delicatessen – now, alas, no more – and caught sight of a sack at the far end of the shop containing peach stones. Or so it seemed until we read the notice WILD DRIED APRICOTS FROM AFGHANISTAN. The owner of the shop, Mr Bedwin, was an enthusiast. He described their mixed flavour of apricot and peach. We left with a large bag full of them. After an hour or two, they filled the warm kitchen with a delicate spiced fragrance, not as strongly as quinces do, but with an equal magic.

It seems that these small fruit, about the size of a mirabelle plum, are not really wild but grow in orchards from Persia to the Hunza valley in Kashmir. And they have been imported into this country for nearly thirty years. How was it that we had never heard of them until 1973? And never read about them in cookery books? 'Well,' said one importer, 'they are sold mainly to Asians; it's in their shops you're most likely to find them'.

In those days that was true. After I wrote about them in the *Observer* magazine in February 1973, they began to appear in good delicatessens as well as in wholefood and vegetarian shops. Whether as a result or a coincidence I do not know. The trouble is variability of supply. It takes a shopkeeper of real passion for good food to keep on and on enquiring about such things, sometimes over severa. years. The last lot I bought came from Suterwallah's cash and carry in Southall. The lot before from a health food shop in Manchester. I buy a large quantity as they keep well.

HOW TO COOK WILD APRICOTS The main thing to remember is that they need no added sugar. Soak them in the same way as ordinary apricots. To cook, add more water to the soaking water, so that the apricots are covered. Bring to the boil and simmer until tender. Remove the fruit, draining the juices back into the pan, extract the stones and crack them to reveal the kernels.

Meanwhile, boil down the cooking liquor to make a syrup.

Put the apricots in a bowl, or individual bowls, pour over the syrup, sprinkle on the kernels, and chill before serving.

WILD APRICOT FOOL OR ICE CREAM

Dried fruit such as prunes make good fools and ices, but these apricots are the best of all.

Soak and cook as above 175 g (6 oz) wild apricots. Chop the kernels. Sieve, liquidize or process the soft fruit, adding enough reduced cooking juice to make 300 ml ($\frac{1}{2}$ pt/1$\frac{1}{4}$ cups) purée. Intensify the flavour with lemon juice. Whisk 300 ml (10 fl oz/1$\frac{1}{4}$ cups) whipping cream with a tablespoon of icing sugar. Taste both mixtures, to make sure that the sweetness is right. Then fold one into the other. Stir in the kernels. Divide between six small bowls and chill, or freeze in the usual way. Serve almond biscuits, p. 459, or keep back the kernels to flavour biscuits.

ARBUTUS

'My love's an arbutus
By the borders of Leane'
we used to sing at school. It sounded mysterious but enviable – what was an arbutus and where was the Leane? We were too shy to ask. The teacher did not explain. She probably did not know herself.

I found the answer twenty years later in the Giardini Hanbury – the Hanbury gardens – near Ventimiglia and the Italian-French border. The gardens were neglected – some interrugnum of guardianship? I do not remember, but they were almost a wilderness, lovely to walk in.

There was a tall bush covered with red fruit, strawberry-red and bristling slightly. 'That's *Arbutus unedo*,' said my husband. 'The strawberry tree.' And I replied, 'But where are the borders of Leane?' He told me that as well as being native to the Mediterranean, there was a patch or district of arbutus in Ireland which was discovered in the seventeenth century, growing in that warm south-west area around Killarney which is on Lough Leane (where Tennyson heard the bugle blowing and its echoes dying, dying, dying).

The thing about that particular arbutus, apart from its prettiness, was that it hung obligingly over the wall so that its fruit dropped down on to the Via Appia, which runs like a wide paved ditch between two high walls holding back the garden. On one wall was a plaque: 'Along this road came Petrarch . . .' The taste of that fruit was amiable rather than decisive, but better than many writers give it credit for. We side with the nineteenth-century Chinese poet who wrote

> Soochow, the good place. The arbutus
> Sweet like heavenly dew,
> Filling my cheeks with juice.

Indeed, such arbutus as you are most likely to come across, unless you live in Killarney or visit the Mediterranean late in the year, are imported in cans from China. They are small and dark red with their characteristic bobbliness of skin, in a beautiful dark red juice. A pretty, if not a remarkable item for taste, in a mixed dish of fruit.

Since I began this book, my husband planted an arbutus, which is not an uncommon shrub from good nurserymen. I anticipate a good summer and a pot or two of *gelée d'arbouses* made, according to a Corsican cookery book, in the Corsican style – which is just the same as everybody else's. Be sure when you strain off the liquid that you press down on the fruit to get every drop of juice; forget about the clarity that prize pots of jelly at village shows are supposed to have. Arbutus jelly is for pleasure not prizes.

As to the name, said Pliny, the reason is that a person will eat only one – *un edo*. Fair enough, if the fruit is not ripe: one taste of its bitterness would be enough. But come November, which is when we visited that good place, the Hanbury gardens, it is mellow and mild on the tongue. Something to bear in mind.

ARBUTUS TARTS
GÂTEAUX D'ARBOUSES

In his *Mémorial historique et géographique de la pâtisserie*, Pierre Lacam gives a cry of distress that I recognize only too well – 'Why is there no yearly register of new cakes, new pastry shops? Why does no one list what is going out of fashion? How many young pastrycooks now know that pain de Gênes and le régent were first made at Chiboust's shop?' Yes, indeed. And how many can recall that Gâteau St Honoré was also invented by Chiboust – in 1879 – and filled with his special light version of *crème patissière*?

About the strawberry tree and its fruits, Monsieur Lacam noted that it will even grow as far north as Tours. The fruits are put into little pastry cases, baked blind and brushed with red currant jelly. Perhaps they were lightly glazed with red currant jelly, too, to give the blandness some zest. A good cake, was his comment. A good cake for readers who live in the warm climate of Killarney.

BANANA

With the great voyages in Tudor and Stuart times, there was inevitably much talk of exotic plants and fruit. Gentlemen with gardening enthusiasm raised oranges, our first exotic, and poets wrote about idyllic islands of the West Atlantic, 'where the remote Bermudas ride'. Such places were a more luscious embodiment of the Golden Age than Virgil had ever imagined. His honey-dew dropping from trees and purple grapes thriving on the untended bramble were far outdone by the bananas and pineapples of reality.

> Such is the mould, that the blessed tenant feeds
> On precious fruits, and pays his rent in weeds.
> With candied plantains, and the juicy pine,
> On choicest melons, and sweet grapes, they dine,
> And with potatoes fat their wanton swine.

By candied plantains, I take it he meant bananas at their full peak of ripeness. Even if Waller had never been to Bermuda – there is argument about this – he could well have seen the first banana in England, at the shop of Thomas Johnson the herbalist, in 1633. It came from Bermuda, and Waller's poem was written in 1638.

Some of the first bananas to be grown here successfully were produced by Sir William Watkins Wynn, the 5th baronet not the original Welsh hero, at Wynnstay, his grand house in Denbighshire. He sent specimens to the Horticultural Society in August 1819. They were 'agreeable, luscious and acid flavoured' – and abundant. Certainly a great stalk, or bunch of bananas that move up like an enormous flower, in regular curving rows or 'hands', is something we do not often see now. My experience of these bunches has been in French groceries, where they sometimes hang from a hook so that customers can choose the 'fingers', the bananas, which they particularly fancy. Some people have claimed that the enormous bunches of grapes brought to Moses from the Promised Land were in reality great bunches of bananas.

Our first shipments of bananas – from the Canaries – docked in 1882. The American trade, from Hawaii, had been going for about twenty-six years by that time thanks to railroad, and to steamship even more as it could take the cargo rapidly to its destination in good state (in schooners, bananas rotted if the ship were becalmed). The next great step was transporting bananas in refrigerated ships. This began in 1901, the year that Elder and Fyffe started in London.

This firm started off with vigorous promotions. I have a little booklet of theirs, undated but an early production to judge by the illustrations. It has some good recipes, recipes that have survived, but as with many similar booklets today it stuns you. Bananas-with-everything rather than the thought that bananas are something delicious, a treat that has now been brought within the reach of most people. Something for the Watkins family in Croydon as well as for Sir William Watkins Wynn with his huge hothouses in Wales. I prefer the poetical chapter written for Edmond Richardin's *L'Art du Bien Manger* (1904, rev. ed. 1913) by the popularizer, Octave Uzanne, who loved what he was writing about, showed his preferences and attracted readers to share his enthusiasm.

Uzanne described the flavour of banana as a mixture of reinette apple and Bon chrétien pear; he sees it as a vegetable butter, ideal and perfect; the smell 'evokes in our senses exotic countries, perfumed Arabia, distant islands and tropical Edens'. He prefers bananas as a savoury dish, with béçhamel, or fried and served with roast or grilled beef: 'a chateaubriant with bananas becomes something exquisite, much to be recommended. Try it.' He loves grilled bananas – slash them in one place first – cooked until they are black from the heat: inside the creamy scented flesh is a real delicacy, to be eaten with a spoon.

The particular point of Octave Uzanne's piece was to try and get the French to take up this paradise fruit as the English had done. He describes a pudding which was all the rage in England – a 2-cm layer of strawberry jam in the bottom of a bowl, then banana slices, then a thick cloud of whipped cream. I remember versions of this at children's parties, and it was disgusting. Perhaps because of the bought jam and the bananas being too firm and pasty. Banana trifle is a more elaborate version, and it has its devotees.

My own impression of bananas is that the tiny ones from the Canaries are particularly delicious and sweet-smelling. Alas, one rarely sees them now. I once stated this preference in the *Observer* magazine, and was severely ticked off by one of the banana firms (not Fyffe's). And I was ticked off not because what I said was wrong, but because these days only five per cent of bananas imported are of this kind. I should not have been inciting the public to demand them. Lie down, public, wag your tongue thankfully for whatever Big Business deigns to give you.

Bananas certainly are big business. Old business, too, possibly the oldest of all as far as the fruit trade is concerned. In his wonderful, specialized book, *Bananas*, N. W. Simmonds gives a time scale of the way bananas spread until they circled the globe. He speculates that the banana was one of the first plants to be domesticated, tens of thousands of years ago, in South-East Asia.

With such antiquity, the naming of bananas, the botanical relationships, the varieties are much confused. N. W. Simmonds abandons the use of the two Latin names, *Musa sapientum* and *Musa paradisiaca*, and deprives us of our amateur division between plantain and banana: 'In East Africa . . . Europeans often use "Plantain" for any banana that is eaten in the cooked state assuming, presumably, that all such bananas must be starchy; in fact the vast majority of them are sweet at maturity and cooking is a matter of custom rather than necessity.' So much for the green banana, which must be one of the dullest things eaten by man.

Pace Mr Simmonds, I value the botanical Latin names for unscientific reasons. They are a shorthand for attractive myths and stories, and were thought up by Linnaeus, the man who named the plants. *Musa* he took from the Arabic *mouz*, banana, deriving either from the Sanskrit *moka* or the coffee town in southern Arabia. In case purists complained that he had strayed from classical restraints, he observed that it honoured the Muses and – but why? – a Roman doctor of the first century BC called Antonius Musa. *Sapientum* he probably took from Pliny, who knew about Alexander and bananas though he had never seen one. He wrote that the wise men of India – *sapientes Indorum* – live on it: 'the leaf is like birds' wings . . . The fruit grows straight from the bark, and is delightful for the sweetness of its juice.' *Paradisiaca*, another reference to the Arabs, who claimed that the banana was the Tree of Paradise, i.e. the Tree of the Knowledge of Good and of Evil. More confusion here, because Europeans – including Pliny – first called bananas figs: *figue d'Adam, figue du Paradis* to the French before they took to *banane, fico d'Adamo* in Italy, rather as the prickly pear is called the *figue de Barbarie* and *fico d'India.*

I have always thought that fig leaves were inadequate for the misuse they have been put to by Popes and Victorians. Perhaps banana leaves would have been a better choice? Young banana leaves, before the wind splits them to fingers.

HOW TO CHOOSE AND PREPARE BANANAS

Bernardin de Saint-Pierre, naturalist, friend of Rousseau, author of *Paul et Virginie*, and briefly head of the Jardin des Plantes in Paris (where

Douanier Rousseau painted his jungle scenes a century later), said that the banana was the most useful of all plants to man. From it he gains shelter and food, clothes and firing, and, finally, a coffin. I do not think that Monsieur Bernardin de Saint-Pierre knew about the banana beer that they make in Uganda.

This scope of usefulness is something we know little about. For us, bananas come in clumps of four, six or eight, and that is that. From the hundreds of different kinds, we can buy three or four. As they are never properly labelled, we never learn the differences between them. Only when the small Canary bananas are about, or deep red bananas in a speciality shop, do we notice anything different. Such red bananas as I have tried have been unexciting. From the streaked dark colour and the slightly orange flesh, I had hoped they would be even more aromatic than the small Canary, but mine weren't. A kind I should like to try is the small rounded lemon bananas that grow on Banana Island in the Nile. Josceline Dimbleby told me about them and I imagine that they belong to the small 'lady's finger' variety. They are a lemon colour, lemon-shaped and have a lemon flavour that is delightful.

The only choice we have, or ever will have it seems, is between stages of ripeness.

Green bananas are for boiling and frying, and serving as a bland vegetable. The same calorie count as potatoes, but less flavour I would say. The peel can be tenacious: slit down the length, along the angles, and cut and pull and persuade the strips away from the fruit inside. For some recipes, you slice the bananas across, soak the slices in salted water, then pop out the bits from the rings of peel.

Yellowish-green bananas can be used as a substitute for the green kind. They can also be kept until riper. In a warm house, they will soon turn the familiar yellow, then yellow patched with brown, which is when they begin to be worth eating. As they turn completely brown, they mellow to a translucent honey, ideal for fools, mousses, ice creams and for sandwiches: shops sometimes like to get rid of them at a keen price, so ask for a reduction. At this late aromatic stage, they are well set off by crushed praline or toasted almonds; walnuts go with them at a slightly earlier stage.

Rum is very much the recommended alcohol for bananas, but I would far rather have kirsch or gin. Coconut is a good companion, so are all kinds of dairy and farmyard things from yoghurt, cream, and fresh soft cheeses to eggs and bacon and chickens. Fruit to go with bananas can be apricots, lemon, lime and orange, strawberries and raspberries. Citrus juice prevents sliced banana going a nasty colour, but it is rarely a good idea to keep banana salads hanging about. Banana trifle is another matter, as the pieces are completely enclosed and protected from the air.

A new preparation of bananas can be bought from wholefood shops. It consists of thin slices of banana that have been deep-fried and rolled in honey. They are quite firm, not exactly crisp but pleasantly chewable. They are nicest when mixed with nuts, coconut slivers and raisins to make a pretty mixture with the name of Tropical Sunrise.

With vegetarian parents, I soon came across dried bananas which I have never managed to like. You just eat them as they are, or soak and cook them, or put them into dried fruit salads. They are very sweet, dark brown, and flabby-tough.

You may think that banana peel is useless, and on the whole I would agree with you. Or at least I would have done before reading *Bombers and Mash* by Raynes Minnis, about the domestic front in the last war. In the food chapter, she writes this: 'Another family who managed to get hold of a banana, after showing it to everyone and meticulously sharing it out, could not bear to part with the skin. They arranged it on the pavement and watched, from behind their curtains, the reactions of passers-by.' Candid camera, wartime style. One thing at least, I am sure that nobody slipped through not noticing that it was there.

FILLETS OF SOLE (WHITING) WITH BANANAS
FILETS DE SOLE AUX BANANES

A friend once told me that she had eaten a delightful dish of fresh haddock with banana cubes, lightly fried in butter. With that on my mind, I tried to find similar recipes for banana with fish, and came at last to a French version. In it, new potatoes and sticks of fresh coconut were deep-fried with the fish fillets. This you can do, but I prefer the method given below.

12 fillets of sole or whiting (six filleted small whiting can be used)
salt, pepper
3 large bananas
seasoned flour
coconut cream, p. 61
18 small new potatoes
3 medium carrots
grated zest and juice of a large orange
1 good teaspoon butter

Season the fish and, if you are using whole filleted fish, cut the whiting up the centre joint so that you end up with 12 triangular pieces. Peel and cut each banana into three across, then into nice little sticks. Sprinkle a tray with seasoned flour, and put the fish on it, skin or skinned side up. At the wider end of each piece, lay a bundle of banana sticks. Roll up and secure with cocktail sticks or tie with thread. Set aside in a cool place.

Prepare the coconut cream, and set it to heat through over the pan in which you boil the potatoes. Grate the carrots coarsely into a pan which already holds the orange zest and juice and the butter. Cover closely and cook briefly, so that the shreds are tender but still a little crisp. Strain off the small amount of carrot juice into the coconut sauce, and add a tablespoon of carrot, swirling it round.

Finally deep-fry the fish until it is nicely browned. Beware of overcooking. Arrange on a warm dish with potatoes and carrots. Pour on a little of the sauce, and serve the rest separately.

NASI GORENG

If you like Indian and Chinese food, you will certainly enjoy this dish from Indonesia (*nasi* means rice, and *goreng* fried). The pleasure of eastern dishes is the way that the different elements are kept separate and retain their identity, while combining into one delicious whole.

As with Chinese fried rice, the important thing is to pre-cook the rice, either the day before or several hours in advance. This is of course no problem in countries where rice is the staple food. Here it needs mild forethought.

Measure out 300 ml (10 fl oz/1 cup) basmati rice and cook it with 600 ml (1 pt/2½ cups) water. When done, fork it up lightly and spread on a plate to cool.

Assemble the following ingredients. Prepare the side dishes first, then the omelette strips, then the fried rice and finally the fried eggs. If you have a *guo* (wok), this is the best utensil for frying the rice: the ingredients fall back into the centre of the pan while you stir, so that everything is kept in an ordered perpetual motion.

SIDE DISHES
cucumber cut into 5 cm (2 in) strips
125 g (4 oz/½ cup) salted peanuts
6 bananas
60 g (2 oz/¼ cup) butter
juice of ½ lemon

OMELETTE
4 eggs
salt, pepper
butter

FRIED RICE
150 g (5 oz/1¼ cups) chopped onion
2 large cloves garlic, chopped
½ teaspoon dried chilli flakes, or a chopped fresh chilli minus seeds
5 anchovy fillets (or ½ teaspoon trasi*)*
1 teaspoon salt
groundnut or sunflower oil
1 tablespoon soya sauce
250–375 g (8–12 oz/1–1½ cups) roast pork/ham, diced
6 eggs (optional)

Put cucumber and peanuts into separate bowls. Peel and slice bananas in half lengthways, place them in a greased ovenproof dish, dot with butter and sprinkle with lemon juice. Bake at gas 7–8, 220–230°C (425–450°F), for 10–12 minutes, turning the pieces over after 5 minutes. Put them in another bowl.

To make the omelette strips, beat the 4 eggs with a little salt and pepper, and cook 3 or 4 thin omelettes in the usual way, keeping them slightly moist. Roll them up and slice across, so that they fall into a tumble of strips. Keep them warm.

For the fried rice, reduce the first five ingredients on the list to a paste in the blender with 5 or 6 tablespoons (½ cup) of oil. Scrape the mixture out into a *guo* or frying pan and cook steadily until the paste begins to bubble and brown. If you do not have a blender, chop the five ingredients as finely as possible, then fry them.

Stir in the cooked rice and fry until lightly browned, adding more oil if required. Keep stirring. Now add the soya sauce and cubes of meat, and cook for another 5 minutes, still stirring, until everything is well heated through.

Spread the rice over a large serving dish and put the omelette strips on top. Quickly fry the six eggs and put them among the omelette strips.

Place all the dishes in the centre of the table, for preference on a dumb waiter. Give each person a spoon and fork, a shallow bowl and lager or beer to drink.

FROG'S LEGS/CHICKEN BREASTS/ SWEETBREADS WITH BANANA RAITA

Kela – i.e. banana – *raita* is one of those delectable yoghurt salads that cool and set off well-spiced Indian dishes, pilaffs (*see* the papaya and ginger pilaff on p. 295) *tandoori* foods, and Mughal braised dishes. While I occasionally enjoy assembling and cooking Indian food, I do not see why certain embellishments such as the various *raita* should not be served with European meals, or eaten in European ways. This particular version makes a fine summer breakfast dish, and it sets off certain mild meats that we often serve with fruit in some form or another. The recipe below is based on Julie Sahni's recipe in *Classic Indian Cooking*, a book of careful instructions and good dishes, published in New York.

An important thing is not to add the banana to the *raita* more than an hour or hour and a half before the meal. The banana loses its bite, and the *raita* becomes very wet with its juice.

Frog's legs in this country – and very often in France, too – are bought deep-frozen. They taste like chicken, but have a delicacy and small size that makes them more appetizing. If you do use chicken – or turkey – breasts, cut the whole breast into its two parts then slice them through into escalopes and remove the skin: this means that three chicken breasts give you a dozen escalopes. Sweetbreads should be blanched, pressed and sliced in the usual way: soak them in salted water for an hour, drain and simmer them in light stock with lemon juice until they lose their pink raw look, drain again and remove knobbly skin and gristle, then press between two plates in the refrigerator for several hours or overnight. Although veal and lamb sweetbreads taste much the same, veal is far superior from the point of view of size and succulence, the consistency is better, too.

RAITA

2 heaped tablespoons (3 tbs) sultanas or seedless raisins
30 g (1 oz/¼ cup) blanched, slivered almonds
200 ml (7 fl oz/good ¾ cup) natural yoghurt
100 ml (3½ fl oz/good ⅓ cup) each soured and whipping cream
1 heaped tablespoon honey
1 medium-sized ripe but firm banana
pinch salt
6 cardamom pods

RICE
375 ml (12 fl oz/1½ cups) brown rice
1 tablespoon oil
1 heaped tablespoon butter

MEAT
30–36 frog's legs
or 1 kg (2 lb) sweetbreads, prepared
or 3 whole chicken breasts, sliced through into 12 escalopes
4 level tablespoons (⅓ cup) flour
2 level teaspoons salt
1 level teaspoon Cayenne pepper
1 level teaspoon curry powder, own mixture or good brand
clarified butter
chopped coriander, parsley, chives

To make the *raita*, pour boiling water over the sultanas or raisins and two thirds of the almonds: leave for 10 minutes. Toast the remaining almonds. Mix yoghurt and creams with the honey.Taste and add more honey if you like, but do not oversweeten. Stir in drained raisins and almonds. Slice the banana thinly and add that, with the salt. Take cardamom seeds from the pods and crush them. Add a little at a time to the *raita* – you may find that half is enough for your taste. Turn into a serving dish and scatter with the toasted almonds. Chill for an hour.

Cook the brown rice in a large pan of salted boiling water, with the oil which prevents the pan boiling over. Simmer for 30 or 40 minutes until tender, drain and rinse with hot water. Mix with the butter and pile up on a warm serving dish. Cover lightly with butter papers and keep warm in the oven.

Pat the meat dry with kitchen paper. Mix flour, salt, pepper and curry powder, and turn the pieces of meat in it. Fry them in clarified butter, not too fast; they should end up a pinkish-gold colour. Frog's legs and chicken escalopes do not take long – turn after 3 or 4 minutes, then test after 7 minutes' cooking time. Sweetbreads can stand a slightly hotter temperature, and longer cooking – time will depend on the state they are in after blanching.

Arrange the pieces up against the pile of rice and round it. Scatter with a little coriander, not everyone likes the taste so go carefully, and then with parsley and chives. Serve with the *kela raita*.

Note When you are frying bananas to serve with chicken, turn the pieces first in the seasoned flour of this recipe. Then fry them in butter, clarified if possible.

BANANA AND BACON ROLLS

A good family supper dish, that can easily be made by children.

For each person allow a medium-sized banana and a good long rasher of streaky bacon (or two short ones). You will also need a slice of bread large enough to take the banana cut in half, and butter to fry it in.

Cut the peeled bananas across the centre. Stretch the rashers of bacon, after cutting the rind away, and divide it into two lengths. Wind a length round each piece of banana, skewering it in place with a cocktail stick.

Grill the bananas for 8–10 minutes, turning them from time to time, until the bacon is brown and crisp at the edges, and the banana soft.

Meanwhile fry the bread. Place the banana rolls on top, remove the cocktail sticks and serve.

BANANA MEAL-IN-A-GLASS

When we were ill as children, or convalescent and not feeling much like a proper meal, my mother used to make us a meal-in-a-glass. When we were little, she would put in ice cream, as we grew up she added a jigger of sherry or rum or brandy. Nowadays with liquidizers and processors, you can make the drink in seconds. It's ideal if you are working and don't want to stop for long.

FOR EACH PERSON
1 large egg
1 rounded tablespoon cottage, curd or full fat soft cheese
½–1 very ripe banana
250 ml (8 fl oz/1 cup) milk
brown sugar or honey
1 scoop of vanilla ice cream or a glass of sherry, rum or brandy

Blend or process the first four ingredients until very smooth. Add brown sugar or honey to taste, and whisk again. Put in a couple of ice cubes and the ice cream or alcohol.

If you want to prepare the drink in advance, and store it in the refrigerator until later in the day, keep back the ice cubes, ice cream or alcohol, and add them just before you serve it.

BANANES BARONNET

In the Béarnaise section of *L'Art du Bien Manger*, 1913 edition, Edmond Richardin sandwiches this delightful banana dish between pigeons with globe artichokes and partridges with ceps and garlic.

Choose ripe yellow bananas patched slightly with brown (this means they are creamy but still firm enough to slice). Peel and slice up one for each person, or a little less than that, into a bowl. Sprinkle with a little lemon juice before adding anything else. Pour over a tablespoon of cream and a teaspoon of the best kirsch from the Vosges for each banana. Sprinkle with sugar and turn the whole thing carefully so that each slice is bathed in the sauce.

If you want to use whipping cream, beat it until it is very thick but still pourable. Not stiff.

BANANAS WITH APRICOT SAUCE

The simplest of all banana recipes – if you have on a hot or moderate oven, you can bake the bananas until they are soft instead of boiling them. If you are lucky enough to find small Canary bananas, buy two per person. One of the giant-sized kind that you usually see is enough for most people at the end of a meal.

4–8 bananas, left unpeeled
250 g (8 oz/¾ cup) apricot jam
4 tablespoons (⅓ cup) water
apricot brandy or orange liqueur to taste

Make the sauce first, and keep it warm. Simmer jam and water until smoothly amalgamated. Sieve it into a sauceboat and stir in the brandy or liqueur to taste. Keep the sauceboat warm in a water bath.

Bring a large pan, half full of water, to the boil. Put in the *unpeeled* bananas and boil for four minutes (just like boiling an egg).

To eat the bananas, each person peels away a strip of skin and opens it out a little – the most delicious smell rises with the steam – to make room for some apricot sauce.

GRILLED BANANAS

Red bananas in particular are often grilled. When they are black and splitting, serve them with lemon juice or apricot sauce.

BANANAS WITH ORANGE AND CARDAMOM SAUCE

Justly or unjustly, I have come to associate certain cookery writers with particular flavourings. Elizabeth David with basil and coriander seed. Claudia Roden with dried mint and cumin (especially in that lovely mixture of toasted sesame seeds, coriander, hazelnut and cumin known as *dukkah*). Cardamom is Josceline Dimbleby's spice, and she uses it to delightful effect in pudding recipes such as this one, and in two others elsewhere in the book.

Peel and halve six bananas lengthways. Fry them gently in 60 g (2 oz/ $\frac{1}{4}$ cup) butter until soft and patched with golden brown. Turn them once. Put the bananas into a serving dish.

Off the heat, put into the buttery juices, 5 crushed pods of cardamom seeds (remove the papery pods first), the juice of 2 large oranges and 1 lemon and 2 rounded tablespoons ($\frac{1}{4}$ cup) of soft brown sugar (I use the light kind). Stir over the heat until bubbling and slightly thickened. Pour in some brandy, rum or orange liqueur to taste. Pour over the bananas and serve with cream.

Note I remove most of the thin rind of one of the oranges with a zester, which reduces the rind to delicate shreds. They go into the sauce with the cardamom etc.

BANANA DUMPLINGS

This way of cooking bananas brings out their quality of aromatic richness that remains hidden and unsuspected in the huge bananas of modern commerce when eaten raw. Although no sauce is necessary as their own juice comes bubbling out of the dumpling crust, the slight sharpness of the lemon and gin sauce on p. 208 sets them off well. As to pastry, puff, flaky or shortcrust are all suitable but I prefer the lightness of the first two: the quantity given below is the final weight not the weight of the flour. Use just ripe bananas that are firm enough to handle easily.

6 bananas
2 heaped tablespoons ($\frac{1}{4}$ cup) sugar
2 heaped teaspoons cinnamon
$\frac{3}{4}$ kg ($1\frac{1}{2}$ lb) pastry
beaten egg or light cream to glaze

Peel the bananas and roll them in the sugar mixed with the cinnamon, spooning it over the inside surface of any curve so that the whole of the fruit is lightly coated.

Cut the pastry into six equal pieces. Roll out the first one into a rectangle and enclose the banana. The easiest way to do this, I find, is to place the banana on the pastry diagonally, then flip a corner over the curved centre part, and roll the banana in the pastry croissant-style. Brush the edges with water and press them firmly together. Make a few slits in the top of the pastry, and lay the whole thing on a baking sheet that has either been lightly greased (shortcrust) or moistened (flaky, puff). Brush over with egg or cream. Repeat with the remaining bananas and pastry. You will have quite a number of pastry trimmings left over by the time you have cut away the surplus thicknesses and odd unnecessary corners: if you are skilful with your fingers, you may find that you can manage with less pastry than I have stipulated above.

Bake at mark 7, 220°C (425°F) for 15 minutes, or until nicely browned. With shortcrust pastry, you may like to lower the heat to mark 5, 190°C (375°F) at this point. These dumplings need far less time than apple dumplings since bananas cook fast.

VARIATION On the island of St Kitts in the Caribbean, they serve these dumplings, made half-size, with the sharp lime sauce on p. 208. Lemons can be substituted for limes, but the flavour is far less good.

BANANA FRITTERS

1) Cut bananas across into three, then halve each piece. Soak in rum and sugar for an hour. Drain and wrap in thinly-rolled puff pastry, to make tiny packages. Deep-fry and serve with the coconut cream below.

2) Cut bananas across into two and halve each piece. Dip into a rum-flavoured batter and deep-fry. Serve with apricot sauce (p. 58) or coconut cream.

3) Mash 2 large bananas with a generous 30 g (1 oz/2 tbs) sugar, and a generous pinch both of cinnamon and nutmeg. Make a thick batter of a large egg yolk, 100 g (3½ oz/scant cup) flour, and equal quantities of water and milk to mix. Stir in the bananas and add a pinch of salt. Fry tablespoonfuls of the mixture in 1 cm (½ inch) of oil, until nicely brown on both sides. Serve hot with chilled coconut cream.

These fritters can be served with bacon, chicken etc. Cut down the sugar and increase the salt slightly. With this kind of fritter, always try out an experimental spoonful first and taste for seasoning and consistency.

More sugar? Or more salt? A little more flour, a little more liquid? A different spice next time?

COCONUT CREAM SAUCE If you can buy blocks of solid coconut cream, all you have to do is to break it up and dilute with water or cream, or both. By heating it or mixing it in the blender, you get a smooth, very white consistency.

If all you can get is desiccated coconut, do not despair. Heat a cupful with a cupful of single cream, to just below boiling. Whizz in blender or liquidizer if possible, then leave to cool. Sieve into a basin, adding a cupful of boiling water towards the end. Add sugar or salt to taste, according to the dish you intend the sauce for.

OCTAVE UZANNE'S BANANA OMELETTE

For two people, allow 2 good-sized bananas and 5 eggs, plus sugar and butter.

Slice the bananas and toss them in granulated sugar. Melt a large knob of butter in an omelette pan, and put in the banana slices. Leave them to cook and caramelize gently – giving an occasional stir – while you separate the eggs. Beat the yolks together and whisk the whites until firm.

When the bananas are golden and buttery, fold the yolks into the whites. Bring the banana slices together into the centre of the pan, and pour round the omelette mixture, raising the banana so that some goes underneath.

When the base is firm and nicely coloured, the top still bubbly and moist, flip the omelette over on to a warm dish and serve at once.

Note If you are nervous about the omelette mixture sticking, cook it in another pan and put the bananas with their juice into the centre when it is almost ready.

PLUM-BLOSSOM AND SNOW

In northern China, plum trees break into flower early while the snow is still resting on their branches. In Prince Genji's Japan, a next-morning letter attached to such a snowy flowering branch was a particularly delicate way for a lover to thank the lady with whom he had spent the night. The image symbolized the coming of spring, the first hint, the idea that in a chilly world love may blossom with delicate joy.

This pudding which I have adapted from a recipe in *Chinese Regional*

Cooking by De-Ta Hsiung reminds me both in title and substance of the French favourite pudding, *oeufs à la neige*.

2 fine eating apples
2 ripe bananas
100 g (3½ oz/scant ½ cup) sugar
3 level tablespoons (¼ cup) cornflour
4 tablespoons (⅓ cup) water
2 large egg yolks
250 ml (8 fl oz/1 cup) milk
2 large egg whites
2 rounded tablespoons (¼ cup) caster sugar

Peel, core and slice an apple very thinly. Arrange on the base of a gratin dish. Quickly peel and slice a banana and arrange the slices firmly on top (speed prevents discolouring). Repeat with the other apple and banana. The apple slices must be very thin, as they do not get much chance to cook through; you could slice them on a wide grater blade, the kind used for cucumber.

Mix sugar and cornflour to a paste with water and yolks. Bring milk to the boil, and whisk into the paste. Return to pan and cook until the mixture is thick and tastes cooked. Pour very hot on to the fruit and smooth it out.

Whisk whites stiff, add half the caster sugar and whisk again, then the rest of the sugar and whisk until satiny. Pile on to the pudding. Bake in the oven at gas 6–7, 200–220°C (400–425°F) until nicely browned and set – about 10 minutes or a little longer. Eat hot or warm.

Enough for 4–6, depending on the rest of the meal.

BANANA TRIFLE

No more than a normal trifle with a layer of bananas through the middle, but it has been a particular favourite at Christmas parties for so many years that I could not leave it out. In summertime, you might substitute cherries, strawberries or raspberries for the bananas, or lightly cooked peach or apricot slices.

When you measure out the ingredients, have extra cake or macaroons in reserve. Your large glass bowl may be bigger or squarer than mine, and need odd corners filling in. There is no reason, for instance, why you should not use both macaroons and cake, or slices of home-made sponge roll with apricot or any other appropriate jam in it. Again, wine can be what you have – we usually bring a small store of Muscat de Frontignan back from France that we save up for Christmas, but other muscat wines can be used instead, or Sauternes.

6 large macaroons or pound of sponge cake slices
kirsch or apricot brandy or fruit eau de vie *of an appropriate kind*
sweet dessert wine
500 ml (15 fl oz/scant 2 cups) single cream
half a vanilla pod, split
2 large eggs
2 large egg yolks
caster sugar from vanilla pod jar
strawberry, raspberry or apricot jam
2–3 ripe, firm bananas
pared rind and juice of a lemon
300 ml (10 fl oz/1¼ cups) double cream
pinch nutmeg
toasted almonds and candied peel for decoration

Make a ground-floor layer of macaroons or cake in the bottom of a large glass bowl. Pour over it 3 tablespoons (scant ¼ cup) of whichever spirit you use, and 150 ml (5 fl oz/⅔ cup) wine. Give this a chance to soak in, then add more wine if you think the macaroons or cake are on the dry side.

Make the custard by boiling the single cream with the vanilla pod. Beat eggs and yolks together; pour on the boiling cream, whisking, then return to pan and stir over a low heat until thickened. Sweeten to taste. Cool, remove the vanilla pod.

While the custard cools, spread a layer of jam over the sponge cake or macaroon layer. Then slice the bananas – after peeling them, of course – and arrange the pieces closely together on top.

Spoon the cooled custard carefully over the banana, starting with the outside, then moving in. This prevents the banana slices being dislodged.

At this stage, it is a good idea to leave the trifle in the refrigerator overnight, for everything to bed down together.

At the same time, put the lemon rind and juice in a bowl with 2 tablespoons (3 tbs) of whichever spirit you used, and 8 tablespoons (⅔ cup) of the wine. Next day, strain the juice into a large bowl, and stir in 60 g (2 oz/¼ cup) caster sugar. Still stirring, pour in the cream slowly, and add the nutmeg. Beat with a whisk until the syllabub holds its shape – do not overbeat or the cream will curdle.

Pile on to the trifle, either right across or in a circle round the edge. Decorate with almonds and peel, or whatever else you like to use. At Christmas, people seem to go wild with glacé cherries, and angelica, in emulation of holly I suppose, but quieter decorations are more appropriate to a real trifle of this kind, made with the best of everything.

BANANA ICE CREAM

A mild but intriguing ice cream, set off by pieces of almond praline.

740 g ($1\frac{1}{2}$ lb) bananas
juice of a lemon
200 g (7 oz/scant cup) sugar
1–2 tablespoons white rum or kirsch
500–600 ml ($\frac{3}{4}$–1 pt/2–$2\frac{1}{2}$ cups) double or whipping cream
almond praline, p. 477

Peel, break up and process the bananas, or liquidize them, with the lemon juice. Dissolve the sugar in 200 ml (7 fl oz/a good $\frac{3}{4}$ cup) of water over a low heat, bring to boiling point and boil hard for 3–5 minutes, until there is 300 ml (10 fl oz/$1\frac{1}{4}$ cups) syrup. Add to it 200 ml(7 fl oz/a good $\frac{3}{4}$ cup) of cold water and add, with the rum or kirsch, to the banana.

Whip the cream until it is very thick but not stiff. Fold into the banana mixture carefully. Freeze in the usual way, with a sorbetière, ice bucket or freezer, following the appropriate instructions; *see also* pp. 473–6.

When you serve the ice cream, scatter it with the praline, broken up into small pieces, not crumbs.

BANANA SANDWICHES

Particularly good if you use walnut bread (*see* p. 462). With wholemeal or plain granary, add chopped walnuts to the mashed bananas. People who are good at cutting bread and like making sandwiches, can serve them instead of biscuits with *coeur à la crème* (*see* p. 470) or vanilla ice cream (p. 473).

250 g (8 oz/1 cup) lightly salted Danish butter
1 heaped tablespoon honey
1 teaspoon lemon juice
2–3 bananas
30 g (1 oz/$\frac{1}{4}$ cup) walnuts, coarsely chopped (optional)

Leave the butter in the kitchen or a cupboard, so that it has a chance to soften, then beat it to a cream adding the honey and the lemon juice drop by drop. Try to keep the mixture creamy, but should it curdle from over-vigorous beating, put it in the refrigerator until it is firm but not hard and then beat it again to make it smooth. Taste and add extra honey or lemon, if you like.

Butter thin slices of bread with this mixture. Then mash the bananas roughly, do not reduce them to a smooth purée. You can leave ripe bananas in slices, but hard firm ones will drop out of the sandwiches when people pick them up.

Add walnuts to the banana mash, if the bread is plain, and use to sandwich the slices of bread together.

BANANA AND WALNUT BREAD

I would say that banana bread came here from the West Indies when bananas returned after the last war. In the six years from 1939–45, they had completely disappeared. Young children were so amazed by this bizarre fruit that when a few began to filter through, they would eat them skin and all. I was away at school during the war, and read in a magazine one day that you could make reasonable ersatz banana by smashing up cooked parsnip and flavouring it with banana essence. We persuaded the cooks to make this for a special tea. The whole school sat down to 'banana' sandwiches in an awed hush. One mouthful and there was a riot. The cooks had forgotten to put in the banana essence. It was reckoned to be all my fault – indeed I should have checked – and I was in trouble for weeks afterwards. Food was a serious matter in those days. Culinary disasters couldn't be laughed off. I tried laughing, but it made things worse. They thought I'd arranged it on purpose, as a joke.

There are many recipes around these days for banana bread, with more and less bits and pieces added to the basic mixture. This simple recipe is my favourite, because it needs no butter when it is newly made, and for a day has a crust that is crisp and a buttery golden brown. Put it in a tin or box and the crust softens.

125 g (4 oz/½ cup) butter
175 g (6 oz/¾ cup) caster sugar
2 large eggs
2–3 large bananas (500 g or 1 lb approx.)
60–90 g (2–3 oz/½ cup) walnuts, coarsely chopped
250 g (8 oz/2 cups) self-raising flour

½ level teaspoon salt
¼ level teaspoon cinnamon or mixed pudding spice
¼ level teaspoon bicarbonate of soda

Cream butter and sugar until fluffy, and add the eggs, beating them in well. Peel and mash the bananas with a fork: do not use an electric chopper or processor, or the bananas will become too liquid. Add them to the mixture, with the walnuts. Sift together the dry ingredients and fold them in, using a metal spoon.

Line a 23-cm (9-inch) loaf tin with Bakewell paper. Brush it with melted butter. Spoon in the mixture, levelling it out. Bake at gas 4, 180°C (350°F) for an hour. Test with a warm skewer, and give it longer if the skewer does not come out clean.

Leave the cake in the tin for 15 minutes when it comes out of the oven, then turn it on to a wire rack to cool completely.

When the cake is too dry to eat with pleasure on its own, spread the slices with brandy butter.

BRANDY BUTTER OR HARD SAUCE With an electric beater or processor, brandy butter takes seconds to make (as long as you remembered to take the butter out of the fridge in good time).

Cream 125 g (4 oz/½ cup) slightly salted butter, add half the weight of caster or icing sugar, and a tablespoon or more of brandy. Some cinnamon is also a good idea.

If you want to make the butter extra light, fold in a stiffly beaten egg white.

Rum butter is made in the same way, using brown sugar and rum. You can also use brown sugar in the banana bread recipe – but I find it much too strong and dominating. The same with rum.

BARBARY FIG, see PRICKLY PEAR

BILBERRY, BLAEBERRY and WORTLEBERRY

I discovered bilberries as a consolation after dragging up Lake District hills in the wake of my parents. After climbing up the steep slope behind the old Langdale Hotel, it was a relief to flop down with sandwiches by the tarn under Pavey Ark and make a dessert of the bilberries that grew up its slopes on the way to the Pikes. That was in wartime. Looking back on those holidays, I am amazed that never in our small hotel, or in any baker's shop, did we see the least sign of a bilberry tart.

As far as polite cookery is concerned, most English people, if they have even heard of bilberries, seem to take the attitude of an eighteenth-century traveller in the Highlands who wrote: 'The only fruit the natives have that I have seen, is the bilberry which is mostly found near springs, in hollows of the heaths. The taste of them, to me, is not very agreeable; but they are much esteemed by the inhabitants, who eat them with their milk.' In other words, bilberries were fit only for semi-starving peasants.

In Ireland, the peasants at least had some fun with bilberries. On the last Sunday of July, or the first of August, they had a great feast, a harvest festival of a kind. Everyone went up to the hills and the young ones in particular picked bilberries into special little rush baskets. To English and Scottish settlers, the Sunday was Lammas Sunday. To the Irish, according to which part of the country they lived in, it was 'Fraughan Sunday, Bilberry Sunday, Blaeberry Sunday, *Domhnach Deireannach* (the Last Sunday, i.e. of summer), Whortleberry Sunday and along the Gaelic-speaking parts of the western seaboard and along the Dingle peninsula (also Irish-speaking) the old pagan name of *Domhnach Cromh Dubh.* According to legend, Cromh Dubh – or the dark bent one – a pagan god was dominant in Ireland until the coming of St Patrick. He was overcome by the saint but his name was still used in living memory.'* I suppose that

* For this information, I am grateful to Brid Mahon, folklorist on the staff of the Department of Irish Folklore, University College, Dublin.

Cromh Dubh might be said to be the patron saint of bilberries, dark with bilberry juice and bent from crouching to pick them.

I wonder if St Columbanus remembered Cromh Dubh when he got to Gaul and founded his three monasteries in the Vosges? He was the greatest of the Irish apostles to Europe, a stern tall man. He first settled with his monks in the ruins of a Roman fort at Annegray. From a recent biography, I gather you can still walk up to his cave and holy well and find bilberry plants along the path which are called the 'brin belu de St Columban and local tradition recalls that when weak with periods of long fasting, he was restored by eating the crimson berries'. They were I suppose the *brimbelles* that people in the Vosges go out to pick on Sundays, in family parties, in the summer.

Bilberries grow on high heaths where the weather can be uncertain, as we have found on our journeys to the source of the Loire. As Gide says in *Les Nourritures Terrestres*: 'I remember picking these mountain bilberries in the snow one extra cold day.'

What either the Irish, the Scottish, the Welsh or the English lacked was a strong enough love for bilberries to invent some implement to make the picking easier, so that they could become a proper feature of summer eating, as they are in France. We were in the Auvergne last spring, at Chaise-Dieu, the highest monastery in France, squatting low – like bilberry plants – on the undulating hills to protect itself from the wind. There in the shops they sell bilberry sweets and liqueurs, and *peignes à myrtilles* (bilberry combs). They are shaped like a shovel, wooden sides and back, with the base made of thick parallel iron wires that curve up at the free end to form a comb. You scoop the shovel through the low bushes with an upward movement, so that berries and leaves fall back and can easily be tipped into your basket. Before you do this, you blow along the shovel so that most of the leaves fly away. There is still some picking over to do, tiny stems, little clusters of leaves, but the whole business is quicker and more efficient than picking the berries one by one.

The liqueur, incidentally, has a new popularity as a substitute for cassis in the aperitif known as *vin blanc-cassis* or Kir (*see* p. 86). No prizes for guessing the name – Myr. This new favourite joins the standard summer offerings of bilberry pies and tarts, bilberry sorbets, bilberry jams and bottled bilberries. Whenever I have bought any of these bilberry preserves, they have been first class, not too sweet, with the right unmistakable taste. I can also recommend the Lenzburg canned bilberries, and bottled Krakus bilberries from Poland.

Do not miss the chance of bilberries in France from the unfamiliarity of the names, which do vary a lot. If you have ever driven down the autoroutes, you will have seen notices for picnic and resting-places with loos, saying Aire de Somewhere or Other. Now *aire* was a threshing floor in a farmyard, and by extension an uncluttered open space or patch of ground. Bilberries which grow on such ground, in their different varieties, are therefore known as *airelles*. However, in general speech and in cookery books, my experience is that the commoner name is *myrtilles* (the botanical name is *Vaccinium myrtillus*).

Other local names are *abrêtier, vacier, raisin des bois, brimbelle* in the Vosges as I have mentioned, and *teint-vin* – perhaps a reference to its use in 'improving' poor wine, and certainly an acknowledgement of its staining power. Our names revolve round the blueness of the fruit – bilberry and blaeberry, of course, but less obviously whortleberry or hurtleberry, the latter from its resemblance to the dark blueish tone of a bruise, i.e. a hurt. In the south-west of England, the word picks up a w – whoam meaning home is another example. It was whortleberry for west-country Coleridge, even when he was in blaeberry Lakeland country.

> At my feet,
> The whortle-berries are bedewed with spray
> Dashed upwards by the furious waterfall.

HOW TO PREPARE BILBERRIES

Pick over the bilberries, removing the stalk and any leaves, then rinse briefly.

BILBERRY SAUCE
SAUCE AUX MYRTILLES

A sauce to serve with Welsh and other hill lamb, or with the tiny, dark-tasting lamb from the Shetlands. It can also be served with duck, especially if you grill the breasts like steaks. Or cook them in a pan without extra fat over a moderate heat: preheat the heavy pan, place the duck breasts skin side down and leave for 10 minutes. Turn them over and cook them for 3 minutes. Keep them hot between two warmed dishes, giving them a chance to relax so that inside they will cut to an even pink. This is Michel Guérard's method, and it works well. Slice the breasts thinly to serve them, season with salt and pepper, and fan them out on separate heated plates or on a serving dish. Put a little of the sauce beside them, with a few berries and a mint leaf. The rest goes into a bowl for people to help themselves. A few new potatoes can go with the meat.

250 g (8 oz) bilberries
150 ml (¼ pt/⅔ cup) dry white wine
150 ml (¼ pt/⅔ cup) duck, giblet or lamb stock
sugar
6 tablespoons (½ cup) white wine vinegar
chopped fresh or dried mint
2 tablespoons (3 tbs) crème de myrtille or cassis
60 g (2 oz/¼ cup) butter
lemon juice, salt, pepper
a few small fresh mint leaves

Cook bilberries with wine, stock and a level tablespoon sugar. When tender, remove a tablespoon of the berries and set aside. Liquidize or process the rest with their liquid until very smooth.

In a tiny pan, heat the vinegar with 2 teaspoons sugar until they boil to a syrupy brownness. Add to the sauce with the mint and the crème de myrtille or cassis. Pour fat from the meat juices in the roasting pan, or from the grill pan if there are any, and add to the sauce. Reheat gently and whisk in the butter. Do not allow to boil from now on. Taste and correct the seasoning with lemon juice, salt and plenty of pepper. Add a little more sugar, but do not make the sauce sweet. Stir in the tablespoon of bilberries. Float the mint leaves on top of the sauce in its bowl.

NOISETTES OF LAMB WITH BILBERRIES
NOISETTES D'AGNEAU AUX MYRTILLES

The combination of two mountain ingredients is, I would have thought, obvious. Only two examples, though, have come my way. The earlier – 1863 – from the Reverend Sabine Baring-Gould's book of his travels in Iceland where he mentions 'hot mutton flavoured with whortleberry jam'. The later – 1979 – from our own experience at the Hôtel de la Marine, which is almost under the Pont Tancarville, right on the edge of the Seine with its barges.

The items of the dish were set out on the plate, new style. Very pretty. Three little *noisettes* of lamb, brown outside, pink within, and bilberry sauce neatly spooned over part of them. At the side, three small new potatoes, and an artichoke base holding julienne strips of lightly cooked carrot and celery.

I thought the artichoke base was the only false note. And how did the chef get hold of bilberries? I found a possible answer, when we reached

Trôo and went shopping in Montoire. Tins of Lenzbourg *Myrtilles au sirop* were on the shelf of the best grocery. Large tins holding 560 g of fruit and 190 ml of syrup, 850 g (1 lb 14 oz) in all. I bought one and tasted the bilberries. They were, I was sure, the same. All that was needed was to heat the fruit in its not-too-sweet syrup, and thicken the juice very lightly with potato flour (*fecule*), arrowroot or cornflour. It needed a little consistency, that was all. If I had had fresh bilberries, 375 g (12 oz) cooked with a little water and the minimum of sugar would have been enough for six.

Noisettes are thick slices cut across a trimmed, boned and rolled piece of loin of lamb. Fry them in clarified butter, fairly fast to brown them, then more slowly to cook them to pale-pinkness.

Strips of celery and carrot should be lightly cooked in boiling salted water, or steamed, so that they are still a little crisp; a tablespoon of each per person is enough. Choose small new potatoes of an even roundness, cooking a few extra, in case some of them come to disaster.

COUPE GERBIER

The high source of the Loire is in a cowshed huddled down below the road and the sugarloaf rock known as Gerbier de Jonc. A climbable, shaly rock with a flat top from which you see a great circle of volcanic peaks and blue Cevennes distances, horizons dropped one behind the other, their sharp edges dissolving down into mist. Solace on the short steep climb is provided by bilberry bushes. The meadow at the foot has bilberries, too, and small bright flowers. The Chalet Hôtel, open in summer, proclaims 'Coupe Gerbier' on a board.

Our first visit was in April and the weather was so bad, so thick with cloud and sleet that we crept slowly up the road. All heights were lost. Meadows crowded with wild daffodils rapidly sloped into the general grey of the blizzard. The Chalet Hôtel was shut, so was the café below, and the cowshed. No stalls selling mountain charcuteries, goat cheeses and little round bilberry sweets. We groped towards the dirty patch of mud that the French Touring Club considers to be the source, and looked up at the Coupe Gerbier board with its three scoops of purple, red and creamy-fawn ice in a stemmed cup with a triangular wafer sticking out of it. We shook the snow out of our eyes and drove away.

On our second visit, we discovered that the three ices, symbol of local fruits of the earth, were bilberry, raspberry and chestnut. The fruit ices were simply made, the bilberry a little gritty from the skins. This is the way I make them:

BILBERRY (SORBET À LA MYRTILLE) Cook 375 g (12 oz) bilberries lightly, with a little water. Liquidize or process, then sieve into a bowl. Add icing sugar to taste, and lemon juice to bring out the flavour. Freeze at the lowest possible temperature, stirring up the ice when it is half-frozen.

The brands of bottled and tinned bilberries that I have tried (Krakus, Lenzbourg or Lenzburg, and bottled bilberries bought at the Gerbier stalls), are not over-sweetened. They can be used successfully. No cooking is required. Just process or liquidize and sieve, adding small doses of sugar as the syrup of the fruit will have added its mild sweetness. Weigh out 500 g (1 lb) approximately of fruit and juice to start with.

RASPBERRY (SORBET À LA FRAMBOISE) Liquidize or process 375 g (12 oz) of the best raspberries you can get hold of. Sieve, then flavour and finish as above.

CHESTNUT (GLACE AU MARRON) Put into a pan 150 ml (5 fl oz/⅔ cup) each of soured and whipping cream. Bring to the boil and whisk into 2 large egg yolks; tip mixture back into the pan and cook for 2 minutes slowly, without boiling. Off the heat, beat in a small 250 g (8 oz) tin of crème de marrons, i.e. sweetened chestnut purée. Taste for sweetness. To bring out the flavour, I add a scant ¼ teaspoon of salt and the juice of a small orange, adding both gradually. Freeze at the lowest possible temperature until almost firm: beat the 2 egg whites until stiff in the bowl of an electric mixer, then add the chestnut ice in spoonfuls, whisking all the time. Return to the freezer to chill until firm.

CHALET TARTS

TARTELETTES AUX MYRTILLES

On our second visit to the source of the Loire, a day of shining beauty and sun, the Chalet Hôtel was open. We were given plates of the good local charcuterie and the triple ice-cream, *coupe gerbier*, the preceding recipe. As I went to pay the bill, the young owner insisted that we try his little bilberry tarts. As he filled the baked pastry cases, he remarked, 'There's no flour in the *crème patissière*, Madame.' This was true. The mixture of crispness, creaminess and the amiable tartness of wild bilberries was just right.

Returning home, I made several experiments and found that a simple mixture of petits Suisses and cream with vanilla sugar was the closest I could get to the Chalet recipe. If you cannot find petits Suisses, you could use Philadelphia cheese, though it can feel too sticky and cloying.

Make 18 tartlet cases of *pâte sucrée* based on 125 g (4 oz/1 cup) flour. Prick them with a fork and bake blind to a crisp golden colour with brown – not burnt – edges.

Beat 6 petits Suisses with 3 generous tablespoons (scant ¼ cup) soured cream and 3 tablespoons (scant ¼ cup) whipping or double cream. When thick, flavour with sugar from the vanilla pod jar.

Sprinkle 250–275 g (8–12 oz) bilberries with vanilla sugar to taste.

Fill the tarts just before serving them if possible. If this makes difficulties, you can put in the cream cheese mixture in advance, but keep the bilberries back until the last moment. You want to avoid their juice making the pastry messy and soft.

BILBERRY TART FROM ALSACE

TARTE AUX MYRTILLES

The Vosges mountains, like the Massif Central, are much visited in August and September by families on a Sunday search for bilberries. When they get home and pick the fruit over, they are often disappointed at the tiny quantity. As we are likely to be, picking by hand in the Lake District or Derbyshire.

The thing to do is to find a tart tin into which the bilberries you have picked fit in a nice close thick layer, coming a third of the way up the tin. Then tip them back into their bowl.

Line the tin with sweet shortcrust pastry. Scatter the base with bread, biscuit or cake crumbs thinly. Then put in the fruit. Bake it for 15 minutes at gas 6–7, 200–220°C (400–425°F).

Meanwhile, for about 500 g (1 lb) bilberries, beat up together 150 ml (5 fl oz/⅔ cup) whipping cream, 100 g (3½ oz/scant ½ cup) sugar and 2 egg yolks. Pour over the bilberries and return to the oven until the cream is set – 10 to 15 minutes.

Note If your quantity is particularly small, make tartlets, and reduce the cream ingredients by half.

BLAEBERRY PIE

In Derbyshire and Yorkshire, bilberry and mint pies are still eaten in some families. The mixture is a good one, but you can of course leave out the mint if you wish.

¾–1 kg (1½–2 lb) bilberries
175–250 g (6–8 oz/¾–1 cup) sugar
1 rounded tablespoon cornflour (3 tbs) (see *recipe)*
1 heaped tablespoon chopped mint leaves
shortcrust or flaky pastry
egg white
caster sugar

Taste the fruit, and reflect on its sweetness and juiciness. Weigh out the sugar according to the first, and decide whether to include or leave out the cornflour according to the second (some bilberries in hot summers can be on the dry side).

Mix the sugar with the cornflour if used and the mint. Put the fruit into a pie dish, layered with the sugar mixture. Cover with pastry. Brush with egg white (or water) and sprinkle evenly with caster sugar. Bake for 15–20 minutes at gas 7, 220°C (425°F) until the pastry is set, then lower the heat to gas 5, 190°C (375°F) and leave for another 20 minutes. Serve hot with a large jug of cream. Or, if you prefer the west-country style, with plenty of clotted cream.

Note This is the standard way of making any fruit pie. The mint addition is worth trying, too, with blueberries and black currants.

HAWORTH BILBERRY PIE

Florence White gives this version of bilberry pie in *Good Things in England*, first published in 1932. She helped to form the English Folk Cookery Association, which had many correspondents, all sending in recipes with the aim of defining a national cookery. Her book is the record of the best things she found. This dish is dated 1867: 'These pies we may be sure were enjoyed by the Brontë sisters, at Haworth Parsonage. Dr Fernie' – presumably the author of *Herbal Simples* and the sender of the recipe – 'tells us they are a feature of Yorkshire "funeral teas".'

½ kg (1 lb) bilberries
125 g (4 oz/½ cup) caster sugar
2 large cooking apples
shortcrust or puff pastry
egg white, lightly beaten, or egg glaze
extra sugar

Mix the bilberries with half the sugar. Core the apples and bake them until tender. Scrape out the pulp and sweeten with the remaining sugar and mix with the bilberries: the point of the apple is to make the pie a little more juicy, which some people like as bilberries can be dry.

Butter and line a pie plate with pastry. Put in the bilberry filling and cover with pastry. Brush with egg white if you are using shortcrust pastry. Sprinkle with a thin layer of sugar. If you are using puff pastry, brush with egg glaze.

Bake in a hot oven first to set the pastry – gas 6–7, 200–220°C (400–425°F) – then lower the heat to moderate, gas 4, 180°C (350°F) to cook the fruit. Serve with cream.

This recipe can be used to make blackberry pie, and many other fruit pies.

BILBERRY CRISP

I am not given to cornflake cookery, but make an exception for this delightful pudding that comes from Mr & Mrs Dan E. Mayers. They intended it for blueberries – it was printed in the leaflet that J. Trehane & Sons Ltd of Dorset supply with their Highbush Blueberry plants – but I find that it is even more successful with the more strongly flavoured wild bilberries. There is no reason why the cornflake mixture should not be used with other fruit.

$\frac{3}{4}$ kg (1$\frac{1}{2}$ lb) bilberries
175 g (6 oz/$\frac{3}{4}$ cup) granulated sugar
2 level tablespoons($\frac{1}{3}$ cup) cornflour
$\frac{1}{4}$ teaspoon salt
1 large pinch cinnamon
$\frac{1}{4}$ teaspoon nutmeg
1 tablespoon lemon juice

TOPPING
90 g (3 oz/scant $\frac{1}{2}$ cup) butter
150 g (5 oz/$\frac{2}{3}$ cup) soft brown sugar
2 level tablespoons ($\frac{1}{3}$ cup) flour
90 g (3 oz) cornflakes

The night before, or several hours before you need to make up the pudding, sprinkle the bilberries with about two-thirds of the sugar and leave in a cool place. When you come to make up the pudding, drain off the juice. Then mix remaining sugar, cornflour, salt and spices in a heavy saucepan. Stir in the fruit juice and lemon and put over a low heat until the mixture is thick and clear. Put in the fruit. Pour into a buttered shallow pan or gratin dish.

To make the topping, melt the butter, stir in the sugar and flour mixed together and stir over a low heat until they are well amalgamated and syrupy. Stir in the cornflakes. When they are well coated scrape them out of the pan and over the fruit. Bake at gas 6, 200°C (400°F) for about 40 minutes. Serve with cream.

See also Blueberry and apple crisp, p. 99.

BLACKBERRY and DEWBERRY

'This being the commonest wild fruit in England is spoken of proverbially as the type of that which is plentiful and little prized.'

That is how the great *Oxford Dictionary* rounds off its definition of blackberry. Unfair, with a touch of class consciousness, as if in manor houses grapes from the hothouse were to be expected and not blackberries from the back lane. Large and very black and shiny and very juicy, blackberries in a good season are as delicious as any of the related soft fruit we cultivate. After all, a basketful costs nothing except the pleasure of blackberrying; and there they are, ready to be eaten with sugar and cream by themselves, ready to add to breakfast muesli, ready to combine with apples, ready for tarts and pasties, for jam, jelly. And ready for some *jeu d'esprit* like the pheasant and bramble recipe later in the section, or a sauce with duck.

Blackberries are world-wide, but it is in the north that they are most plentiful, where there are not so many juicy sweet wild fruits. So for thousands and thousands of summers, blackberrying must have been going on. And indeed, when a neolithic burial was excavated at the beginning of this century on the Essex coast, there was about a pint of seeds found in the area of the stomach – with blackberry seeds predominating.

What can be done with blackberries can be done with the best of their wild relatives, dewberries, the fruit of *Rubus caesius*, which are blue-grey instead of black. The flavour of dewberries is delicious and individual, it is more delicate than blackberry flavour. But dewberry brambles are not so common, and it is often disappointing that each berry is made up of only a few drupelets.

HOW TO PREPARE AND USE BLACKBERRIES

This is something most people know all about, from their earliest years. But there are one or two points that are perhaps not so commonly thought of.

When picking blackberries, make a point of picking some red ones as well. They are not ripe, of course, but they fortify the flavour and help the set if you make jelly.

When making blackberry jelly, you will get a better set by adding a proportion of apple; sour windfalls are best, up to equal weight with the fruit. I find a proportion of 2 parts blackberries with 1 part apple gives a good result. Strain off the liquid if you want to make a clear jewel-like jelly in an abundant year. Strain off the liquid and push through as much of the pulp as you feel inclined for a thicker but equally delicious jelly in a meagre year.

Blackberry jelly can be used in apple pies in place of liquid and sugar; it makes a pleasant surprise in February. In one family I know, February is a sort of carnival month: the precious store cupboard, with all its jars that have been carefully guarded, can be raided without thought of tomorrow. This way the shelves are clear for the new season.

BLACKBERRY SAUCE

Several years ago now, a farmer's wife in Devon, Gwen Troake, won fame in a cookery competition, then sympathy, admiration and a wider fame on a television programme. One of her special dishes was duck served with blackberry sauce. Versions of the sauce have been appearing ever since, but I find that it goes better with puddings, steamed apple pudding for instance, or an apple plate tart. If you want to try them with meat, use them to make the French black currant sauce on p. 84.

250 g (8 oz) blackberries
30 g (1 oz/2 tbs) light soft brown sugar
1½ level teaspoons arrowroot
5 tablespoons (⅓ cup) red wine, or 2 tablespoons (3 tbs) port or other fortified wine
1 cinnamon stick 5–8 cm (2–3 in) long
2 cloves

Bring blackberries and sugar to boiling point with 100 ml (4 fl oz/½ cup) water. Have ready the arrowroot mixed to a smooth cream with a further 50 g (1 fl oz or 2 tablespoons/3 tbs) water and stir it into the pan. The moment everything is well mixed, put in the wine and spices. Simmer about 5 minutes. Taste and adjust sugar if necessary, and remove the cinnamon stick.

If you do want to serve this quick and easy sauce with meat, use 2 teaspoons arrowroot and pour in the meat juices from the roasting pan, after removing the fat. Taste and see if extra sugar is needed, but do not oversweeten the sauce. Grind in plenty of black pepper.

AMULREE GROUSE

Catherine Brown of the Scottish Tourist Board, who published her own *Scottish Regional Recipes* in 1980, told me about this recipe. It comes from the Board's *Taste of Scotland Manual*, and was a prize-winning recipe in a student competition. Wild raspberries and grouse I had come across before, but this combination never. And why not? They are in season together and the flavours marry well. The little choux nests are fun to make: if you are not used to such things, do it in advance and see how you get on. Choux buns can be stored in an air-tight box and reheated.

Incidentally, in the north of England and in Scotland, blackberries are called brambles.

This recipe is enough for 10 people.

10 young grouse
1¼ kg (2½ lb) brambles
400 ml (12 fl oz/1½ cups) malt whisky
275 g (10 oz) streaky bacon
275 ml (½ pt/1¼ cups) game stock
salt and pepper

CHOUX
300 g (10 oz/2½ cups) plain flour
250 g (8 oz/1 cup) butter
600 ml (1 pt/2½ cups) water
8 eggs

Season the grouse inside. Marinate the brambles in whisky for an hour.

Make the choux nests. Sift the flour. Bring butter, water and a pinch of salt to the boil. Remove from the heat and mix in the flour. Put back on the stove and cook, stirring all the time, until the mixture leaves the side of the pan. Cool slightly, then add the eggs one by one, to make a dropping consistency. Pipe into fluted patty or brioche tins, to make nests. Bake 30–35 minutes at gas 6, 200°C (400°F).

Fill the grouse with three-quarters of the brambles. Truss for roasting, and pour the whisky from the marinade into the cavities. Cover breasts with bacon, roast in a hot oven for 12–15 minutes.

Remove the bacon jackets and arrange the grouse on a serving dish and keep them warm. Deglaze the pan with the game stock, and boil it down vigorously. Meanwhile, reheat the choux nests. Fill them with the last of the brambles and put them round the birds.

Strain the small amount of sauce over the birds, pepper and salt their breasts, and serve.

BLACKBERRY MUESLI

I like blackberries and milk together very much better than blackberries and cream, and blackberries with a milky muesli best of all. Turn to p. 301 for the basic muesli recipe, and substitute blackberries. Do not omit the grated apple.

LA MÛRÉE

In Flanders, they cook blackberries – *mûres* – to a thick jammy sauce – *la mûrée* – and serve it tepid with a grainy cake such as semolina or ground rice cake. It makes a good tea after a blackberrying expedition (if you had the forethought to bake the cake in advance, that is).

Rinse ½ kg (1 lb) of blackberries. Put them into a pan with 250 g (½ lb/ 1 cup) sugar, the juice of half a lemon and 25 g (scant ounce/¼ cup) of flour. Stir for about 15 minutes until cooked, over a slow heat at first until the juices run.

BLACKBERRY AND APPLE PIE

When you pick blackberries in the autumn, and gather windfall apples to make this pie, quantity and variety of fruit do not much come into it. You make the best of what you have. This is the way it should be. This is how regional dishes once developed. People used what their garden and neighbourhood could provide. Sometimes a suggestion from a visitor or the arrival of a new ingredient with the development of trade and manufacture might give a new aspect to an old dish, with luck a new refinement, most likely – at least in recent times – a short cut or substitute ingredient that did not improve it. I cannot help feeling that before the great Bramley conquest – it was introduced in 1876 – blackberry and apple pie made with the windfalls of pippin or reinette varieties tasted much better. Nowadays, I use tart windfall Blenheim orange apples, and then after Christmas Belle de Boskoop. This is a real treat as we do not grow it in Wiltshire, and depend on the kindness of January visitors from France for an occasional supply.

With apples of this quality, you do not need so many blackberries (cookery books often give equal weights of each). If you are making the pie soon after picking blackberries, rather than from a store in the freezer, weigh them and add up to double their weight in apples.

Assuming you start with 1 kg (2 lb) apples and $\frac{1}{2}$ kg (1 lb) of blackberries, put half the blackberries into a large saucepan. Peel, core and slice the apples, sprinkling them with lemon or orange juice to prevent them darkening: put the peel and cores into the blackberry pan and cover with water. Cook slowly at first, then faster as the juices run, so that you end up with about 150 ml ($\frac{1}{4}$ pt/$\frac{2}{3}$ cup) of strained liquid. Dissolve 200–250 g (6–8 oz/1 cup) sugar in the liquid, the quantity according to the tartness of the fruit.

Pack the sliced apples and remaining blackberries in layers in a deep pie dish. Mound up the fruit in a curve above the rim of the dish. Pour the sweetened, cool juice over the whole thing. Cover in the usual way with shortcrust pastry. Brush over the top with egg white and sprinkle with caster sugar.

Put into the oven at gas 6, 200°C (400°F) for 45–60 minutes depending on the depth of the dish. Turn the heat down slightly once the pastry is set firm and lightly coloured. Protect it with butter papers from becoming too brown. Test the apples with a thin pointed knife through the central hole in the pastry: when it goes through them easily, they are done. Serve with custard sauce or cream, preferably cream, clotted cream above all.

BLACKBERRY AND APPLE PLATE TART

With small quantities of fruit and a family to feed, blackberry and apple plate tarts are a good solution. Use a tin pie plate of about 23 cm (9 in) diameter for the best result.

Cook half the blackberries and apple trimmings as in the recipe above, but use the liquid to mix with cream for a sauce. There is enough juice in the fruit to make a moist double-crust tart: no need for the more abundant liquid of a deep-dish pie.

Line the pie plate with plain shortcrust pastry. Brush the rim with water. Pile in the fruit, mounding it well up so that you will get a nicely curved lid. Sprinkle the layers with sugar. Cover with pastry, decorate and finish with egg white and sugar. Bake as above, but test the fruit after 30 minutes. It may well be done.

Beat up the reduced blackberry liquor with double cream and serve very cold with the hot pie. Or heat the liquor with single cream, and thicken very slightly with arrowroot or cornflour. Serve hot. Sweeten the sauces to taste.

If you want to spice these pies, add some ground cinnamon to the sugar.

BLACK, RED and WHITE CURRANT

Black currants, red currants and white currants are northern natives (white currants are varieties of red) which the English seem to have added to cooking, after starting to cultivate them early in the seventeenth century.

Black currants, the French *cassis*, don't appeal much to the French except for the drink or drink basis they have given their name to. I think there is an explanation. *Cassis*, berry by berry, were swallowed as a substitute for imported *cassia* pods, senna pods, that is to say. One French writer in 1722 wrote that black currants had a savour of bed bugs.

Red currants – and white currants, too – have had more success. They gleam delicately on a bush or in a bowl, catching the light of candles. They do not look like bed bugs, they haven't that strange scent. They have found themselves in poems –

> Here the currants red and white
> In yon green bush at her sight
> Peep through their shady leaves, and cry
> Come eat me, as she passes by.
>
> Robert Heath in *Clarastella*, 1650

They began to gleam too in Dutch and Flemish still-life paintings of fruit.

The French derived their currants from England and Holland. Versailles became a commercial centre for them. There, in the eighteen forties, an English nurseryman named John Jaffers, became famous for growing currants (including a variety called Queen Victoria) arranged in groups according to the colours of their fruits. Currants were soon grown all round Paris, as well as in every private garden throughout France.

Before the days of lorry transport, it was some sight to see great wicker panniers of red currants being unloaded from the trains – always the last possible trains so that the newly picked currants would look their freshest next morning in markets and shops.

Red currants are good to eat on their own when they are very ripe, provided there is plenty of sugar with them. Some of Clarastella's currants would certainly have graced the dinner table in the display of fruit that began to be the pride of seventeenth-century gentlemen, but most of them would have been preserved according to recipes that we still use today. As a sideline on preserving days, cooks would make little red currant creams – a type of syllabub without wine, or a fool made with fruit juice.

SAUCE CASSIS

I have adapted this sauce to serve with roast duck from the elegant original of the Troisgros brothers in Roanne. They are the bright spot – opposite the station – in an unappealing city whose ancient charms and bustle as a Loire port have been superseded by lorries and summer traffic jams. Once it was famous for velvet. It has always been famous for food, cherishing the travellers up and down the Loire on their way to Lyons and Italy, or to Paris, Nantes and the Americas.

In 1646 John Evelyn stayed there with his friend the poet, Edmund Waller. They enjoyed 'the best that all France affords', as you still can, then came down the Loire, the watery main road of France, singing and picnicking on its banks, in the summer weather.

This sauce they could not have had, as black currants were not about until the following century. I have altered the recipe in the interests of the family cook, substituting giblet stock and duck juices for the broken carcase of duckling and demi-glace that are specified in the original. For this you must consult *The Nouvelle Cuisine of Jean and Pierre Troisgros*, published by Macmillan in 1980.

300 g (10 oz) black currants
1 level tablespoon icing or caster sugar
5 tablespoons (6 tbs) wine vinegar
1 tablespoon red currant jelly or black currant jam
2 tablespoons (3 tbs) crème de cassis
250 ml (8 fl oz/scant 1¼ cups) giblet stock
juice from roasting the duck
90 g (3 oz/6 tbs) lightly salted butter

Cook a heaped tablespoon of the black currants in a little water. Drain off the liquor into a larger pan and set aside the black currants for the final garnishing. Bring the liquor up to 3 tablespoons with water. Put in the rest of the black currants and the sugar. Cook until the fruit is tender.

In a small pan, reduce vinegar and jelly to a caramel and deglaze with the crème de cassis. Stir in the stock. When all is smooth, add to the pan of black currants and simmer for 20 minutes. At this point, add the juices from the duck roasting pan, making sure that all the fat has been poured off. Simmer another 5 minutes. Strain the sauce into a clean pan, pushing through the juices: be careful not to spoil the texture of the sauce by working through too much of the debris of the fruit. Reheat with the tablespoon of black currants, and whisk in the butter just before serving.

BLACK CURRANT EXTRACT AND SAUCE

The strength of black currants makes them suitable for economical flavourings. The extract described below makes a base for mousse or ice cream: it can also be added in very small quantities to beef stews as a substitute for wine or crème de cassis, because the vigorous flavour enhances the general richness without being identifiable. The sauce, made from the extract debris, can be served on its own with hot puddings or with a cream cheese dish like the *coeur à la crème* on p. 471; it can also be used as the start of a sorbet or a simple bavaroise.

EXTRACT Put 1½ kg (3 lb.) black currants into a pan with ½ litre (generous ¾ pt/2 cups) water. Cover and cook slowly until the fruit is soft. Tie up in a cloth or jelly bag and hang over a basin to drip. Leave several hours or overnight. You will end up with about ½ litre (good ¾ pt/2 cups) strong black currant extract.

Freeze in 150 ml (5 fl oz/⅔ cup) plastic tubs, and in ice cube trays. You then have a supply for puddings, and small quantities to use more as an essence.

SAUCE Put the debris from the cloth or bag into a clean saucepan and cover with water. Simmer for 20 minutes, or slightly longer. Process or liquidize to a sludge. Pour through a sieve, pressing through the juice with a pestle or wooden spoon, into a measuring jug. You will have about 1½ litres (2¼ pt/5⅔ cups): if you have less, don't worry, just reduce the next ingredients.

Now stir in the juice of 2 oranges, sugar to taste and red wine to bring the quantity up to 2 litres (3½ pt/8¾ cups). The mixture will be thick and smooth.

Freeze in 250-ml (8-oz) 1-cup cream pots, or in the kind of quantities you find useful. The sauce can be reheated, but keep it below simmering point and whisk it well together.

CRÈME DE CASSIS

A friend in France produces this *crème de cassis* every summer, to a recipe that came from a Burgundian great-aunt. Black currants are one of the great crops around Dijon, so great that it is sometimes difficult to find a market. French mothers do not feel the need to pour Ribena into their children, neither do they have our addiction to summer pudding. The only hope of shifting the crop is to turn it into an alcoholic beverage. Canon Kir, now dead but once the mayor of Dijon and a Resistance hero, had the idea of popularizing the local *rince-cochon* of *crème de cassis* and white wine (usually aligoté). It was renamed Kir in his honour. In the summer now you drink it all over France. Everyone drinks it. Either to help down grocery plonk, or to vary the boredom of daily champagne, or to colour up a decent local *vin blanc* with an air of festivity. It's refreshing and pretty in the heat, an affable drink so long as you do not overdo the *cassis*.

So famous is the drink these days that there is now a rhyming Myr, with a bilberry liquor (*myrtille* is French for bilberry) and white wine, that you can drink in the hilly districts of Central France, or in knowing bars in large towns. An even newer drink is Cardinale, red wine and cassis.

Crème de cassis is not easy to find here, so this recipe is worth knowing. Try it in half quantity the first time. Remember that it makes a good flavouring for sweet dishes, and is useful to heighten meat stews and sauces. There is often a tablespoon of cassis in *nouvelle cuisine* dishes.

Two awkward parts of the recipe are the brute strength needed to squeeze the black currants through a cloth, and the patience for cooking the brew so slowly that it thickens without boiling.

The first problem is solved with a processor or liquidizer, or solved to the extent of making it bearable. The second with a kitchen timer.

1 kg (2 lb) black currants
1 litre (scant 2 pt/scant 5 cups) reasonably good red wine
about 1½ kg (3 lb) sugar
about ¾ litre (1½ pt/3¾ cups) brandy, gin, or vodka

Soak black currants and wine together in a bowl for 48 hours. Put a large piece of old sheeting into a large basin. Gradually feed the currants and wine into liquidizer or processor and tip the mush into the cloth-lined basin as you go. Pull the cloth together and twist it so as to squeeze out all the liquid.

Measure it, put it into a preserving pan and to each litre (5 cups) add 1 kg (2 lb) sugar. Note the level. Stand over a low to moderate heat and stir until the sugar is dissolved. Regulate the heat so that the liquid keeps above blood heat but well below simmering and boiling points: use a thermometer to make sure. Set the kitchen timer for 15 minutes. Then check the temperature of the liquid and give it a thorough stirring. Do this two or three times more, then set the timer at longer intervals. In about 2 hours, the level will have gone down slightly and the liquid will be slightly syrupy. It is difficult to describe what it should look like, because it really does not look all that different while it is hot. Leave to cool.

Into a clean bowl, pour a mug of whichever spirit you are using, then add 3 mugs of black currant syrup. Repeat until all the syrup has gone, adjusting your quantities appropriately towards the end in the ratio of 1:3. Pour through a funnel into bottles – I use spirit bottles or glass mineral water bottles recently emptied and then kept tightly closed in order to save the business of washing out narrow-necked bottles.

Leave for two days before broaching. I should also point out that *cassis*, ice and Perrier water make an excellent long drink: this is the way I like it best. *Crème de cassis* keeps well.

PEAR AND BLACK CURRANT TART
TARTE AUX POIRES ET AU CASSIS

This is a recipe I came across in France that sounds odd. You might think the black currant would swamp the pears, but it doesn't. The two blend well together. An advantage of the recipe is that most of the work can be done in advance. Final assembly and the cooking of the meringue should be done at the last moment.

Make the creamed pastry mixture on p. 456, and line a 25 cm (10 inch) tart tin. Bake blind in the usual way, but do not allow the pastry to brown. Aim for a natural blond colour.

FILLING
1 kg (2 lb) pears
150 g (5 oz) black currants or black currant extract, p. 85
150 g (5 oz/⅔ cup) sugar
2 egg whites
100 g (3½ oz/7 tbs) caster sugar

Peel, core and slice the pears. Put the black currants or extract into a pan with the sugar, bring to simmering point. After a minute or two, put in the pears and leave them to cook gently for about 20 minutes. Turn them occasionally so that they colour evenly. You should end up with a thick mass of dark purple, containing large pieces of pear that have not entirely lost their identity.

Spread this mixture into the pastry case. Whip the whites until very stiff, add half the sugar and whip again. Fold in the last of the sugar. Spread over the tart, right to the pastry edges. Bake in the oven heated to gas 4, 180°C (350°F) until the meringue is nicely browned – about 15 minutes. Eat warm, with cream if you like.

Note If you do not wish to bother with meringue, or do not like it, pre-bake the pastry case until it is just firm, put in the pear-black currant mixture and criss-cross with a pastry lattice made from the trimmings. Brush with milk, cream or egg glaze and bake in the oven, moderate to hot, until the pastry is nicely coloured.

BURNT TART WITH BLACK CURRANTS

As black currants have such power, they should be used in smaller quantities than you might think normal for a fruit tart. Because of this, use a pie dish of about 3 cm (generous inch) depth or a *moule à manquer* rather than the usual shallow tart tin. This way, you get more cream filling to fewer black currants. Apricots, pineapple, red currants, white currants can all be used instead of black currants. The recipe can be adapted to milder fruit, by using the normal shallow tart tin of about 25 cm (10 inch) diameter and more fruit in relation to the cream.

Using a sweet shortcrust or nut pastry – *see* end section – line the pie dish. Bake blind until firm but not coloured.

Put enough black currants into it to make a single layer, then tip them into a pan. Add two or three tablespoons of water plus a level teaspoon of cinnamon and cook them lightly. They should not be allowed to collapse. Sweeten to taste with soft light brown sugar or demerara. When tepid or cool, spread over the cooled pastry.

Mix together 2 large eggs, 2 large egg yolks, 250 ml (8 fl oz/scant cup) whipping cream and 4 tablespoons ($\frac{1}{3}$ cup) *crème de cassis* or kirsch. Add just enough sugar to bring out the flavour, without really sweetening the cream. Pour over the fruit. Bake at gas 4, 180°C (350°F) until barely firm – if the centre is slightly wobbly it does not matter; it will continue to cook in its own heat. Leave the tart to cool.

Protect the pastry edges with strips of foil pressed into place. Sprinkle the filling with a thin layer of granulated sugar. Put under a pre-heated, very hot grill so that the sugar caramelizes. It will not turn into a sheet of brown glass as in the traditional *crème brulée*. Rather the sugar turns to a thin crisp layer that sets off the cream and black currants well, and the different crispness of the pastry.

Serve straightaway while still warm from the grill, having removed the foil, of course.

BLACK CURRANT PUDDING AND PIE

Black currants are an ideal fruit for family puddings. An unusual north-country flavouring is chopped mint, which is also added to bilberry pies, or to currants to make a Yorkshire mint pasty.

The general system is the same for both pudding and pie. Only the pastry and the method of cooking are different.

Make a suet crust or a plain shortcrust pastry.

Choose the size basin or dish appropriate to the family's appetite and fill it two-thirds full of fruit. Tip the fruit on to the scales. To each ½ kg or 1 lb, add about 175 g (6 oz/¾ cup) sugar and a tablespoon of chopped mint leaves. Line basin or dish with pastry, fill with the fruit, then cover with pastry.

Finish and cook the pudding or pie (or plate tart) in the usual way. Pudding instructions are to be found under the suet crust recipe in the end section.

Serve with cream or custard sauce, or with the soft *coeur à la crème* on p. 470.

CHESTER STEAMED PUDDING BLACK CURRANT JAM SAUCE

Black currants grow well in the north, and are popular in family puddings and pies. This is very much a winter pudding from the time before freezers, when summer fruit had to be turned into jam for storing.

PUDDING
125 g (4 oz/1 cup) white breadcrumbs
125 g (4 oz/1 cup) self-raising flour
125 g (4 oz/scant cup) chopped suet
60 g (2 oz/¼ cup) soft light brown sugar or caster sugar
125 g (4 oz/scant ½ cup) black currant jam
a large egg, beaten
milk

SAUCE

4 large tablespoons ($\frac{1}{3}$ cup) black currant jam
300 ml ($\frac{1}{2}$ pt/$1\frac{1}{4}$ cups) water
60 g (2 oz/$\frac{1}{4}$ cup) sugar
lemon juice or orange juice

Butter a generous 1 litre (2 pt/5 cup) pudding basin. Mix pudding ingredients in order given, using the milk to mix to a soft dough. Put into the basin, tie up as for a suet pudding and boil or steam for 3 hours in the usual way (*see* suet crust recipe on p. 457 for cooking instructions).

Meanwhile make the sauce. Bring jam, water and sugar to the boil. Cook 2 or 3 minutes. Add lemon juice to taste, or orange juice if you prefer a milder note, and push through a sieve.

CURRANT CREAMS

This seventeenth-century cream, half a syllabub, half a fool, can be made when you are in the process of straining off the juices intended for jelly, *see* recipes at end of this chapter. It works best with red and white currants, but black ones can perfectly well be used.

Take off 250 ml (8 fl oz/1 cup) of the juice, and if necessary sweeten it. Simmer $\frac{1}{2}$ litre (16 fl oz/2 cups) whipping cream with a blade of mace and 60 g (2 oz/$\frac{1}{2}$ cup) of coarsely chopped unblanched almonds. Keep a lid on the pan, for minimum evaporation and, after 5 minutes' simmering, turn off the heat and leave the cream to cool. Put two-thirds of the juice into a basin, and strain on the cream. Whisk until foamy, taste and add extra sugar if necessary, and a drop or two of natural almond essence or German bitter almond essence. Divide the remaining juice between six glasses, and pile the cream on top. Serve with almond biscuits.

Do not waste the almond debris. Remove the mace and add the almonds to a pound cake, or add them to the biscuit mixture to serve with the creams.

BLACK CURRANT LEAF CREAM

We do not make enough use of the flavour of some fruit tree leaves. I was most surprised the first time I made a peach leaf aperitif at how much they contributed an almond-like taste. This recipe comes from an Australian collection, *Early Settlers' Household Lore*, made by Mrs Pescott and first published in 1977. It is full of echoes of European food of the past, with American colonial recipes and occasionally dishes such as slippery bob, which was kangaroo brain fritters cooked in emu fat, or the way to cook witchetty grubs (in honey – delicious) and make bush cake (long keeping). This particular recipe belongs more to genteel houses, settled places, with their vegetable and flower gardens.

½ kg (1 lb) sugar
¼ litre (8 fl oz/1 cup) young black currant leaves
2 egg whites
juice of a lemon
150 ml (5 fl oz/⅔ cup) double cream, whipped

Dissolve sugar into 300 ml (½ pt/1¼ cups) water over a low heat. As the liquid becomes clear, put in the leaves (well washed first and shaken dry). Boil 15 minutes – do not stir. Beat the egg whites electrically until stiff, then strain on the boiling syrup. Continue beating all the time, until the mixture is very thick. Stir in the lemon juice, then the cream. Divide between individual glasses. The mixture can also be frozen.

In Russia, ices and jellies are made on the same principle.

RED CURRANT CAKE
RIBISELKUCHEN

The best of red currant desserts, and this is an Austrian speciality in the version given to me by a friend who lived many years in Vienna. She makes it now for her dinner parties in Paris, cutting it into smallish squares. The combination of almond pastry, soft and crisp meringue, sharp red freshness, makes it the ideal end to a meal.

160 g (5 oz/generous ½ cup) butter
375 g (12 oz/1½ cups) caster sugar
4 egg yolks
100 g (3 oz/¾ cup) ground almonds
300 g (9 oz/2¼ cups) flour
½ kg (1 lb) red currants, free of stalks
4 egg whites

Cream butter and one-third of the sugar, add yolks, almonds and flour. Or else mix to a dough in the processor. Roll out into a circle to fit a 23 cm (9 inch) tart tin, preferably one with a removable base. Bake at gas 5–6, 190–200°C (375–400°F) for 30 minutes or until cooked.

Meanwhile mix red currants with half the remaining caster sugar (some people like to cook them gently, then strain off and reduce the liquor, to make a jammy mixture, but this is not necessary except in wet summers). Whip egg whites until stiff, again add half remaining sugar and whip once more until thick and soft. Finally fold in the last of the sugar.

Take the almond base from the oven. Raise the heat to gas 7, 220°C (425°F). Spread red currants and sugar (or their thick syrup) over the pastry, leaving a narrow rim of cake free. Pile on the meringue, right to the edge of the cake. Fork it up and put into the oven for 15–20 minutes until the meringue is nicely caught with brown. Be guided entirely by appearance, rather than precise timing. Serve warm or cold.

Note If you want to get ahead, make the whole thing earlier in the day. Or just make the base the day before and complete with red currants and meringue nearer the time of serving.

You can use other sharp fruit, raspberries, gooseberries – use young green ones and cook them first – or in winter a purée of dried apricots cooked in orange juice.

BLACK CURRANT JELLY

Remove the stalks of the currants, and measure them in a 1 litre jug (1¾ pt/scant 4½ cups). Tip them into the pan and to every 5 litres (8¾ pt/scant 22 cups), add ½ litre (¾ pt/scant 2 cups) of water. Cover and put closely into a cool oven for 2 hours. By standing the pan in a bain-marie, you will get more juice.

Strain off the liquid. To every litre (1½ pt/4½ cups) of juice add 750 g (1½ lb) sugar. Stir over a low heat, then when the two are dissolved together, raise the heat and boil hard until setting point is reached.

This method can be used for red and white currants, too.

Sieve the debris, and boil it up with an equal weight of sugar to make a black currant cheese (*see* p. 362 for method) or use it to make a second liquid for syrup for fruit salads.

UNCOOKED RED CURRANT JELLY

This recipe is given by Joseph Favre in his *Dictionnaire Universel de cuisine pratique* which began to come out in parts in 1883.

'The nomads of Arabia and Indo-China have their own way of making a currant jelly without fire which is as good as any made in Europe, both for its transparency and its delicacy.

'They remove the stalks from the currants, and put them between two stones to extract the juice. This juice goes into an earthenware pot in the proportion of one-third juice to two-thirds sugar. This preparation is put out in the burning sun; they pour it off gently into another pot two or three times, and when it begins to coagulate, they put it in little cups. Next day they put it out again in the sun. Then they cover it with parchment and store it in the cool. These jellies are exquisite, but they do not keep as long as those which have been cooked on the fire.'

Urbain Dubois gave a practical interpretation: for every litre ($1\frac{3}{4}$ pt/scant $4\frac{1}{2}$ cups) of juice, warm 1 kg (2 lb) of sugar in a heavy bowl in the oven until it is really hot. Put the bowl on a table and stir in the juice until the sugar is dissolved. Strain and pour into little jars. Leave them in an airy place for 12 hours, and then cover them. Best stored in the freezer.

Uncooked raspberry jelly can be made in the same way.

RED CURRANT JELLY

Run a thin layer of water over the base of the preserving pan, then put in an equal weight of red currants – no need to remove the stalks – and sugar. Stir and heat slowly until the sugar has dissolved, then raise the heat and boil hard for 8 minutes. Tip out on to a sieve set over a bowl, or into a jelly bag, and pour the resulting liquid into small pots.

This jelly, which is a version of Eliza Acton's, gives a strong jelly of fine flavour.

BAR-LE-DUC RED CURRANT JAM

Bar-le-Duc was famous for its jams centuries ago. Mary Stuart said that they were potted sunshine. Here is a recipe.

'With a quill pen cut from a goose's feather, remove the pips from the finest large red or white currants: put them on plates as you go. When you have 1 kg, bring 3 kg (6 lb) of sugar to 111°C (232°F) with 750 ml ($1\frac{1}{4}$ pt/generous 3 cups) water.

'Add a glass of red currant juice and the currants. Give the whole thing a gentle boil, and when it reaches 105°C (221°F), pour into a basin.

'When the jam has cooled a little, put it into jars and cover as usual.'

Another recipe from Joseph Favre.

Do not forget to stir up the jam before potting it, so as to get an even distribution of fruit. A splendid jam for serving with delicate cream cheese mixtures.

THE ORIGINS OF CUMBERLAND SAUCE

In 1726, John Nott gave a collection of thirteen red currant recipes in his *Cook's and Confectioner's Dictionary* (facsimile reprint, 1980, Rivington, with an introduction and glossary by Elizabeth David). They are mostly preserves, from jelly and marmalade to sugared currants in bunches and currant wine. He gives a cream and a red currant water either for drinking or turning into a water ice. What he does not give is the simple sauce for venison made by melting down red currant jelly. Twenty years later, Hannah Glasse does and it continued with an alternative port wine or claret sauce, the two together side by side. Under German influence it seems, they were amalgamated into simple versions of Cumberland sauce. Alexis Soyer, as Elizabeth David pointed out in her discussion of Cumberland sauce, acknowledges the German origin of his sauce and said it was to go with boar's head. Francatelli has a similar version in 1861, described in the English style as venison sauce. This is the way we think of red currant jelly now – as a flavouring for meat sauces, or as an accompaniment to winter lamb. I would recommend a return to some of the red currant desserts of the past, for their freshness.

1747 – Hannah Glasse: '. . . you may take either of these sauces for venison – currant jelly warmed; or half a pint of red wine with a quarter of a pound of sugar, simmered over a clear fire for five or six minutes; or half a pint of vinegar, and a quarter of a pound of sugar simmered till it is a syrup.'

1817 – Dr Kitchiner: *Wine sauce for venison or hare.* 'A quarter of a pint of claret or port wine, the same quantity of plain unflavoured mutton gravy' – i.e. stock not Bisto – 'and a tablespoon of currant jelly; let it just boil up, and send it to table in a sauce boat.'

1846 – Alexis Soyer: sauce for a boar's head, described as German. 'cut the rind, free from pith, of two Seville oranges into very thin strips half an inch in length, which blanch in boiling water, drain them upon a sieve and put them into a basin, with a spoonful of mixed English mustard, four of currant jelly, a little pepper, salt (mix well together) and half a pint of good port wine.'

1861 – Francatelli: *Venison sauce.* 'Two tablespoons of port wine, a small stick of cinnamon bruised, the thin rind of a lemon, half a pound of red currant jelly; boil for five minutes, and strain into a sauceboat.'

1970 – Elizabeth David: *Cumberland Sauce*, for ham, pressed beef, tongue, venison, boar's head or pork brawn. The rind from 2 large oranges, cut into strips and blanched as in the Soyer recipe. Put them in a basin with 4 tablespoons red currant jelly, a heaped teaspoon of Dijon mustard, pepper, pinch salt and sprinkling of ground ginger. Heat until smooth over simmering water. Add 7–8 tablespoons of medium tawny port. Simmer another 5 minutes. Serve cold. Enough for four. Keeps well in the refrigerator.

CHERRY AND RED CURRANT SAUCES

Can be served hot or warm with duck, tongue, guineafowl or venison, or cold with tongue especially.

Grate the zest of an orange. Put its juice into a pan with 250 g (8 oz) stoned cherries, and a very little sugar. Simmer until cooked. In another pan, a wide pan, reduce 200 ml (7 fl oz/a good ¾ cup) port by half. Lower the heat, add the grated zest, 250 g (8 oz) red currant jelly, and a pinch of mixed spice. Stir until the jelly has dissolved, without allowing it to boil. Strain on to the cherries, and add pepper.

This sauce is derived from Escoffier, and is typical of cherry sauces which were popular at the beginning of the century. Elizabeth David has a delicious cold variation of her own:

Melt 250 g (8 oz) red currant jelly over boiling water. Stir in 2 teaspoons French yellow mustard, then 3 tablespoons wine vinegar and some pepper. When the mixture is smooth, put in 250 g (8 oz) stoned cherries and a teaspoon of cinnamon. Serve cold.

The cherries should be of the dark kind. Tinned or frozen ones can be used: drain them well.

OTHER FRUIT AND RED CURRANTS

Red currants are quite often used to reinforce the flavour of other fruit, as if they were a kind of lemon. The best summer pudding is made from raspberries and red currants in the proportion of 4:1. Similar quantities should be used when making raspberry jelly – pudding or potted as a preserve – and the following sauces:

CHERRY Crush ½ kg (1 lb) of cherries (dark red Kentish cherries are the best) to crack the stones which will help to release their almond flavour. Put with a generous handful of red currants in a pan with 175 g (6 oz/¾ cup) sugar, bring to the boil and boil for 10 minutes gently. Keep stirring with a wooden spoon. Strain through a sieve, pushing through enough pulp to make a good consistency for the pudding it is destined to accompany: for ice cream make it fairly thick, for a baked or steamed pudding make it thinner and more pourable.

RASPBERRY Cook ½ kg (1 lb) raspberries and 150 g (5 oz) red currants as above, with extra sugar to taste.

RED CURRANT WATER

A refreshing drink on a hot day. Processor liquidize ½ kg (1 lb) red currants with ½ litre of water, adding 2 good tablespoons (3 tbs) of sugar. Put through a sieve, pressing on the pulp but not putting too much through. Taste and add extra sugar if you like: the flavour should be strong. Store in the refrigerator. Serve with ice cubes and soda water, adding gin or vodka if you like, as a cooling long drink.

Why we do not make more of these waters in the summer I do not know; cherry water and raspberry water do very well, especially if you add red currants to the fruit to sharpen it. The thing is to drink them fresh, within a couple of days. They can be stored in the freezer for longer periods, of course, but they take up quite a lot of room – if you have this in mind, reduce the quantity of water in the recipe.

BLAEBERRY, see BILBERRY

BLUEBERRY

In America, blueberries have been declared 'the aristocrat of soft fruit' – with some justification, too, whether we use them fresh, frozen, or canned. Our wild European bilberries are fine, as I have said, but these North American relatives are bigger, sweeter and more succulent. Try them, and you understand a poem* about blueberries, or blueberry picking, which Robert Frost wrote in England some seventy years ago, when he was homesick for the farm he had owned in New Hampshire and for the burnt-off hills where blueberries flourished mysteriously:

> Blueberries as big as the end of your thumb,
> Real sky-blue, and heavy, and ready to drum
> In the cavernous pail of the first one to come!
> All ripe together, none of them green.

Frost was pining for wild 'lowbush' blueberries (*Vaccinium angustifolium*). He marvels at their sudden appearance in the cleared and burned-off woods:

> . . . you may burn
> The pasture all over until not a fern
> Or grass-blade is left, not to mention a stick,
> And presto, they're up all around you as thick
> And hard to explain as a conjuror's trick.
>
> It must be on charcoal they fatten their fruit.
> I taste in them sometimes the flavour of soot.
> And after all, really they're ebony skinned:
> The blue's but a mist from the breath of the wind,
> A tarnish that goes at a touch of the hand . . .

These wild blueberries are frozen and canned by the ton. They characterize the blueberry barrens, like the burned ground in the poem; but they are also cultivated in various forms. So are 'highbush' blueberries, fruit of *Vaccinium corymbosum*, which are easier to harvest from their tall heavy-cropping plants. It is these highbush blueberries which a few

**The Poetry of Robert Frost* edited by Edward Connery Lathem

English growers are raising now in Dorset and Devon. Some of the recipes following are from – or rather are based on – a leaflet of blueberry dishes supplied by J. Trehane & Sons Ltd, nurserymen at Hampreston in Dorset (you can get the bushes from them, too). Some of the recipes are American, from Mr and Mrs Dan E. Mayers of Wadhurst, Sussex. Blueberries can be used for bilberry recipes, and vice versa.

BLUEBERRY SAUCE

Blueberries made into a sauce seasoned with spices and lemon juice go particularly well with pancakes, the firm *coeur à la crème* on p. 470, or cream cheese *papanas* on p. 464. They can be sandwiched between crisp fried layers of puff pastry (*see* recipe later in the section). Americans like blueberry sauce with ginger, ice cream and shortcake (p. 423).

60 g (2 oz/¼ cup) caster sugar
¼ teaspoon each nutmeg and cinnamon
2 level teaspoons cornflour
good pinch salt
½ kg (1 lb) blueberries
grated zest and juice of a lemon

Put sugar, spices, cornflour and salt into a heavy saucepan, and mix together with 150 ml (5 fl oz/⅔ cup) water. When smooth, put in the blueberries and set over a moderate heat. Stir until the liquid clears and thickens.

Taste to see if more sugar is needed, and add extra water if you want a runnier consistency. Stir in the zest, then the lemon juice gradually to taste.

To make a sauce for serving with game, use half blueberries and half peeled, sliced cooking apples. Keep the sugar down, and grind plenty of coarse black pepper into the sauce just before serving.

BLUEBERRY PIE

Elisabeth Lambert Ortiz's recipe in the mango section can be neatly adapted to good, ripe blueberries.

Fill a baked pastry case with uncooked berries. Cover them with a glaze made from a 250 g (8 oz) blueberries, cooked and thickened according to the instructions on p. 234.

LIGHT BLUEBERRY CHEESECAKE

Finding many cheesecakes too cloying on account of the full fat soft cheese they are made with, I now make a light version with *fromage frais*. Some supermarkets now stock it under the French Jockey or German Quark labels; you should also be able to get it at specialized delicatessens or health food shops. If you cannot buy it, make a fuss – it really is the best of the soft cheeses to use with fruit, the fat content is medium (in the brand we import anyway), the consistency just right.

Macaroons can also be a problem. People in the bakery trade assure me that no ground rice is used in commercial macaroons. What therefore makes them taste so nasty? I suspect that synthetic almond essence – or 'flavouring' – is added to the mix that bakers buy from their wholesalers. To get round the problem, I buy French packet macaroons, but this is expensive. Sometimes shortbread or other crisp biscuits are used instead.

Line a tart tin of 20 cm (8 inches) with sweet shortcrust pastry. Assemble the following:

150–175 g (5–6 oz) blueberries
50 g (1½ oz/3 tbs) caster sugar
60 g (2 oz/½ cup) macaroons, crushed
250 g (8 oz) fromage frais
2 good tablespoons (3 tbs) cream, whipping or double
2 large egg yolks
about a tablespoon granulated sugar
1 large egg white, whisked

Mix blueberries with the sugar. Sprinkle a generous half of the macaroon crumbs over the pastry and put the blueberry mixture on top. Beat the *fromage frais* with the cream and yolks, then put in the sugar. Taste and add a little more if you like, but do not oversweeten. Fold in the egg white. Spread on top of the fruit and sprinkle with the remaining macaroon crumbs.

Bake at gas 7, 220°C (425°F) for 15 minutes, then at gas 4, 180°C (350°F) for half an hour longer, or until the filling is puffed up and just set. It will sink as it cools and become firmer.

BLUEBERRY AND APPLE CRISP

The addition of apple, sour apple, gives an extra edge to the blueberries, without being identifiable. It melts into the purple fruit. The superior crumble mixture is so good that it is worth using with other fruit: pecans are difficult to find at other times of the year than Christmas, but walnuts, almonds or hazelnuts – whichever is best suited to the fruit – can take their place.

300–325 g (10–11 oz) blueberries
250 g (8 oz) peeled, sliced cooking apples
juice and grated rind of a lemon
175 g (6 oz/¾ cup) soft brown sugar

TOPPING
150 g (5 oz/1¼ cups) flour
125 g (4 oz/½ cup) granulated sugar
1 level teaspoon baking powder
¾ level teaspoon salt
50 g (1½ oz/3 tbs) butter
1 large egg, lightly beaten
8 pecan nuts or walnuts, shelled and chopped
½ teaspoon cinnamon

Mix the blueberries and apples with the lemon rind and juice and the brown sugar. Spread out in a gratin dish of 1½-litre (2½-pt/6¼-cup) capacity, which has been lightly buttered.

Mix the first four topping ingredients, rub in the butter, then add egg and nuts. Spread over the fruit. Sprinkle cinnamon over the whole thing and put into a hot oven, gas 6, 200°C (400°F) for half an hour or until the top is nicely browned and the apples are soft. If the top browns too fast, lower the temperature.

Serve hot with cream, or vanilla ice cream.

Note: see also Bilberry crisp, p. 75.

BLUEBERRY CAKE OR PUDDING

If you have only a few blueberries, say between 150 g and 250 g (5–8 oz), this is a good way of making them go round half a dozen people.

Make up a pound cake or genoese mixture, based on 125 g (4 oz/1 cup) flour. Add a heaped tablespoon ground almonds and the juice of a lemon.

Put half into a buttered cake tin or baking dish. Spread on the fruit and sprinkle it with sugar. Cover the fruit with the remaining cake mixture.

Scatter blanched split almonds over the top, and bake in a moderate oven in the usual way.

BLUEBERRY YEAST CAKE
HEIDELBEERKUCHEN

Make a brioche dough, *see* page 463, and leave it to rise until doubled in bulk. Spread it out on a greased baking tray until it is about 1 cm thick (scant ½ inch), either with your hands or a rolling pin.

Mix 1 kg (2 lb) blueberries with sugar to taste from the vanilla pod jar and a level teaspoon of cinnamon. Add the grated zest of a lemon, then its juice. Spread over the dough. Sprinkle with extra sugar and leave to rise for a further 30 or 40 minutes.

Bake at gas 5, 190°C (375°F) for about 40 minutes, or until cooked. Serve hot or warm straight from the oven, cut in slices. This kind of cake loses its charm as it cools.

Note Other fruit can be used instead of blueberries – red currants, or red and white currants mixed, or black currants. Apricots and plums, too.

> 'The mountain tops of New Hampshire, often lifted above the clouds, are covered with this beautiful blue fruit in greater profusion than any garden.'
>
> Thoreau, *Journal*
> 30 December 1860

CAPE GOOSEBERRY, see PHYSALIS

CARAMBOLA

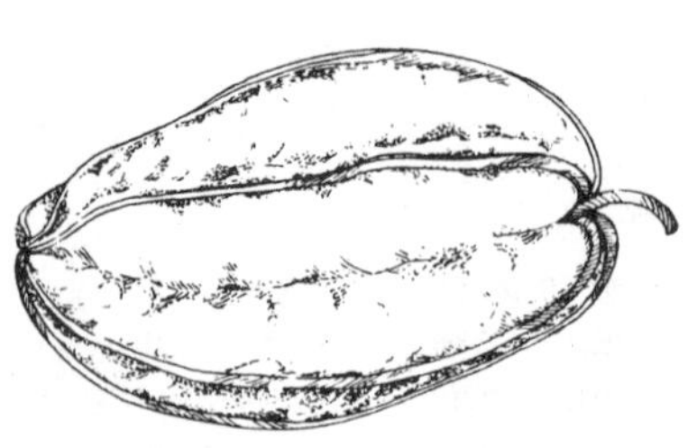

An amusing fruit. At first glance, it looks like a small banana gone mad, bright yellow, waxy and shining, with five fluted sides. When you cut it across into slices, you end up with a series of yellow stars. Peeling it would be impossible. Luckily you do not have to, though the fine sharp lines down the apex of the flutings sometimes need to be peeled thinly away as they can be tough.

The name – nothing to do with billiards – comes from the north Indian *kamranga*, which goes back to Sanskrit *karmara* meaning food-appetizer, an excellent description of its pleasure. What does it taste like? Difficult to describe, but here is part of an instructive conversation of 1563, from a Portuguese dialogue on medical matters and fruit, published first in Goa:

O. Antonia, pluck me from that tree a *carambola* or two . . .
A. Here they are.
R. They are beautiful; a sort of sour-sweet, not *very* acid.
O. . . . They make with sugar a very pleasant conserve . . . Antonia! bring hither a preserved *carambola*.

I wish Antonia lived in England to pluck us a few carambolas from her tree; they are a useful and pleasant addition to the cook's repertoire, because of this sour-sweetness. You can eat them raw, a bit sharp – at any rate by the time they got to France where I encountered them – but they are best cooked.

They made an unusual addition to a sauté of chicken: some of the slices are cooked in with the meat, and sieved into the final sauce, and the rest of the slices are lightly cooked by themselves to serve as garnish, with a very little sugar. Two carambolas for six people is about the right quantity.

Another good way to use a couple of the fruit is to put them into a fruit salad. Make a light syrup (300 ml water to 600 ml water (10 fl oz to a pint/$1\frac{1}{4}$–$2\frac{1}{2}$ cups), and when it simmers, slip in the sliced carambolas. Cook them gently until tender, allowing the syrup to reduce slightly. Transfer the carambolas to the serving bowl, and use the slightly lemonish syrup to cook any other fruit that needs it, then use as the syrup for the salad. When the whole thing is assembled, bring up the starry lemon-coloured slices to the top of the fruit, or enough of them to catch people's eye.

Carambola slices are made into sweetmeats in India, and into chutneys. I should like to experiment with a sweet-sour pickle, but this would be tricky; the fruit being sharp on its own account, you would need to avoid drowning the natural flavour with vinegar. I have never been able to buy more than one or two carambolas, in Paris, so experimentation on a vast scale has been impossible.

Bilimbi is a similarly shaped fruit to the carambola, and the only other member of the genus Averrhoa (*A. carambola* and *A. bilimbi*). It is altogether more sour than the carambola and is used particularly with fish in south-east Asia. Both fruits were so anciently cultivated that neither is known in its wild form. Travellers might look out for slices of dried bilimbi to bring home – they are sold in markets – as well as the fresh bilimbi. Both fruits are misleadingly – from the point of view of the cook – called star fruit. You need to look at the recipe to judge which one is required.

PADANG SOUR-SHARP FISH

PANGEK IKAN

This recipe gives an idea of how the sharp bilimbi or star fruit is used in south-east Asian cookery. Lemon and lemon grass are used as well, giving three tones of sharpness. Lemon grass can be bought at oriental stores, so can macadamia nuts which are also to be found in good groceries (their texture is quite different to any other nut's, but closest to the brazil).

I have taken the recipe from Yohanni Johns' *Dishes from Indonesia* (Chilton Book Company).

5 macadamia nuts
2 sliced onions
2 cloves garlic
8 fresh red chillies, seeded, or 1 tablespoon chopped chilli
2 slices green ginger, peeled
¼ teaspoon tumeric
½ kg (1 lb) sliced tunny or bonito
2 tablespoons lemon juice
1 stalk lemon grass, bruised
salt
6 tiny star fruits (or ¼ lemon, cut in thin wedges)

Mince or process the first six ingredients to a smooth paste. Season the fish with the lemon juice and salt, and leave to marinade.

Put the chilli paste into a heavy saucepan, with lemon grass, salt and 4 tablespoons of water. Bring to simmering point and slip in the star fruit or lemon. Stir together and leave to cook for 5 minutes, giving an occasional stir to prevent sticking. Put in the fish cutlets, turning them over so that they are covered in the chilli paste. Cook the fish gently until done, moving the pan with a restrained shake, to make sure they are not stuck Serve hot, or reheated next day when it tastes even better.

CHERIMOYA, CUSTARD APPLE and SOUR-SOP

These fruits 'rarely appear in the markets of the temperate world', an authority wrote in 1969. In England, in London, it wasn't quite so true a dozen years later, partly because West Indian immigrants like to have the sweet custard apples and the rather less sweet sour-sops they were used to back in the Caribbean.

All three belong to the American tropics by origin, all three belong to the same *Anona* family, all three with their white pulp suggest 'sops', as bread soaked in milk used to be called. Open them up and there, with a minimum of kitchen assistance and possibly an addition of double cream to their creaminess, nature provides a kind of 'fool' of its own, a natural fool as delicious as gooseberry fool or raspberry fool.

The aristocrat of the three is the cherimoya. Cherimoya trees began their world history in the uplands of Peru and Ecuador, a wild tree taken into cultivation by the Incas ('cherimoya' comes from the name of the tree and the fruit in Quechua, the Inca language).

Custard apples and sour-sops are fruits of the lowlands – rather humbler, I was tempted to say, though no doubt Montezuma ate his custard apples and his sour-sops, along with tomatoes and avocados and melons and maize. The Israelis are among today's pioneer growers and marketers of the cherimoya. All three fruits will be more familiar to our children, I am sure.

HOW TO CHOOSE, PREPARE AND EAT ANONA FRUITS

The cherimoya looks like a leathery fat green pine-cone, but do not be fooled. The skin is easily broken. When you buy some, put them carefully on top of your basket. The larger fruits have a less firm look, and they are even more delicate. I say this with feeling, having brought a very large sour-sop home on the train one day, when there were many other packages to wrestle with. Later I had to spend a long time separating skin from sweet white flesh.

With small fruit, just cut a lid from the top and dig in with a spoon, adding a little cream from time to time (not necessary, but nice). The lovely smooth perfumed custard inside also contains the smoothest of black seeds, which feel delightful as they slide against your teeth and on your tongue but which are not eaten.

The pulp, sieved to get rid of the seeds, can be added to all manner of drinks and desserts – just add cream, sugar and a hint of lime or lemon, to make an ice cream mixture, or the basis of a chilled soufflé. The sieved pulp can also be used as a sauce, being so convenient a natural custard.

CHERRY

In the Middle Ages – and until recently in some parts – the cherry fair was a great festival. People wandered about the orchards; the fruit was picked and sold; there was dancing, drinking and making love (a few years ago there were still some old people in our Wiltshire village with birthdays nine months after Clyffe feast, which took place every year at cherry time). The poignancy of colour and glory in lives which were normally brutish had by the thirteenth century turned the fair into a symbol of the passing moment:

> This life, I see, is but a cherry fair.
> All things pass and so must I algate.

Which is also, I suppose, the feeling behind Japanese cherry blossom festivities, when everyone goes out to picnic beneath the flowers – and reflect on the *neiges d'antan*.

The precious unkeepable cherry was the fruit of paradise, the glimpse and symbol of perfection. No wonder Mary longed for cherries when she was walking with Joseph and saw a laden cherry tree. She asked Joseph to reach some down, but he felt irritable and refused 'with words so unkind, Let them get thee cherries who got thee with child!' Then Jesus cried out from Mary's womb to the cherry tree. It leaned down one of its branches right to her hand. She called out like any girl at a cherry fair, 'See, Joseph, there are cherries for me!'

I wonder if Memling knew that carol when he painted his 'Mother and Child with two angels', now in the Uffizi. Jesus sits on Mary's knee, ivory pale against her rust-coloured robe, with a cherry clutched in one hand, symbol of the Heaven he had left behind. With the other hand, he stretches towards a fine red and yellow apple, held up by one of the angels and the symbol of man's fall by which death came to the world. Jesus leans towards his own death, taking on man's fate and redeeming him. The paradise cherry orchard is for a time forgotten.

Oddly enough in a world where seasons of fruit and vegetables have broken down completely, where many are available much of the time, the season of cherries always seems very short, gone before we have made the most of it. When we return to France in the spring, fat tight *bigarreaux* cherries from Provence are already on sale in Normandy. We buy some as a great luxury, as a sign that winter is over, but we have to wait a month before our true season begins in Loir et Cher. Then in the lane below the house, children skip and sing:

Un, deux, trois,
Dans les bois,
Quatre, cinq, six,
Cueillant des cerises.

It's time for everyone to go out and pick the wild red and black geans to make the best cherry brandy. The sharp taste is good. And not just in France either. 'I remember,' wrote an Austrian friend, 'that when I was a child, everybody had his or her secret wild quince, crab apple, morello, bird-cherry, chestnut tree in the woods. Once I upset an aunt in the country when she discovered that we children climbed over the churchyard wall to feast on wild cherries growing between the graves (definitely the best ones I ever ate).'

The pleasure of cherries does not belong only to country children. My mother, when we lived in a distinctly fruitless northern town, used to tell us about the splendid white heart cherry tree in their Croydon garden when she was little. It was so huge that she could hide with a book, high up in a fork of the branches, reading and stuffing handfuls of cherries into her mouth. I used to think that that was paradise, and I still do. We planted a white heart cherry when our daughter was small. It grows too slowly. The birds get the fruit before we do, just before it is ripe. I suspect that the first children to hide in it will be our great-grandchildren. Perhaps it should be enough that we have planted for the future, somebody's future?

Anyway the blossom is a delight. So is the name cherry, or *cerise* in French. It carries a great deal, not just the idea of summer and love and their passing, or the thought of paradise, but the whole history of the fruit. Both come, via the old French word *cherise*, and Latin *cerasus*, from *karsu*, the Accadian word used by the Assyrians and Babylonians who first cultivated the fruit. I like Pliny's tale, even if it is not certainly true, that the great general Lucullus – the elegant gourmet who committed suicide when he was down to his last few thousand, at the thought of frugal suppers – brought the cherry to Italy from Pontus. He had gone to those parts to fight against Mithridates in 74 BC, and came back to Rome eleven years later. In just over a century 'they have crossed the ocean and got as far as Britain'. Pliny must have been talking about a superior variety of cherry. Wild cherries are native in Europe, and there must surely have been some cultivated cherries in Italy before 63 BC? 'The cherry is one of the earliest fruits to repay its yearly gratitude to the farmer . . . moreover it can be dried in the sun and stored in casks like olives.'

Dried cherries can be bought in Greek markets sometimes. They are small and black and wizened; the stone occupies too much room. The magic has all gone, leaving nothing but a dry curranty taste. They improve with soaking and can be used to make jam (there is a recipe in Claudia Roden's *Book of Middle Eastern Food*).

I would imagine that Pliny – and the churchmen who worked out the symbolism of fruit – preferred to ignore the flirtatious games that must be as old as cherry orchards. Girls teasing young men by dangling bunches of cherries over their ears, daring them to come and nibble the fruit in the sheltering warmth of their long hair. Or two people eating a couple of joined cherries, laughing in a first embarrassment and in the end kissing – and then starting on another pair.

Do people, I wonder, play these games any more? I imagine they do, if with a more open and amused excitement, less pressure. Given our freer times, cherries cannot be the same strong key to recollected pleasure that they once were for everybody, even for kings, judging by Charles V, the king of France who got rid of the English (we were left with a few Channel ports, the cherry-pits of the great Plantagenet Empire that once went from the Scottish border down to the Pyrenees). About 1364 he turned his thoughts, for a time at least, from arms to fruit-growing. At his two gardens of Tournelles and St-Paul, he had planted, all in one season, 115 apple trees, 100 pear trees, 150 plum trees – and 1125, which is to say *one thousand one hundred and twenty-five*, cherry trees. There's passion for you, at the royal level. At the court again, but later in the time of Louis XIV, when Madame de Sevigné wanted to express a very particular delight in the fables of La Fontaine, she compared them to a basket of cherries, 'You pick out the best – and the basket is empty!'

Our great lover of cherries was the poet and priest, Robert Herrick. Down in Devon in the seventeenth century, feeling exiled from London, he consoled himself with writing poems to unattainable girls. Julia was the name he gave to a special favourite, 'prime of all', the girl of cherry-time. In gloom one day he sat with Julia eating cherries, and chucking the stones into a small hole. This was the game of cherry-pit, a country version, or inversion, of quoits. Herrick thought it summed up his hopeless situation:

She threw; I cast; and having thrown,
I got the pit, and she the stone.

A few poems later, he bounces up again with a joyful embroidery of the cherry-seller's street cry:

Cherry-ripe, ripe, ripe, I cry,
Full and fair ones; come and buy:
If so be, you ask me where
They do grow? I answer, There,
Where Julia's lips do smile;
There's the land, or cherry-isle,
Whose plantations fully show
All the year where cherries grow.

KIRSCH, CHERRY BRANDY AND MARASCHINO

Kirsch, correctly known as *eau de vie de kirsch* (France) or *kirschwasser* (Germany), is a white brandy distilled from cherries – wild cherries are used on account of their acidity and the bitter almond flavour of their stones which are crushed in with the fruit.

There are a number of other brandies of this kind. When they are made with soft stoneless fruit, raspberries, wood strawberries, or the soft service tree berries known in France as *alisier* (*see* Sorb section), the fruit is soaked with alcohol for some time before distillation. The Germans make a difference between the two in the names: the first kind is *-wasser*, -water, the second *-geist*, -ghost. The French call them all *eau de vie*. I like the second German name. When you pour any of these brandies over a sorbet, it seems to be the fiery spirit of the fruit itself; when you drink one of them, the essence of the fruit is first smelled and seems more magical at that moment than when you taste it.

Each fruit brandy has a place in exalting the flavours of their fresh original: Poire William poured over a pear sorbet, quince brandy added to apple or quince desserts, and so on. Kirsch, though, goes with many other fruit. It is by far the most adaptable. With a bottle of kirsch and a bottle of Cointreau, you can cover most fruits of the year, but the kirsch has the dryer strength. Avoid bottles labelled *kirsch fantaisie* when you are on holiday in France. It is not at all the real thing. You would do better to have a bottle of sharp cherries covered with vodka on the shelf, as you might have a jar of sugar with vanilla pods in it. Top up with more vodka, until all the cherry strength has gone.

In spite of its name, cherry brandy is, like Cointreau, a liqueur, or cordial as such drinks were called in the past (and still are in America). The fruit is soaked in an appropriate alcohol, then sweetened and flavoured in other ways as well, sometimes by adding a small quantity of kirsch. Danish Cherry Heering is one of the best-known brands in this country. A pity that it is not easier to find Grant's Morello Cherry Brandy which is made in Kent. It is no problem finding a good cherry brandy of one make or another, and you can produce a homely version yourself if you turn to the end of this section.

Maraschino tastes quite different, and for many dishes it is the next best to kirsch, being lighter than cherry brandy. The Amarasca cherry is the favoured variety for making maraschino. It grows in Dalmatia, around Zara, which was part of the Venetian Republic when Francesco Drioli first made maraschino there over two centuries ago. Other firms set up in rivalry. They all had to leave at the end of the Second World War when Zara was included in Yugoslavia; they re-settled in the Veneto where new amarasca orchards were planted. Drioli, now in Venice, is still the best-known firm in this country. Maraschino cherries may be false (read the small print on the jar) but true Maraschino liqueur can be bought for trifles, and ices, and fruit salads – and for drinking.

BUYING CHERRIES

For practical purposes, there are three main groups of cherries. The sweet dessert cherry (*Prunus avium*), the acid cooking cherry (*Prunus cerasum*), and the hybrid Dukes which are sharp enough to cook well but which also include some dessert varieties.

Very light coloured, 'white' cherries are sweet. Dark cherries can be sweet or acid or in between. You need to ask and taste before buying. Acid cherries are often sold as morellos, a name which covers several varieties with dark juice, or as amarelles which are acid cherries with light juice. In France the name Montmorency also covers acid cherries: Dukes are called *royales*.

In fact, you can cook with sweet cherries, no law against it, but the acid varieties and Dukes have more character, especially if you are cooking a meat dish or making jam. It may sound odd but, in spite of all the sugar, morello cherry jam can be useful in cooking if you cannot buy fresh morello cherries or canned ones either. Add some lemon juice, and if the recipe lists sugar in the ingredients, leave it out, or reduce the quantity.

Some frozen cherries are of good quality. They are on the sweet side, but cook reasonably well.

Stoning cherries is a trying business. If you cannot avoid it, buy one of the little gadgets that will also do for stoning olives. Avoid pampering your family in this respect. Half the fun of a cherry pie is putting the stones on the side of the plate, and counting them out later.

Remember the almond *noyau* flavour that the stones give to cherry brandy: almonds make a good partner to the fruit, almond pastry for cherry tarts, for example, or an almond cake mixture on top of the cherries (p. 123).

HUNGARIAN CHERRY SOUP

MEGGY LEVES

I've seen a number of recipes for cherry soup, but none to equal this, which comes from Victor Sassie, proprietor of the Gay Hussar restaurant in Soho. It is unimaginably delicious, a wonderful refreshment in late July and August when morello cherries are about in the shops.

½ kg (1 lb) wild sour cherries or morellos
60 g (2 oz/¼ cup) sugar
1 bottle Riesling
good pinch powdered cinnamon
grated rind 1 lemon
juice of 2 lemons
1 double brandy
600 ml (1 pt/2½ cups) soured cream

Stone cherries over a pan so that the juice is not wasted, and reserve fruit. Add stones and stalks to the pan, with sugar and wine. Bring to the boil: simmer for 5 minutes. Strain liquid into a clean pan. Add cinnamon, rind and lemon juice. Bring to the boil, then tip in the cherries and any more juice that has collected. Bring back to a rolling boil, then remove immediately from the heat. Allow to cool slightly. Stir in the brandy.

Put cream into a bowl. Whisk in the soup, holding back the cherries until the mixture begins to make a smooth blend. Then tip in the rest of the soup. If you like a thicker soup, liquidize or process some of the cherries. Serve well chilled.

ENGLISH CHERRY SAUCE

A sauce to be served hot with ham, tongue and duck, from the group of red currant and wine sauces that became popular in the nineteenth century (*see* p. 84). There is no reason why it should not be served with venison and hare, too.

3 oranges
1 lemon
1 generous tablespoon red currant jelly
150 ml (¼ pt/⅔ cup) each port and red wine
2 tablespoons (3 tbs) of boiling stock, or juices from a roast, as appropriate
black pepper
125 g (4 oz) stoned black cherries

Cut the zest thinly from one orange and the lemon. Slice it into matchstick-long shreds, simmer them in water until they are soft. Drain and reserve.

Bring the juice of all the oranges and of the lemon to the boil with the jelly, wines and stock or roasting juices. Whisk to dissolve the jelly. Boil down to half quantity.

If slices of tongue or ham need to be reheated, put them into the reduced sauce until thoroughly warmed through, then arrange them on a serving dish and keep them warm (remove the slices with a pierced skimmer, so that the sauce falls back into the pan).

Add plenty of black pepper to the sauce, put in the cherries and simmer until they are cooked (frozen cherries will not take as long as fresh ones). Add any juices that emerged when you stoned the cherries. Put in the shreds of peel to heat through.

Pour some of the sauce over sliced meat, or round roast meat, and serve the rest in a sauceboat.

RUSSIAN CHERRY SAUCE

Poultry stuffed with apples, served with roast potatoes and cherry sauce – a dashing combination of flavours that works well. A well-hung turkey or a guineafowl is likely to give a better result than most chickens. Fill the cavity with peeled, cored and quartered cooking apples. Butter and roast in the usual way, along with a pan of potatoes.

½ kg (1 lb) stoned morello or dark, tart cherries
sugar
3 cloves
¼ teaspoon cinnamon

1 level tablespoon arrowroot or cornflour
100 ml (3½ fl oz/½ cup) Madeira or port
1 tablespoon lemon juice

Put cherries with 600 ml (1 pt/2½ cups) water into a pan with 2 level tablespoons (¼ cup) sugar and the spices. Simmer until cherries are soft. Slake arrowroot or cornflour with a little of the cooking liquor, and stir into the sauce. Cook, stirring, until thickened. Add wine and lemon. Simmer another two minutes, adding more sugar to taste, and a shade more cinnamon if you like (but don't overdo it). Aim for a rich, slightly sweet flavour. Pour off the meat juices in the roasting pan, skim off the fat and add the juices to the sauce. Heat through and pour into a hot bowl to serve.

FRENCH CHERRY SAUCE

A sauce to serve with brains or sweetbreads – it is milder than English or Russian versions – that have been poached in a *court bouillon*, sliced and fried in butter. Use some of the *court bouillon* to moisten the sauce. If you want to serve it with guineafowl or turkey, use a giblet or appropriate poultry stock.

1 good tablespoon butter
1 tablespoon arrowroot or cornflour
court bouillon *or stock*
100 ml (3½ fl oz/½ cup) dry white vermouth, e.g. *Noilly Prat*
2 generous tablespoons whipping, sour or double cream
lemon juice
250 g (8 oz) sharp black cherries
salt, pepper, sugar

Melt the butter. Slake arrowroot or cornflour with 100 ml (3½ oz/good ½ cup) bouillon or stock and stir into the butter with the vermouth. Cook for a minute or two, then add cream and more stock if the sauce is very thick. Taste and put in a little lemon juice, but go carefully as the cherries will add sharpness. Put in the cherries and simmer gently until they are tender.

Taste and season as required.

Note Brown rice goes particularly well with brains or sweetbreads and cherry sauce, providing a slight chewiness which so soft a dish requires.

DUCK/DUCKLING MONTMORENCY
CANARD/CANETON MONTMORENCY

Fruit can be the duck's amiable companion. More often, it's the vehicle of disaster from too much sweetness. Apples are fairly safe, but when it comes to cherries – or to peaches or orange – people forget that it's acidity that makes for success. Caneton Montmorency reminds us that the cherries to go with duck should be the acid kind. Montmorency, the acid French cherry, is called after the district north of Paris where it was first developed. Some people in this country grow them – for instance, F. P. and C. P. Norbury Ltd of Sherridge, near Malvern in Worcestershire – and they are worth pursuing for their special flavour which does not deteriorate in freezing. Otherwise, morellos or Kentish Red can be used.

For 4 to 6 people, you need one very large duck or two smaller duckling, complete with giblets, plus:

1 bouquet garni
1 medium onion stuck with 3 cloves
1 small carrot quartered
salt, pepper
½ kg (1 lb) Montmorency cherries, stoned
2 rounded tablespoons (¼ cup) sugar
grated rind and juice of an orange
¼ teaspoon each cinnamon and nutmeg
150 ml (¼ pt/good ½ cup) port
1 tablespoon red currant jelly
60 g (2 oz/¼ cup) butter
bunch of watercress

Put the giblets, minus the liver, into a large pan with the *bouquet*, onion and carrot. Cover with a litre (1¾ pt/4½ cups) water. Bring to the boil and leave to simmer down, uncovered, for about 3 hours. This can be done in advance.

Roast the duckling in the usual way, or follow the method given in the Orange chapter, p. 264. Pour off fat from time to time, and at the end remove the last of the fat from the pan and keep the juices for the sauce.

Strain off the giblet stock, pour it into a wide shallow pan and reduce to 250 ml (8 fl oz/1 cup). Season lightly. Put in the stoned cherries with any juice, and simmer them in the stock until just cooked. Remove two-thirds of the cherries to serve as a garnish to the duck later on. Meanwhile, dissolve the sugar with two tablespoons (3 tbs) of water over a low heat, then raise the heat and boil the syrup to a golden brown caramel. Off the heat, stir in the orange rind and juice, going carefully as the caramel will splutter. If some of it clings in a lump to the spoon, put the pan back on to the stove and stir until it dissolves. Tip this mixture into the cherry pan. Add the spices, then the port and red currant jelly gradually in small quantities to taste – you may not need the full amount.

Remove the duck(s) from the roasting pan when cooked and put them on a dish. Having poured off the fat from the pan juices, add them to the cherry sauce. Raise the heat under it and leave it to boil down to about 375 ml (12 fl oz/1⅓ cups).

Carve or cut up the duck and arrange it on a warm serving dish. Scatter with the garnishing cherries, lay butter papers over the top and keep warm in a very low oven.

When the sauce is well reduced, liquidize or process, then sieve it into a clean pan. Do not worry if you cannot get all the cherry pulp through. Taste the sauce and make any adjustments to the flavour. Salt or plenty of pepper, and a little sugar might be required. The sauce will be strong at this stage.

Just before serving the dish, bring the sauce to boiling point, give it a fast bubble or two and remove from the stove. Whisk in the butter, cut into little bits. This thickens and softens the sauce slightly, and improves the flavour. Pour a little over the duck and serve the rest in a warm sauceboat. Tuck sprigs of watercress round the edge of the dish and serve.

QUAILS ON THEIR NESTS

Quails in the corn say 'Wet my lips! Wet my lips!' in England, '*Paie tes dettes! Paie tes dettes!*' in France. Ours in Wiltshire, however, seem to have gone for ever. In Trôo, walking down the slopes of the village to the field called Rome, we sometimes hear them in the evening. A case of drinkers go under, debt collectors – or moralists – survive?

However, in both countries, quails in the large food shops come from quail farms. The quails that sit in Fauchon's window with grapes on their breasts, or that lie in leggy packages on a Selfridge meat counter, have led uneventful lives. And by comparison with wild ones, they taste like it. Even so, they have flavour; delicious morsels that you pick up finally with your fingers so as not to waste the last scraps.

This combination of cherry, carrot, orange peel and red wine sets them off well. Quantities are for six.

12 quail
500 g (1 lb) large, dark cherries (morello if possible), stoned
4 large sections of candied orange peel
1 tablespoon sugar
½ litre (generous ¾ pt/2 cups) red wine
375 g (¾ lb) medium to large carrots
175 g (6 oz) onions
butter
pepper, salt

Into each of the quails, put 2 stoned cherries, then bend their legs back, close to the body, and tie them into shape with button thread.

Cut the orange peel into thin shreds and put into a wide pan with the sugar, and with the remaining cherries in a single layer. Pour on the wine, bring to simmering point, cover and cook for a few minutes until the cherries are just done.

Meanwhile, top and tail and scrape the carrots, then shred them coarsely (use processor or mandoline). Skin and slice the onions thinly. Melt a large tablespoon of butter in a wide heavy pan, put in carrots and onions, cover and cook for a few minutes until they are tender but still a little crisp. Season them, then arrange them in 6 'nests' on a large serving dish and keep warm.

Brown the quails in as little butter as possible in a wide sauté pan. Drain off surplus butter when they are nicely coloured and strain the cherry liquor into the pan. Cover and leave them to finish cooking – another 10 or 15 minutes – turning them over from one side to the other occasionally. When they are tender, add cherries and orange peel to heat through.

With a slotted spoon, remove the quails, arranging them two by two on the nests. Salt them lightly. Then put the cherries and peel round them. If there is the likelihood of the dish having to wait – for instance, while you eat a first course – cover the whole thing lightly with butter papers and leave in a very low oven to keep warm.

Just before serving, finish the sauce. The liquor in the quail pan will have reduced to a strong dark taste. Soften and freshen it by whisking in 100–125 g (3–4 oz/½ cup) butter, cut in cubes. As you do this, keep raising the pan from the heat so that the sauce keeps hot but well below boiling point, and the butter amalgamates with the liquor smoothly. Taste and add pepper and salt, and a pinch or two more of sugar. Pour carefully round the nests of quail and serve.

Note If you cannot get large pieces of peel, the kind that is sold with other candied peels by good grocers and delicatessens, use three good tablespoons of home-made coarse cut marmalade and no sugar.

Cream can also be used, with the butter, to smooth out the sauce.

CHERRIES WITH GOAT CHEESE

A Canadian friend living in the South of France discovered that cherries with fresh goat cheese is a fine combination – 'people fall about when I describe it. A curiosity for English cooks because I don't suppose you can get fresh goat cheese in England, but the taste of it with fresh cherries is marvellous. If you stone the cherries and mix them with the goat cheese, or perhaps half goat and half ordinary *fromage frais*, plus sugar to taste, it makes a delicious, somewhat lurid dessert.'

Fresh goat cheese is becoming easier to get; try your local wholefood or vegetarian shop. *Fromage frais* or its German equivalent, *quark*, is on sale in many supermarkets.

CHERRY RATAFIA CREAM

A seventeenth-century pudding, simple to make, with the right feeling for cherry time. I think you get the best result with morellos or Montmorency cherries because their sharpness invigorates the cream, but rather than go without the pudding, I make it with whatever cherries I can find. For look's sake, they should be red.

To me, the point of the cream is the bay leaf flavouring which gives it an almond air – though I would not go so far as to say that it tastes like an almond cream exactly. Seventeenth-century cooks might also have used orange-flower water or rose-water (triple strength from Greek shops or good delicatessens). You could do the same if you can't get hold of fresh bay leaves.

500 g (1 lb) lightly cooked cherries, drained
150 ml (5 fl oz/⅔ cup) soured cream
450 ml (15 fl oz/2 cups) whipping cream
2 large egg yolks
1 large egg
3 large fresh bay leaves
sugar
30 g (1 oz/¼ cup) blanched, split and toasted almonds

Put the cherries into a glass serving bowl, or into 6–8 individual glasses. There is no need to stone them.

To make the cream, first mix the creams together, in a saucepan. Take out a couple of tablespoons and beat them up with the yolks and egg. Put the bay leaves into the remaining cream and bring it slowly to the boil. Keep an eye on it, and keep tasting. When the flavour is strong enough – remember it will weaken as the cream cools – remove the bay leaves.

Pour some of the boiling cream on to the egg mixture, whisking vigorously. Tip it into the pan, taken from the heat, whisking again. Add sugar to taste. Put the pan back over a low to moderate heat, stirring until it thickens. Do not let it come near boiling point. Taste again, and add more sugar if necessary, but do not overdo the sweetness. When cool, spoon carefully over the cherries. Decorate with the almonds and chill.

SWISS CHERRY BUNCHES

Very much a family pudding, for lunch or a snack in the kitchen. Everyone can help by tying up the cherries, while you prepare the batter and see to the oil.

BATTER
125 g (4 oz/1 cup) flour
3 tablespoons (scant ¼ cup) oil
¼ teaspoon salt
1 egg white, stiffly beaten
cinnamon sugar

Mix flour, oil, salt and about 150 ml (¼ pt/½ cup) tepid water to make a smooth batter. Leave to stand, adding the egg white at the very end.

Allowing 10–15 cherries for each person, according to age and capacity, tie them into bundles of 5 with button thread, leaving a nice tail.

Dip them into the batter, then lower them by the tail of thread into a pan half-full of hot oil (better still, into an electric deep fryer, thermostatically controlled). Do three or four bunches at a time, no more, or they will be overcrowded and spoil.

Eat straightaway sprinkled with cinnamon sugar (one part cinnamon to four of sugar).

CHERRIES JUBILEE

A dish that children love for its drama, and for the contrast of heat and icy cold. Flaming burns off the alcohol, making it perfectly innocuous.

Freeze some vanilla ice cream in a tall round mould (a 1 lb coffee tin, with the sharp edge knocked flat, saves the expense of buying a special tin). Turn out the ice and cut it across into the appropriate number of slices. Lay them on a piece of greaseproof paper on a baking sheet and put into the freezer: the ice cream should be really hard.

For the sauce, assemble:

500 g (1 lb) sweet red cherries
150 g (5 oz/$\frac{2}{3}$ cup) sugar
thinly-cut peel of half a lemon
1 level tablespoon arrowroot or cornflour
4–5 tablespoons cherry brandy or kirsch

Stone the cherries while the sugar, lemon peel and 300 ml ($\frac{1}{2}$ pt/1$\frac{1}{4}$ cups) water are boiling to a syrup. Fish out the strips of peel after 4–5 minutes, then put in the cherries to cook at simmering point. When they are tender, remove them with a perforated spoon or skimmer and boil down the liquor to the original syrup level.

Slake the arrowroot or cornflour with a little cold water, then stir in some of the cherry liquor. Stir this mixture back into the pan, and go on stirring until you have a dark, clear sauce which is lightly thickened but not gummy. It should have no floury taste. Put back the cherries. Pour the alcohol into a tiny pan. Put a serving dish into the refrigerator.

Now you are ready for the finale. Transfer the ice cream slices to the chilled serving dish. Bring the cherry sauce to boiling point and warm the alcohol. Pour the boiling sauce on to the ice cream (the dish should be standing on the table which lessens the risk of it cracking). Set light to the warm alcohol and pour it flaming over the sauce. Take rapidly to the table where people are eating.

CHERRY WATER ICE

SORBET AUX CERISES

Any cherries can be used to make a water ice, but the richer the flavour of the fruit the better the ice will be. You may need to adjust the quantity of syrup added to the fruit, less for very sweet cherries of a light flavour, a shade more for sharp morellos. The recipe works quite well with commercial frozen cherries which are sweet and dark red, making a refreshing sorbet for winter parties.

500 g (1 lb) cherries
300 g (10 oz/1$\frac{1}{4}$ cups) sugar

thinly cut peel of a lemon
juice of a lemon
4–6 tablespoons (⅓–½ cup) cherry brandy or kirsch
German bitter almond essence or Langdale's natural almond essence

Stone the cherries into the bowl of a liquidizer or processor, being careful to keep the juice. Stir the sugar in a pan with 600 ml (1 pt/2½ cups) water and the lemon peel over a low heat until dissolved. Raise the heat and boil for 5 minutes. Strain and cool.

Add the lemon juice to the cherries and 300 ml (½ pt/1¼ cups) syrup. Liquidize until smooth, adding more syrup to taste. Freeze to a slush, then beat and add the alcohol and a few drops of almond essence. Go carefully with the German essence, it is particularly strong – 2 drops may well be enough.

Refreeze, keeping the mixture on the slushy side, like an Italian *granita*. Serve in glasses, topped with a whirl of whipped cream.

CHERRY ICE CREAM

Make the above recipe in half quantity. When the mixture is half frozen, beat and fold in 300 ml (10 fl oz/1¼ cups) whipping or double cream whisked until firm with the alcohol and almond essence. You can amalgamate the two mixtures completely, or swirl them together in a rippled effect. Refreeze until firm.

CHERRY CREAM TARTS

For 24 little tarts, made with sweet shortcrust or oatmeal pastry, you will need:

1 kg (2 lb) large black cherries of acid or Duke varieties
1 egg white
200 ml (7 fl oz/¾ cup) soured cream
2 eggs
1 egg yolk
125 g (4 oz/½ cup) caster sugar

Stone the cherries. Beat egg white to break it up a little and brush it over the pastry cases (this diminishes the risk of sogginess). Put the cherries into the pastry cases, closely together. Mix any cherry juice with the cream and remaining ingredients. Pour over the cherries, being careful not to fill the pastry brimful, as the filling will rise.

Bake at gas 7, 220°C (425°F) for 15 minutes. Check and leave a little longer until the filling is puffed up and nicely browned. The filling will collapse unless you serve the tarts straight from the oven. If you cannot do this, don't worry; many people will agree with me that such things are best eaten warm anyway, rather than hot. Compared with this, the puff doesn't matter.

Note You can use one large pastry case. If you do, scatter the base, after brushing with egg white, with a layer of crumbs – macaroons, boudoir biscuits, plain cake, bread, chopped raw pastry can all be used. The crumbs can be flavoured with cardamom seeds (see recipe for lemon meringue pie, p. 214).

DUCLAIR CHERRY TART
TARTE AUX CERISES DE DUCLAIR

Duclair is on the Seine upstream from Tancarville, not far from Rouen. Ducks are a speciality, so is this unusual cherry tart. The tricky bit is cooking the cherries so that they caramelize without becoming over-cooked: you need a high heat once the butter, sugar and Calvados are blended, but take care the cherry juices do not burn – this kind of thing is much easier to do on a gas burner than an electric ring.

You need a puff pastry case, baked blind in a 23–25 cm (9–10 inch) tart tin. This can be done in advance, and the case reheated to freshen it if necessary. The cream and cheese can also be mixed earlier in the day. But the cooking of the cherries and the final assembly should take place not too long before the meal.

500 g (1 lb) firm red cherries, stoned weight
60 g (2 oz/¼ cup) butter
100 g (3½ oz/scant ½ cup) granulated sugar
3 tablespoons (scant ¼ cup) Calvados or malt whisky
250 g (8 oz) fromage frais
caster sugar
150 ml (5 fl oz/⅔ cup) double cream

Drain the cherries in a sieve. Melt the butter, stir in the granulated sugar and Calvados, or whisky. When they are blended, raise the heat and put in the cherries. They should caramelize lightly, so keep them moving and do not take your eye off them. Drain and cool.

Beat the *fromage frais* with caster sugar to taste. Whip the cream and fold it in, check again for sweetness and taste a cherry to make sure that you do not over-sweeten the mixture.

To assemble, spread the cream cheese and cream on to the pastry. Put the cherries on top.

FRANZIPAN CAKE

When I first knew my husband, he took me to lunch occasionally at the Maison Sagnes, in Marylebone High Street, for a special treat. The bay tree by the door, the small tables in the plain room, the unambitious first courses were all a prelude to the cakes. Wonderful cakes they were, in that grey time after the war. No doubt the cakes are still wonderful. Perhaps they still do this cherry franzipan cake. I cut the recipe from a magazine a few years after our last visit and now, settled in Wiltshire, I produce the cake from time to time as a souvenir. And also because I like it.

Why franzipan? First, it was the name of a perfume for gloves made by the Marchese Frangipani from the red jasmine of Mexico in the sixteenth century. In the eighteenth century, the name was used for almond- and orange-flower water fillings for tarts, and in the nineteenth century it was used for the Marchese's original Mexican flower (three centuries late).

creamed pâte sucrée, *p. 456*
60 g (2 oz/½ cup) cake or biscuit crumbs
60 g (2 oz/¼ cup) butter
60 g (2 oz/¼ cup) caster sugar
1 large egg
30 g (1 oz/¼ cup) plain flour
90 g (3 oz/¾ cup) almonds, blanched and ground
1 can of morello cherries, or 375 g (12 oz) fresh morellos, lightly cooked
2 tablespoons apricot jam
4 rounded tablespoons (scant cup) sieved icing sugar
1 teaspoon lemon juice

Line a 20 cm (8 inch) tart tin with the pastry and scatter over it the crumbs. Cream butter and caster sugar, add the egg, then fold in the flour and ground almonds. Drain the cherries well and spread closely in a single layer over the base of the tart. Dab the almond mixture over the cherries, then spread into a smooth layer with a knife.

From the pastry trimmings, cut long thin strips and lay them parallel to each other across the cake, about 2 cm (¾ inch) apart. Bake at gas 5, 190°C (375°F) for 30–40 minutes. Remove when nicely browned, and put on a rack.

Heat the apricot jam with a dessertspoon of water and, when it boils, sieve into a bowl. Brush over the cake. Let it cool a little further, then brush over another layer.

To make the icing, mix the icing sugar with the lemon juice and 2 teaspoons of water in a small pan. Bring to just under boiling point. Brush a coating over the top of the cake. Let it cool, then rewarm the rest of the icing to a spreadable consistency and brush on a second coating. Leave to dry and cool completely.

Note When you line the tart tin, make a straight edge to the pastry, so that it will eventually lie flush with the top of the baked filling.

ANTON MOSIMANN'S CHERRY STRUDEL

KIRSCHENSTRUDEL

STRUDEL AUX CERISES

Turn to page 16 in the apple section for full instructions for making strudel dough, and the method of rolling and baking it. Use this cherry filling instead of the apple version given there.

50 g (2 oz/1 cup) white breadcrumbs
20 g (¾ oz/scant ¼ cup) pine kernels
50 g (2 oz/⅜ cup) hazelnuts, toasted and skinned, then ground finely
1 kg (2 lb) stoned cherries
125 g (4 oz/½ cup) caster sugar
25 g (1 oz/¼ cup) cinnamon
100 g (4 oz/¾ cup) sultanas
a little kirsch

Do not fry the breadcrumbs, but mix them with the nuts. Put the cherries with the sugar, cinnamon and sultanas.

Spread the crumb and nut mixture over the dough, after brushing it with butter. Then put the cherries and sultanas on top. Sprinkle with a little kirsch (I use 2 tablespoons (3 tbs)).

Roll up and bake as for apple strudel.

CHOCOLATE AND CHERRY CAKE

Real Black Forest cherry cake – *Schwarzwalder Kirschtorte* – is an extravaganza of chocolate whirls and curls, cream and cherries. People without the skill of a German pastrycook, or the time to acquire it, may prefer to join me in a simpler version no less rich and wicked, if less glorious to contemplate. A warning – this is not a cake for those who suffer indigestion: serve it in very small slices in any case. Lighter versions can be made with cocoa-flavoured sponge cakes.

CAKE
100 g (3½ oz/½ cup) bitter or semi-sweet chocolate
2 tablespoons (3 tbs) milk
200 g (7 oz/⅞ cup) slightly salted or unsalted butter
a pinch of salt, or two, according to the butter
200 g (7 oz/⅞ cup) caster sugar
4 large egg yolks
4 rounded tablespoons (1 cup) flour
5 large egg whites

FILLING
500 g (1 lb) black cherries, sour if possible
300 ml (10 fl oz/1¼ cups) double cream
4 tablespoons (⅓ cup) single cream
2–4 tablespoons (3–5 tbs) kirsch (optional)
icing sugar

Break up and melt the chocolate with the milk in a large pudding basin, over a pan of simmering water. Stir as it begins to melt and remove from the pan as soon as possible. Cut up the butter and whisk it into the chocolate; put in salt and sugar. Beat well, then add the yolks gradually, one by one. Sift the flour into the basin and fold it in.

Beat the whites until they are stiff, but not too dry. Fold into the chocolate mixture gently, with a metal spoon.

Have the oven preheated to gas 6, 200°C (400°F). Brush three shallow sandwich tins, diameter 20–23 cm (8–9 ins), with melted butter. Dredge in flour, shake the tins about until they are completely coated; tip out surplus flour. Divide the cake mixture between the tins, smoothing it out. Bake for 20–25 minutes, or until the cakes are risen. If you are baking on two shelves, swop the tins around after 15 minutes. Put the tins to cool for

5 minutes on a wire rack. The cakes will shrink slightly from the side; they will also subside, but not tragically. Turn out the cakes and leave them on the rack to cool down.

The filling: stone the cherries, working over a large basin that will catch the juice that falls. Set the cherries aside. Pour the creams and the kirsch into a bowl and add a tablespoon of icing sugar: whisk until very thick, tasting as you go and sweetening further, or adding extra kirsch. Bear in mind the sourness or sweetness of the cherries.

To assemble: put a wide strip of foil across a wire rack (this helps when you transfer the cake to a dish for serving). On it place the second-best cake, upside down. Cover with half the cherries, then with half the cream. Put on the worst cake, upside down again, and cover with cherries and cream. Top with the best cake, right side up. Dredge evenly with icing sugar. Put in the refrigerator to give it a chance to settle down.

If you want to keep the cake overnight, cover the whole thing with plastic film and put it back in the refrigerator. If you are serving it later the same day, remove it to a plate or dish with the aid of the foil and a fish slice. Take away the foil carefully.

These instructions may sound fussy, but this kind of rich moist chocolate cake is extremely tender to handle.

VARIATIONS In winter, use either frozen cherries (well thawed and drained), or black cherry jam and barely sweetened cream.

For the cherries, you can substitute slices of pear poached in syrup.

For the cherries and cream, you can substitute a mousse of apricots, p. 35.

PICKLED CHERRIES
CERISES AU VINAIGRE
CERISES À L'AIGRE-DOUX

A paragraph I like very much in Elizabeth David's *French Provincial Cooking* is the one in which she describes her first acquaintance with pickled cherries. She was at Mère Germaine's restaurant at Châteauneuf du Pape. 'The hors d'oeuvre are beautifully served here, in small oval *raviers* of glowing deep yellow Provençal pottery. Brought on a basket tray which was left on the table so that we could help ourselves, they consisted only of olives, marinaded mushrooms, a rice and prawn salad, anchovies, these pickled cherries and the *tapénade* already described . . . the visual effect had been skilfully thought out, and against the yellow background these few simple things made a dazzling display of colour.'

This French recipe is similar to Mrs David's. Simplicity itself, cherries put to soak with a sweet spiced vinegar and left for at least a month:

1 kg (2 lb) morello, Kentish Red or Montmorency cherries
1 litre ($1\frac{3}{4}$ pt/scant $4\frac{1}{2}$ cups) white wine vinegar
500 g (1 lb) soft light brown sugar, or granulated
6 cloves
6 juniper berries, lightly crushed
thinly cut peel of a lemon
5 cm (2 inch) stick of cinnamon

Pick over the cherries, cutting the stalks down to a centimetre ($\frac{1}{4}$ inch) and throwing out any damaged fruit. Rinse and dry them thoroughly, then pack them into small preserving jars. Jam jars can be used, so long as you have close-fitting plastic lids for them: metal screw-top lids turn nasty in contact with vinegar.

Stir the remaining ingredients together in a stainless steel or enamelled pan. Bring to the boil and simmer for 10 minutes. Cover and leave overnight to cool and flavour the vinegar. Strain over the fruit, making sure it is completely covered. Keep in a dark, dry cupboard

Note In times past, cherries were kept in vinegar alone, with savory between the layers.

CHERRY BRANDY

To make this in France, people go looking for deserted cottage gardens to find the old-fashioned *guignes*, the black and red cherries that we call geans. They make a superb liqueur, but the cherries are not much fun to eat. My grandmother, who was a great cherry brandy-maker, used Kentish Reds which are a Duke cherry: and her cherries were worth eating (except when they were twenty five years old, by which time they were rather tasteless having given everything to the liqueur).

Whether in England or France or anywhere else, the method is the same. You fill a clean bottle or bottling jar almost to the top with cherries. Cut their stalks down to a tiny length first, then prick them several times with a darning needle and put them into the bottle as you go.

Pour in caster sugar to come about a third of the way up, then enough brandy to cover the fruit. Cork it up tightly, or close the bottling jar in the usual way.

Leave in a cool, dark place for as many years as you can bear to, or at least until Christmas. In time, the sugar dissolves. You can help it along by giving the bottle a gentle turn from time to time. Bottling jars can be turned upside down.

CHINESE GOOSEBERRY or KIWI

Chinese gooseberries have turned out to be a puzzle. They have suddenly descended on us, publicity, leaflets, all the paraphernalia of a press campaign, and yet they seem to be a fruit without a history. I remember being given one just before the war, a special treat – this beautiful fruit with its clear green colour and tiny fine black seeds. But where had it come from?

Chinese it certainly is. *Hortus Third*, the great American plant dictionary, says so. It even gives the Chinese name, *Yang t'ao*. But I have hunted in all my Chinese cookery books, in scholarly books on Chinese food, and there is no mention of it. It does not even creep in, unindexed, as a little garnish, an item in a general chop-up. Nothing at all. You would expect it to have been a favourite in the courts of T'ang and Sung, the pleasure of a beautiful concubine as the lichee was to Yang Kuei-fei. I have never seen it in books about Chinese painting, or sliding round the edge of a decorated porcelain bowl.

Yet New Zealand, where it now flourishes in excelsis, got the seeds from China. A Mr James McGregor brought some back and gave them to Mr Alexander Allison who grew them at Wanganui. His vines first fruited in 1910 (they take a while to get to that stage) and all the varieties now in New Zealand have been developed from those plants of *Actinidia chinensis* from Yangtze valley seeds.

Success came by chance, when Bruno Just, a horticulturalist, began selling bundles of Chinese gooseberry plants to dairy farmers in the Bay of Plenty, at Te Puke. Here the soil and climate were exactly right. The plants flourished and in 1937 one brave farmer, Jim MacLoughlin, staked out an acre in Te Puke. Another farmer 150 miles away did the same thing and wisely chose a particularly good variety called Haywards which had been developed by a famous nurseryman who specialized in exotics and unusual plants and was called Hayward Wright. This fine-flavoured fruit is the one that now dominates the enormous export trade. Areas of the Bay of Plenty are draped with Chinese gooseberry vines hanging over pergolas, stretching over long lines of T-bars. Their large pale yellow flowers, the five petals wide open to show a thick circle of gold stamens, hang down and turn into the plain egg-sized brown-skinned fruit that are beginning to be so familiar to us.

The first shipment abroad was to London in 1953, to the London firm of T. J. Poupart (who helped to popularize the courgette). In that year, 2,500 trays were exported. In 1981, six and a half million were sent out – 200,000 to this country. In France, the little fruit huddled in their trays in neat rows came to be known as *souris vegetales*, vegetable mice, which I think is a better name than the recently-adopted 'kiwi'. But I can see that if the American market was squeamish about Chinese gooseberry as a name – on account of politics, or dislike of the ordinary gooseberry? – it would take even less kindly to 'vegetable mouse'.

So it seems we have to put up with the name of kiwi, like it or not. And as the New Zealanders have done the work, they have the right to choose. One day, though, I shall find out why the Chinese have done nothing about the fruit beyond remarking that it was thirst-quenching, and greatly relished by the monkeys in the mountain districts of Shensi.

HOW TO CHOOSE AND PREPARE CHINESE GOOSEBERRIES

A big reason for the success of the New Zealand export trade is the obliging way in which this fruit puts up with storage. It is picked when hard, like pears, and allowed to develop slowly. In a cool place it will remain in reasonable condition for up to eight weeks. If it is kept in a chilled atmosphere, wrapped in plastic to promote humidity, it is good for up to six months. This explains the long season we have. The fruit is picked over six weeks only, starting at the beginning of May. By the end of July it is over. But we go on eating the fruit until after Christmas, into the New Year.

So the standard is high. The greengrocer has no problems of waste which makes avocados and persimmons difficult to stock. This must mean that we shall see more and more Chinese gooseberries. At the moment it is still in the exotic stage. Chefs in France have made it almost a badge of the *nouvelle cuisine, salads aux kiwis, sorbets aux kiwis, clafoutis aux kiwis*, are all a sign of chic. In a year or two when everyone knows them, they may be demoted, but at the moment they are at the top of the culinary pyramid.

The brown hairy peel is just tender enough to be eaten, and just tough enough to provide a shell so that the halved fruit can be scooped out handily. But mostly the fruit is peeled.

Chinese gooseberries have become a New Zealand and Australian tradition with passion fruit cream in a soft meringue shell (the Pavlova). With strawberries, they can be the star of a mixed fruit salad. With oranges, cucumber and salted cashew nuts, they can be served as a first or main course salad, especially with cold chicken and guineafowl at a summer lunch or buffet.

Puréed and mixed with the juice and zest of one lemon, they can be used to flavour a mousse (*see* the lemon section): decorate the top with slices of the fruit and chopped toasted hazelnuts.

If you enjoy sautés of chicken or pork in the Chinese manner, add slices of Chinese gooseberry at the end, with strips of green, red or yellow sweet peppers. Season with honey and light soya sauce, and serve with rice. They could be substituted for lychees in the sauce on p. 226.

Try fish with Chinese gooseberries, thinking of them as a substitute for grapes. A little lemon juice brings out their flavour, and clarified butter should be used to heat them through so that nothing spoils their delicacy. A number of recipes in the grape section could be made with Chinese gooseberries.

ESCALOPES OF CHICKEN WITH CHINESE GOOSEBERRIES

POULET AUX KIWIS

Here is a simple way of trying out kiwis in a salted dish for 2 people to see how you like the idea.

Remove the complete breast from a chicken, and skin it. Cut it in half down the centre, to separate the two sides. Then slice carefully through the two pieces, so that you end up with four escalopes. Beat them out, not too vigorously, between sheets of damp greaseproof paper. Turn them in seasoned flour to which you have added enough Cayenne pepper to turn it slightly pink.

Fry the escalopes in clarified butter until they are a pinkish-golden brown on each side. Remove them from the pan and keep them warm. Put in one large or two small Chinese gooseberries, peeled and sliced, and heat them through. Keep the stove low. When they are thoroughly hot, arrange them on the chicken pieces. If you have a few strips of sweet pepper, red or yellow for preference, they can now be added to the pan. Deglaze it with some muscat wine (Cypriot, Frontignan, Beaumes de Venise), about 6 tablespoons (½ cup), and 150 ml (¼ pt/⅔ cup) chicken stock. Reduce this liquid. Fish out the strips of pepper and put them on the chicken pieces. Add either some cream or a large knob of butter to the pan sauce, stir it in well, and then strain around the chicken. There should not be a lot of sauce, just a few spoonfuls.

PAVLOVA

A pudding made for Pavlova when she visited Australia in the Thirties. The soft and chewy-centred meringue is not, I think, original to Australia, but comes from the European *pâtisserie* tradition. What is original is the shape, and the idea of mixing the cream with passion fruit and Chinese gooseberries. Other fruit can be used – strawberries and passion fruit is another popular combination.

3 large egg whites
175 g (6 oz/¾ cup) vanilla caster sugar
1 teaspoon cornflour
1 teaspoon vinegar
250 ml (8 fl oz/1 cup) whipping cream
2 passion fruit
extra sugar
2–4 Chinese gooseberries, peeled, sliced

Use an electric beater if you can and whisk the whites until stiff. Add the sugar, beating until the meringue looks smooth and silky. Then mix in the cornflour and vinegar. Pile or pipe on to a baking sheet lined with vegetable cooking parchment, hollowing out the centre to form a nest.

Bake 1¼–1½ hours at mark ½, 130°C (250°F) until firm. When cool remove to a plate.

Pour the cream into a basin. Scoop out the pulp of the passion fruit. If you wish to sieve out the seeds, simmer with a little sugar just to warm the pulp which will then go easily through a sieve: it should in no way be cooked. Whisk the cream, flavouring it to taste with the passion pulp. Pile into the nest, then arrange the slices of Chinese gooseberry on top.

KIWI MERINGUE PUDDING

Here is a good way of serving fruit which should not be cooked – or which there is no need to cook – but which does benefit from a little warmth. This applies particularly to imported, hot climate fruit. I came across the idea first in an article by Sheila Hutchins in the *Daily Express*, with a recipe for a meringue-topped grape flan. She packed the fruit and sugar into a pastry case, pre-baked, and topped it with an unusual meringue mixture which only needed a very low temperature to cook it. The snag of the recipe was that the fruit gave out large quantities of very liquid juice: now I leave out the pastry and bake the pudding in a gratin dish.

Cover the base of the dish with sliced kiwi fruit and sugar them as required. Whisk the whites of two large eggs stiffly, then fold in one of the yolks, 6 tablespoons ($\frac{3}{4}$ cup) of caster sugar and 1 tablespoon ($\frac{1}{4}$ cup) of sifted flour. Do this with great care – it is easier if you beat the yolk first to make it more liquid.

Spread over the fruit. Bake at gas 1, 140°C (275°F) for 40 minutes or a little longer. The top should have a crunchy dry look, delicately browned.

BURNT CREAM WITH CHINESE GOOSEBERRIES
CRÈME BRULÉE AUX KIWIS

Burnt cream has been made in England since the seventeenth century, but it gained a new reputation when a chef at Trinity College, Cambridge, took it up at the end of the last century. Being a rich pudding, it was usually served with fruit. Not long ago, someone had the idea of putting grapes underneath the pudding. And why not? Chinese gooseberries taste even better than grapes, so do Sharon persimmons which can be cut into slices and eaten while they are still firm (unlike other persimmons which have to be very soft). Raspberries and poached sliced peaches and pears do well, too.

The success of burnt cream depends partly on the flavour of the cream you use. Loseley or other farm brands or *crème fraîche* are ideal. Indeed, if you can buy one of them, and are in a hurry, you can use it whipped, instead of making a custard.

about $\frac{3}{4}$ kg (1$\frac{1}{2}$ lb) Chinese gooseberries
1 litre (1$\frac{3}{4}$ pt/4$\frac{1}{2}$ cups) cream
thinly cut peel of a lemon
5-cm (2-inch) cinnamon stick
4 large eggs
4 large egg yolks
sugar

Choose a large gratin dish that will hold 1$\frac{1}{2}$ litres (2$\frac{1}{2}$ pt/6$\frac{1}{4}$ cups) at least Cover the base with peeled and sliced Chinese gooseberries.

Bring cream, peel and cinnamon slowly to just under boiling point. Strain on to eggs and yolks that have been beaten together in a large pudding basin: whisk together vigorously at first, then occasionally until you have a smooth cream. Put the basin over a pan of barely simmering water, and stir with a wooden spoon until very thick – the back of the spoon should be coated. Should the custard begin to show a hint of graininess, rapidly pour it into a processor or liquidizer and whizz at top speed for a few minutes. Pour over the fruit and chill for several hours.

A couple of hours before serving, sprinkle enough granulated sugar over the whole thing to make a ½-centimetre (scant ¼-inch) depth. Preheat the grill to make it as red as possible. Slip the dish underneath. Do not turn your back on it, but be patient and watch as the sugar melts to a marbled brown glassiness. If the grill heat is uneven, you will have to turn the dish.

Some people stand the gratin dish in a tray of crushed ice before grilling the sugar. This is a good idea if your grill is not very hot, as it prevents the custard overheating underneath and bubbling up through the sugar. But I have never found it necessary.

The quantities above make enough for ten or even a dozen people, depending on the rest of the meal.

MINI GRAND DESSERT

At Les Eyzies in the Dordogne, we stayed at the Centenaire for a night; at the end of the meal – mainly fish, and excellent – we had this delightful plate of ice cream and fruit.

In the middle, a scoop of coconut ice cream, surrounded with strawberries and raspberries like a necklace, carefully placed on top of a raspberry purée (sauce Melba, sieved raspberries with icing sugar to taste). And on top was a slice of kiwi fruit, worn a little like a badge. I laughed when I saw it and wished at the end that there had been more kiwi and a little less raspberry.

175 g (6 oz/2 cups) desiccated coconut
250 ml (8 fl oz/1 cup) single cream
caster sugar
300 ml (10 fl oz/1¼ cups) double cream
300 ml (10 fl oz/1¼ cups) Melba sauce or sweetened strawberry purée
24 magnificent strawberries, or raspberries
3 Chinese gooseberries, peeled and sliced

Bring the coconut and cream very slowly to the boil. Turn off the heat, cover and leave to infuse for about 10 minutes. Then whizz in the processor or liquidizer and sieve, adding 150 ml ($\frac{1}{4}$ pt/$\frac{2}{3}$ cup) very hot water to help get all the coconut liquid through. You should be left with rather a dry mass of coconut shreds. Sweeten the coconut cream to taste, and put in the freezer. When it is firm round the edges, but still slightly slushy in the middle, whisk the double cream and add sugar to taste. Then fold into the coconut ice, and return to the freezer.

To serve, put a scoop of ice on six plates. Carefully pour round the sauce or purée, so that the flat of the plate is covered. Arrange either strawberries or raspberries – whichever you have not used for the sauce – round the edge of the ice, with halved kiwi slices, reserving some to put on the top.

Note To help the smoothness of coconut ice, which is made without eggs, you can dissolve a dessertspoon of gelatine in the hot water before pouring it through the sieve.

KIWI AND TAMARILLO SALAD

A reader, Mrs Hayman, sent me this recipe which includes tamarillo, sometimes to be seen in our shops. Scoop out the flesh of two ripe soft tamarillos and chop into bite-sized pieces. Peel and slice two kiwis. Mix and add about 2 teaspoons sugar. Leave an hour at least. Tamarillo is slightly tart, a good foil for kiwi fruit.

CHINESE GOOSEBERRY CAKES

1) Make a lime cake (p. 219), without the syrup, using the lime juice to marinade Chinese gooseberry slices. Sandwich the cake with the slices, and whipped cream flavoured with coconut cream; and keep some of each back to decorate the top.
2) Use Chinese gooseberry slices for an upside-down pudding (hot) or cake (cold) (p. 347). Add 60 g (2 oz) desiccated coconut, and a little coconut cream if you have any.
3) Use Chinese gooseberries and whipped cream to layer in with rich chocolate cakes. Dredge the top with icing sugar.
4) If you want to add crispness to dishes with Chinese gooseberries, use toasted chopped hazelnuts, cashews or macadamia nuts.

CITRON, see GRAPEFRUIT

CORNEL or CORNELIAN CHERRY

Many people grow a bush of *Cornus mas*, the cornel or cornelian cherry, just as the Romans did, for the sake of its flowering in February and March. No leaves, just hard twisting branches of a dull flat brown, veiled with a greenish-yellow mist of tiny flowers growing straight from the branch. Later come the leaves, then the green 'cherries' which brighten to red in the autumn.

Owners of *Cornus mas* appreciate the fruit, too – for as long as the birds leave it to be admired – but only for its beauty. As food, they would be of Circe's opinion, if they thought about it at all. When Odysseus' men explored the island of Aeaea, and took refuge from terrifying beasts in the great porch of Circe's castle, she invited them in and turned them into pigs. Their minds she could not change. They shed tears in their pig-sties. 'But Circe flung them some acorns, beech-mast and cornel cherries, and left them to eat this pigs' fodder and wallow in the mud.'

To the ancients, it seems, cornel cherries were for wild picking and eating, by children perhaps, and not fit food for man. In the *Aeneid*, Virgil recounts that the marooned Achaemenides complained that he survived on a miserable diet of berries and stony cornels. Though if one only had wild strawberries to live on, it would not seem much of a way to survive either. 'Pliny, indeed, had no great fondness for cornels, for he says that they were dried in the sun, like prunes just to show that there was nothing not created for man's belly.'*

In fact the belly – or its ills, such as dysentery and looseness of the stomach – was regarded as the main beneficiary of cornels. They were eaten with meat, from which pairing the following recipes have developed. The physician Dioscorides, of the first century AD, wrote of their efficacy, commented on the olive-like appearance of the fruit and said they were sometimes pickled like olives. His herbal was translated into English by John Goodyer (who introduced Jerusalem artichokes into this country) in the early 1650s. It was never published, but perhaps John Evelyn saw it and pinched from it the idea of pickling olives. 'I have some

***The Trees, Shrubs, and Plants of Virgil* John Sargeaunt, 1920.

years since invented the pickling of Cornelians, and have frequently made them pass for olives of Verona.'

Evelyn gathered them when they were beginning to blush. Then they went into brine – dissolve sea salt in water until an egg will float in it – with fennel and bay leaves. The jar was stoppered tightly, and left for a month. If the pickle was too strong, it could be diluted with water before the jar was put on the table. (I imagine in exactly the way we put a jar of gherkins on the table with pâté.)

Unfortunately, we are never in Wiltshire when our cornelian cherries are at this stage, so I have never been able to try out the pickling. By the time we get home, the berries are ripening to red, and the race with the birds is on. If I win, I make them into a jelly to go with Christmas turkey, a jelly of the most beautiful pink colour. Rarely are there enough to give me more than two small pots. For the interest of those whose harvest is more lavish, I give a recipe from Richard Bradley's *The Country Housewife and Lady's Director* of 1732. Again, I have not tried it, but the idea of white wine seems a good one, worth trying in small quantity.

CORNELIAN CHERRY PRESERVE

Richard Bradley was given this recipe – and a number of others for preserves – by a 'Barbadoes lady'. The idea of pickling cornels in syrup must have seemed more attractive then Evelyn's olive treatment, especially with sugar from the West Indies plantations becoming more abundant and therefore cheaper. In earlier times, sugar had been a sign of wealth: the blacker your teeth, the richer you were. During the eighteenth century, it began to swing the other way, until now I would say that the richer you are the less sugar you buy.

This style of preserve is eaten in Russia on its own, in little dishes, with tea – glasses of tea. The writer Konstantin Paustovsky was in Sukhum in the early 1920s, in the tobacco republic of Abkhazia, and describes how he would sip his tea while eating spoonfuls of cornel jam, 'the sourest in the world . . . the syrup of this jam was as blood-red as a mountain sunset'.

Weigh the cherries, then weigh out an equal quantity of sugar and warm it. Layer fruit and sugar in a preserving pan and pour on enough white wine to wet the sugar and prevent it burning as it melts. Set on a low heat, and give an occasional stir once the panful heats up. Boil vigorously until the fruit is cooked. Skim and pour into sterilized jars or bottling jars and seal.

These cherries could be drained of their syrup and used as an edible decoration for meat and game, when you are serving cornel jelly (*see* below).

CORNEL JELLY

A delicately sharp jelly that is ideal to serve with meat such as turkey, duck, lamb, pork – think of it as taking the place of red currant or cranberry jelly.

Having picked your cornels, weigh them and weigh out an equal quantity of windfall apples – or cooking apples, of course. Cut out any blemishes before you weigh them, then chop them roughly into a pan. peel, core and all. Put the cherries in as well, and cover with water. Bring to the boil, put on the lid and lower the heat so that fruit simmers slowly until very tender and pulpy.

Strain out the juice through a jelly bag. Boil it with sugar in the proportion of 600 ml juice to 500 g sugar (in other words a pint to a pound/2½ cups to 2 cups). When setting point is reached, pot in small sterilized glass jars of a size suitable for one meal's serving.

VARIATION The debris can be heated through with a little more water, then sieved through, so that you get a purée free of peel and pips etc. Weigh it, and cook it with an equal weight of sugar until it is very thick indeed. This gives you a cornel cheese, which can be served in slices with grilled meat.

NORWEGIAN VENISON WITH GJETOST SAUCE

Reindeer is the original intention of this recipe, which came to me via an American friend, Evan Jones, but it works well with venison and beef, or with pheasant and guineafowl.

The sharp sweetness of the sauce comes from any tart fruit jelly. I find cornel jelly, with red currant as an alternative, the best. The other main flavouring ingredient is *gjetost*, the brown caramelized cheese that is usually sold here in small, foil-wrapped bricks. Some people say they dislike it. I suspect this may be because they eat it in wedges and find it sickly. *Gjetost* should be shaved from the block in transparent slivers. On dark rye bread it is good, and in this sauce, excellent.

1½ kg (3 lb) venison or beef roasting joint
butter, salt, pepper

generous ½ litre (¾–1 pt/2–2½ cups) venison or beef stock
heaped tablespoon (¼ cup) flour
150 ml (5fl oz/⅔ cup) soured cream
60 ml (2 fl oz/⅓ cup) double cream
30 g (1 oz) gjetost, *in slivers*
cornel or red currant jelly

Put meat on a rack and butter it generously all over (take the precaution of larding it, if you like). Season. Pour three-quarters of the stock into the roasting pan underneath. Cook 15 minutes at gas 7, 220°C (425°F) then lower the heat to gas 4, 180°C (350°F) and leave for an hour. The meat should be pink, whether venison or beef.

Meanwhile brown a generous tablespoon of butter in a heavy pan, stir in the flour and add the rest of the stock and the creams. When smooth and very thick, set aside until the meat is cooked. Then degrease the pan juices and add them to the sauce, stirring until smooth again. Taste and boil down if necessary to strengthen the flavour. Add seasoning.

Add *gjetost* and half-teaspoons of jelly gradually until you get a balance of flavours that appeals to you. Keep the sauce under boiling point as you do this. Pepper it well. Surround the meat with potatoes and watercress. Pour over some of the sauce, serve the rest separately.

Note With pheasant and guineafowl, treat the cooking as a braise and leave the bird until completely cooked.

CRANBERRY

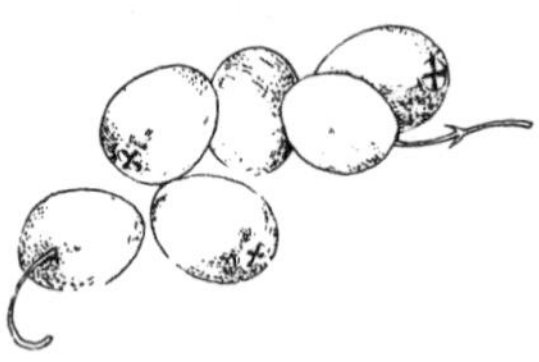

William Cobbett's experience in America made him quite sure that cranberries were 'the finest fruit for tarts that ever grew'. Whether you agree with him or no (I think I don't agree), cranberries do have their special history.

Why are they called 'cranberries'? Americans, who like them as much as Cobbett did, say because the flowers are like a crane's head. More probably and practically, if less poetically (why bother about the flowers), it is because cranes (which used to be common in Britain) and cranberries are both of them at home in bogs.

When the early settlers reached Plymouth and Cape Cod, they were pleased to find what they supposed were bigger and better cranberries than the ones they had known at home – fruit, in fact, of an American species *Oxycoccus macrocarpus*.

At home, in the boggy lands of the north of England and of Scotland, cranberry tarts were made from the smaller fruit of a different species, *Oxycoccus palustris* – tarts you won't find mentioned in our early cookery books because they belonged to the unwritten tradition of peasant cookery. Peasants brought their cranberry tarts – and their bilberry tarts – in to market to sell. 'Cranberry', too, is a word not to be found in polite speech of the sixteenth and earlier centuries, but it probably existed as a dialect word: in Scotland and the North they still talk of *Oxycoccus palustris* berries as 'crones' or 'cranes' as well as cranberries.

Certainly it is true – about the only altogether certain thing in this matter of the name – that 'cranberry' very quickly became a standard English word once the larger fruit of this American *Oxycoccus macrocarpus* began to be imported – which happened soon.

Cranberries are among the good fruits (like the related bilberries and blueberries) which grow by nature in what would otherwise be waste land – for cranberries, it is acid sandy bog, waste land and water. They were a gift, to be had for the picking; and what made the New Englanders value them so much, in their first straitened existence, was that cranberries keep fresh so long – from autumn into winter – because of their waxy skin.

The taste for cranberries grew. Some two hundred years after the settlers came, cranberry cultivation began, to supplement the huge wild harvest of every fall. Ingenious Americans found out how to manage cranberry bogs – how to devise them, too; and eventually they began raising modern improved strains.

Cranberry growing, canning, freezing and the rest of it, is business now for a world market, carried on in the old cranberry lands of Massachusetts, and in New Jersey, in Wisconsin between the Lakes, in Oregon and Washington, in British Columbia, and between Quebec and Montreal.

One of the very splendid sights of the American fall in some districts is the harvesting of the cranberries by floating them. Water is let into the bogs, the cranberries are mechanically detached, and float, in acres of scarlet. Then they are rafted into islets of scarlet – scarlet on blue water – which are towed to collecting points on the water's edge. Be careful with cranberry recipes in publicity leaflets and magazine articles: quantities of sugar are often far too high.

Remember that cranberries cooked with sugar have tough skins: always sweeten when they are tender.

BASIC CRANBERRY SAUCE

Weigh a packet of cranberries if the measure is not given on the wrapping, then tip them into a saucepan. Add water or orange juice to come almost to the top but not quite. The liquid should show through the berries without floating them from the base of the pan.

Simmer without a lid until the berries start to pop open – this takes place quite rapidly. The more berries burst, the softer and thicker the sauce will be. Four minutes gives the consistency we like.

Remove from the heat and stir in half the cranberries' weight in sugar. Most people will find this sweet enough; too much sugar brings out the bitterness.

I find that a 175-g (6-oz) packet of cranberries makes enough sauce for 6 to 8 people. If you make more, it can be kept in covered jars in the refrigerator.

CRANBERRY AND ORANGE SAUCE

Can be served with venison and hare, as well as with turkey.

Squeeze oranges until you get 375 ml (12 fl oz/1½ cups) orange juice – do not be tempted to use concentrated or canned juice. Put it into a pan and add 500 g (1 lb) cranberries. Cook, uncovered, as above.

Remove from the heat and sweeten to taste – you will need from 200–250 g (7–8 oz/¾–1 cup) sugar.

Cranberry sauce can be potted into rinsed jam jars, dried upside down on a rack in a low oven. It keeps well in the refrigerator.

TWO CRANBERRY TARTS

My two favourite cranberry tarts. If I want to serve a crisp sort of pudding, after a main course such as a *blanquette* or a hare stew, I make the first one. If the main course is a grill or a roast joint, I make the meringue version with its melting softness.

The cheese mixtures balance the strong cranberry taste in a most successful way. At other times of the year, mashed raspberries, or lightly-cooked red or black currants or bilberries, could be used instead of cranberries.

1) PASTRY

125 g (4 oz/1 cup) flour
125 g (4 oz/1 cup) unblanched almonds, ground
125 g (4 oz/½ cup) butter
about 2 tablespoons (3 tbs) single cream
extra cream or beaten egg for glazing

FILLING

250 g (8 oz/1 cup) curd or medium fat soft cheese, drained weight
375 ml (12 fl oz/1½ cups) cranberry sauce, p. 141
about 3 tablespoons (⅜ cup) sugar
60 g (2 oz/½ cup) toasted, split almonds

Make pastry in the usual way. Line a tart tin of 23–25 cm (9–10 in). Roll out the trimmings and cut enough strips to make a lattice when the tart is filled.

Mix the cheese with half the sauce, and add sugar gradually to taste. Spread over the pastry. Put the remaining sauce over the top. Decorate with the pastry strips. Brush them with single cream or egg to glaze. Put the toasted almonds into the gaps, pushing them in lightly so that the tart bristles with little spikes. Bake at gas 5, 190°C (375°F) for about 35 minutes, or until pastry is nicely browned. Serve warm with cream, clotted cream is best.

2) Assemble the almond pastry ingredients, as in the previous recipe, but blanch the almonds before grinding them. Make up the paste and line a 20–23 cm (8–9 inch) tart tin.

FILLING
250 g (8 oz/1 cup) full fat (cream) soft cheese, or Philadelphia cheese
375 ml (12 fl oz/1½ cups) cranberry sauce, p. 141
2 large egg yolks
caster sugar
60 g (2 oz/½ cup) toasted, slivered almonds
2 large egg whites

Mix the cheese, half the cranberry sauce and the yolks together. Sweeten to taste. Spread on the pastry, with the remaining cranberry sauce on top. Scatter on half the almonds. Bake for about 30 minutes at gas 5, 190°C (375°F), and remove from the oven. Raise the heat to gas 6, 200°C (400°F).

Meanwhile, whisk the egg whites until stiff, stir in 30 g (1 oz/2 tbs) sugar, and whisk again until thick and soft-looking. Fold in a further 30 g (1 oz/2 tbs) sugar. Heap on to the tart, right to the pastry edges. Sprinkle with the remaining almonds and a rounded tablespoon (2 tbs) of sugar. Put back into the oven for 15 minutes, or until the top is nicely browned. Serve warm.

Cream is not really necessary, but some people like a little with the tart – use whipping cream beaten until very thick and just pourable.

BAKED CRANBERRY PUDDING

By adding rather a large amount of sugar to the usual genoese type of sponge mixture, you get a crisp meringue-like brown top which is particularly successful with the sharp cranberries.

Make it in a pie plate of almost 1-litre capacity (1½–1¾ pt/3¾–4¼ cups) and 3–4 cm (1¼–1½ inch) depth.

Butter the plate generously. Sprinkle on 250 g (8 oz) roughly chopped cranberries, 100 g (3½ oz/scant ½ cup) demerara or light brown sugar, and 100 g (3½ oz/scant cup) chopped walnuts.

Beat together 1 large egg and 1 large egg yolk with 175 g (6 oz/¾ cup) granulated sugar; when they are thick and foamy, fold in 100 g (3½ oz/scant cup) flour. Melt 125 g (4 oz/½ cup) butter, pour it down the sides of the mixing bowl, then fold into the mixture. Spread over the fruit and nuts.

Bake at gas 3, 160–170°C (325°F) for 45 minutes, or until the top is brown and crisp.

Serve with cream whipped until thick, but not stiff.

WINTER PUDDING

A winter version of summer pudding, with a lively taste – which means that it goes quite a long way, enough for eight people, and that it should be served with plenty of cream.

Start the pudding two days in advance, or more if you like as it keeps well. Use good bread, unbleached or white. The problem with this style of dish can be a shortage of juice – the bread must take on the colour of the fruit, or it looks pallid and uninviting. Cook the cranberries slowly, the lid of the pan jammed on tightly, so that as little liquid as possible evaporates. Counsel of despair: boil up a few extra cranberries or raspberries or both in some water with sugar to taste, to make more juice.

½ kg (1 lb) frozen raspberries
175 g (6 oz) cranberries
caster sugar
about half a large loaf of bread

Thaw raspberries overnight in a nylon sieve with a bowl to catch the juice. Next day, cook the cranberries in the juice, plus 300 ml (½ pt/1¼ cups) water. Cover the pan and simmer until they pop and look mushy. Stir in the raspberries, turning and mixing them to heat through. No more than a minute. Take pan from the stove, sweeten to taste and tip the fruit back into the sieve. Leave to cool.

Cut enough wedge-shaped pieces of bread, and a circle, to line a pudding basin of generous 1-litre (2-pint/5-cup) capacity. Starting with the circle, dip one side of each piece into the juice from the cooked fruit, and line the basin. Fill up any tiny chinks in the lining, with small bits of bread. Spoon the fruit into the cavity, and put another circle of bread on top as a lid. Pour remaining fruit juice over the lid and round the edge.

Weight the pudding down with a plate that sits inside the rim of the basin. Chill until next day if possible. Slide a knife blade between basin and pudding before turning it out. If there is to be any delay in serving, put the basin back over the turned-out pudding to prevent collapse. This is a useful tip – you can turn out the pudding before the meal, put on a pale dress and not worry about stray purple splashes.

CRANBERRY STREUSEL CAKE

Fruit crumbles are often attributed to America. I suppose that the idea of the crisp topping came from Austrian streusel cakes, via Jewish immigrants. A clever idea to lighten rather solid yeast cakes. And another clever idea to turn it into a pastry substitute for fruit pies. This recipe combines the virtue of both ideas. Be careful not to overdo the cranberries, and remember that the cake can be served hot or warm as a pudding – with cream – as well as cold with a cup of coffee.

CAKE
1 large egg, size 1 or 2
60 g (2 oz/¼ cup) soft butter
90 g (3 oz/scant ½ cup) demerara sugar
90 g (3 oz/¾ cup) self-raising flour
½ level teaspoon baking powder
2 tablespoons (3 tbs) cranberry or orange juice
or milk
or orange liqueur

TOP
60–75 g (2–2½ oz) cranberries, coarsely chopped
90 g (3 oz/scant ½ cup) demerara sugar
90 g (3 oz/¾ cup) plain flour
60 g (2 oz/¼ cup) butter
½ level teaspoon cinnamon

Either mix ingredients together in order given, beating well, or tip everything together into an electric beater bowl or processor and whizz until smooth. Line 2 sides and the base of an 18-cm (7-inch) square tin with a long strip of Bakewell paper or foil. Butter the whole thing generously and put in the cake mixture, spreading it level with a knife.

Scatter the cranberries evenly over the top. Mix remaining ingredients together to a crumble consistency – this is the streusel – and put over the cranberries.

Bake at gas 4, 180°C (350°F) for about 50 minutes, or until cooked. Cool for a couple of minutes in the tin, then lift the cake on to a wire tray with the aid of the ends of the paper strip, or on to a serving dish if the cake is to be eaten warm.

To serve, cut across twice in each direction, making 9 pieces.

DECEMBER KRINGLE

This Danish cake is a winner, not too sweet, cheap and easy to make. Normally, it is just filled with butter, sugar and cinnamon, but I like the tartness of cranberries in the middle. The icing becomes rather a hectic pink with the cranberry juice: you might prefer to make half white, and half pink.

250 g (8 oz/2 cups) strong plain bread flour
1 packet Harvest Gold yeast
or 15 g (½ oz/½ cake compressed yeast) fresh yeast
5 tablespoons (6 tbs) warm milk
100 g (3½ oz/scant ½ cup) slightly salted butter
1 large egg, beaten

FILLING
250 g (8 oz) chopped cranberries
125 g (4 oz/½ cup) sugar
60 g (2 oz/½ cup) toasted, chopped hazelnuts

TOP
1 beaten egg
2 heaped tablespoons(3 tbs) chopped hazelnuts
1 heaped tablespoon sugar
½ quantity glacé icing on p. 468, made with cranberry juice (see *below)*

To make the kringle dough, mix flour with Harvest Gold yeast, or with fresh yeast mashed with the milk and left 10 minutes to froth. Melt the butter and add to the flour, plus the milk if not used with the yeast. Finally bind to a dough with the egg. Knead lightly on a floured surface. Rinse out the bowl if necessary, dry it and brush with oil. Put the dough back into it. It rises better in a close, steamy atmosphere – tie into a carrier bag and leave in a warm place until doubled, 50 minutes, or more depending on temperature.

Meanwhile cook the cranberries with ¼ litre (8 fl oz/1 cup) water until they begin to soften. Drain off and keep the liquid to make the glacé icing. Stir sugar and nuts into berries.

Break down the dough. Roll into a strip about 75 cm (30 in) long and 14 cm (5½ in) wide. Spread tepid or cooled cranberry filling along the centre, then flip over the two long edges to overlap each other and enclose it. Use a little beaten egg from the top ingredients to brush the underneath edge, so that the two layers stick together. Bend the long strip in three so that you can lift it to a parchment or greaseproof-lined baking tray. Shape the two ends round to make a giant pretzel (as you might fold your arms)

Envelop in a plastic carrier, loosely bagged above it to avoid the dough sticking. Leave to prove for 20–30 minutes.

Brush over with the beaten egg, sprinkle with chopped hazelnuts and the tablespoon of sugar. Bake at gas 6, 200°C (400°F) until golden brown – this takes at least 20 minutes.

Allow to cool to tepid, then run trails of icing over the top. Best eaten the same day, while the dough is still tender, and if possible slightly warm.

CRANBERRY TEA BREAD

A pretty-looking bread with its mixture of red berries, orange-coloured peel and nuts. Leave for a day before slicing and buttering, as it improves with keeping. I match the oil to the nuts, almond oil to almonds, walnut oil to walnuts and hazelnut oil to hazelnuts. These can be bought from Justin de Blank groceries, by post from Robin Yapp the wine merchant of Mere, Wiltshire, and from shops such as Neal's Yard or Culpeper's, and shops supplied by Taylor and Lake.

½ kg (1 lb/4 cups) sifted flour
225 g (7 oz/scant cup) sugar
4 level teaspoons baking powder
1 level teaspoon salt
1 large egg, beaten
375 ml (12 fl oz/1½ cups) milk
2 tablespoons (3 tbs) nut oil or melted butter
125 g (4 oz) cranberries, coarsely chopped
60 g (2 oz/⅓ cup) chopped mixed candied fruit and peel
60 g (2 oz/½ cup) chopped almonds, walnuts or hazelnuts
extra flour

Sift dry ingredients together. Add next three ingredients. Toss cranberries, candied fruit and nuts in a little extra flour, shaking off any surplus, then stir into the mixture. Put into a large 23-cm (9-inch) long loaf tin, or into two smaller tins. Bake at gas 4, 180°C (350°F), until a skewer pushed into the loaf comes out clean – about 70 minutes for one large loaf, 50 minutes for two smaller ones.

Keep at least 24 hours, wrapped in foil after it has cooled down, or stored in an air-tight plastic box. Serve sliced and buttered.

CUSTARD APPLE, see CHERIMOYA

DAMSON, see PLUM

DATE

A new era for dates in cookery and as table fruit has come in with fresh dates from Israel, in place of the over-sweet sticky box dates in their long round-ended boxes (*deglet nour*, incidentally, means 'beautiful maiden'). A few years ago now, I read in a vegetarian leaflet that these box dates were washed in detergent and then dipped in syrup (p. 156): how true this still is I do not know, but it is a story to match the one I was brought up on, about compacted dates in solid packages – they had been crushed together by sandy unwashed feet. One tip I was and remain glad of, always split dates and remove the stone before eating them: sometimes an insect has lain its eggs inside, and although I doubt one would come to much harm, I feel such dates are to be avoided.

A favourite book of mine, rather a batty one, is Dr Fernie's *Herbal Simples*, and after saying in its solemn way that dates 'are the most wholesome and nourishing of all our imported fruit', adds (writing before 1914) that 'a wholesome, sustaining and palatable meal' of dates may be had for one penny with a little bread; also that dates are the best portable food for jurymen, 'preventing exhaustion and keeping active the energies of mind and body'. (And for walkers or students on long train journeys, I can recommend the date nougat given later in this section.)

Dr Fernie is not surprised by this because the date palm was the Tree of Life in the Garden of Eden.

Certainly, it was the tree of life to Bedouin moving with their black tents through the deserts of Arabia. Though, as Tom Stobart points out in his *Cook's Encyclopaedia*, this was in pre-oil days. He goes on to quote a story about a sheik's banquet 'which included bowls of *leben* (buttermilk) with lumps of rather dirty-looking butter floating in it and dates. After the meal of a whole sheep or lamb (with the eyes and tail fat as the guest's particular right), the guest drinks the buttermilk and eats the dates with the bits of butter. If you care to try, you will find that dates actually do go well with butter and with yoghurt.'

Agreement with this opinion comes from France: 'If you love elegance, replace the date stone with a small bit of butter' – unsalted butter. And from India, as you will see from the fried date recipe later, which also includes clotted cream. Dates and chopped toasted nuts are delightful in home-made yoghurt, the kind with a crust on top that comes from adding extra cream to the milk.

The name 'date' comes in a roundabout way from the Greek *dactylon* which meant a toe (as well as a finger). Dates were dusky Arab toes. And more than toes, I would guess from the Shakespearean jokes about date and quinces in pies.

HOW TO CHOOSE AND PREPARE DATES

By convention, the impacted blocks of dried dates are kept for cooking, the sticky box dates for eating as a fruit. Now with the fresh dates, we have a general purpose kind that can be used successfully in both ways. The date harvest around Jericho takes place in the late autumn, but fresh dates freeze well which means they can be bought all the year round. Store them in the refrigerator.

Some people do not care much for their tough papery skins. This is easily removed. Nip off the stalk end and squeeze from the opposite end. The date will pop out.

As I remarked in the introduction, it is a good idea to slit open dates and remove the stones, so that you can make sure there are no insect eggs inside.

The cavity of dates can be filled with mixtures based on soft cheese, either the full fat cream type, or the medium fat curd cheese. Cottage cheese can be used, too, but requires a little cream to bind it: even then it is difficult to handle, although it tastes perfectly all right with dates. With such dry cheese, you would do better to mix it with chopped dates and a little cream and grated lemon peel, and use it spread on toast.

Some of the most delicious stuffed dates I ever ate, had a little orange liqueur mixed in with the cream cheese (Sabra liqueur from Israel). The dates were skinned and rolled in coconut and chopped nuts. They looked attractive in long rows on a tray, alternating with other sweetmeats. Flavourings for cream cheese might also include chopped drained ginger, walnuts or toasted hazelnuts, raisins and sultanas soaked in liqueur, or tiny bits of crushed crisp bacon.

CHICKEN (OR LAMB) TAJINE WITH DATES

The charm of a Moroccan fruit *tajine*, according to Madame Guinaudeau, in *Fez: Traditional Moroccan Cooking*, 'consists in the strange mixture of sweet fruits combined with the sharp shock of ginger'. Often the quantities of fruit – apples, pears, quinces – are equal to the weight of the meat.

Prunes being dried are added in half quantity, unsoaked. To our taste this can be too much, especially with quince or dates.

large chicken, jointed, or 1½–2 kg (3–4 lb) lamb cut in large pieces
3 large chopped onions
3 tablespoons (⅓ cup) butter
1 teaspoon ginger
pepper, salt
3–4 tablespoons (¼–⅓ cup) chopped coriander or parsley
pinch saffron
375 g (12 oz/1¾ cups) fresh dates, stoned

Put meat, one onion, the butter, half the ginger, pepper, pinch of salt and coriander or parsley into a wide sauté pan. Sizzle over a high heat for a minute or two, turning everything about.

Dissolve saffron in a little warm water, then tip it into the pan with more water almost to cover the meat. Simmer until nearly cooked, without a lid on the pan so that the liquid reduces by about three-quarters. Put in the remaining onion.

When the meat is tender, remove it to a serving dish and keep it warm. Add the dates to what is now an onion purée. Continue to simmer for 20–30 minutes: the dates should not break up. Taste from time to time and add the rest of the ginger if you like. Put in more salt if needed. Pour off fat, then cover the meat with the purée-sauce.

Serve with couscous or boiled rice or pitta bread.

FRIED DATES

I came across this recipe in Madhur Jaffrey's *Invitation to Indian Cooking*. For a quick sweet end to an impromptu meal, it is much to be recommended. Mrs Jaffrey would not approve I am sure, but oatcakes go well with the sweetness and the richness: they have a calming effect. Three fresh dates are enough for most people, when presented in this way, although you might do a couple more – say twenty for six people – for those who like sticky sweet things particularly.

Stone the dates. Put 250 ml (8 fl oz/1 cup) clotted cream into a bowl (or half-whipped double cream) and mix in a generous tablespoon of slivered pistachios.

Fry the dates in clarified butter quickly – 20 seconds is the maximum time. Keep them moving. Divide among individual warmed dishes. Serve immediately with the cream.

JEFFERSON DAVIS TART

Jefferson Davis was head of the Confederacy of the Southern States of America, until defeat in 1865 by the North at the end of the Civil War. It seems inappropriate that anyone who shared in that experience of carnage should have a sweet frothy-topped tart named after him – Pavlova cake, yes, but gâteau Petain, Montgomery sponge, Rommelstrudel, Haig fairy cakes? I think not.

Like a number of other English dishes, transparent pie or tart has survived in America when it has disappeared here. The holding mixture of eggs, butter and sugar – in other words, a custard with melted butter taking the place of cream or milk – sets to a semi-transparent firmness. Flavouring items differentiate one version from another. Two favourites here were sweetmeat pudding and Duke of Cambridge pudding (p. 199) and I reckon they are due for revival. In the States, especially in the South, chess tarts, Kentucky pie, pecan or black walnut pie and this date and nut tart are still very much alive.

For me, the tart is improved by using all dates, rather than half dates, half raisins, and I halve the quantity of sugar to the amount listed below (it is still very sweet). A good pudding for a family lunch party.

For a 23-cm (9-inch) tart tin, lined with plain shortcrust pastry, mix together the following ingredients in the order given:

125 g (4 oz/½ cup) softened butter
150 g (5 oz/⅔ cup) light brown sugar
3 large egg yolks
250 ml (8 fl oz/1 cup) whipping cream
1 level teaspoon cinnamon
½ teaspoon allspice
⅓ nutmeg finely grated
6–8, stoned, chopped dates
100 g (3½ oz/½ cup) raisins
60 g (2 oz/½ cup) roughly broken pecans, or walnuts

Bake at gas 6, 200°C (400°F) for 10–15 minutes, then down to gas 3, 160°C (325°F) for a further 20 minutes.

Top with 3 large egg whites, whisked and sweetened with 125 g (4 oz/

½ cup) caster sugar, to make a meringue. Put back into the oven until nicely browned.

Note Kentucky pie has no nuts or spices; chess tarts are made small with raisins and nuts, no dates or spices.

DATE KICKSHAWS

Kickshaws, as their name implies, should be served in small quantity. They are light, short and fattening. They make a delightful snack with coffee, when everyone is sitting round in the kitchen and talking, and can be eaten straight from the pan. If you have an electric deep-frier so that you can fry the kickshaws at the last moment with the minimum of fuss and the maximum of safety, they are a good way to end a dinner party.

Allow three dates – enough for three fritters – for each person. Use fresh dates if possible. Stone and chop them coarsely. Add the finely-grated zest of an orange for each 9 dates, and some chopped nuts if you like.

I use shortcrust pastry, as it fries to an unusual sandy texture. Puff pastry gives a less friable result. Roll out the pastry thinly and cut two circles for each fritter, using a large scone cutter of about 6-cm (2½-in) diameter. Brush the rim of half the circles with water, and put a teaspoonful of filling in each centre. Put on the remaining circles, pressing and twisting the edges together to make a firm seal. Chill until required for cooking, then deep-fry at 180°C (350°F) or a little higher, until a light golden-brown. Keep warm on a thick layer of kitchen paper, and serve as quickly as possible.

Note Of course you can use fila pastry, brushed with melted butter and cut into strips: put some filling at one end, fold a corner down over it, then continue to turn the filling over and over until you end with a triangular cushion. Repeat with more strips.

DATE MERINGUE CAKE

An elegant cake with contrasting flavours of date, almond, apricot and chocolate. They work well together, none being too dominant. The cake may seem small, but it is quite enough for 8 people at the end of a dinner. If you prefer, you can always bake the meringue in two larger tins. An unusual and very good cake which tastes even better if kept until the following day.

250 g (8 oz/1¼ cups) fresh dates
2 level teaspoons flour
200 g (6½ oz/1½ cups) blanched whole or ground almonds
150 g (5 oz/⅔ cup) sugar
4 egg whites

FILLING
150 ml (5 fl oz/⅔ cup) whipping cream
3 tablespoons (scant ¼ cup) apricot jam
100 g (3½ oz/good ½ cup) good plain chocolate
3 tablespoons (¾ cup) icing sugar
2 teaspoons butter
5 fresh dates

First prepare the cake. Roll the dates in flour, so that they don't stick to your fingers. Remove the stones and cut in small bits. Put into a bowl. If you use whole almonds, grind them and mix with the dates: put ground almonds straight in. Add the sugar and mix well. Finally fold in the whites stiffly beaten.

Line three 20-cm (8-inch) sandwich tins with Bakewell paper and brush them with melted butter. Spread the cake into the tins, and cook at gas 4, 180°C (350°F) for 45 minutes. Keep an eye on the cake after 30 minutes. Cool on a wire rack.

Whip the cream. Spread two of the cake layers with jam, then with cream, reserving some for decorating the top. Sandwich together.

Melt the chocolate and sugar with a tablespoon of water in a basin over simmering water. When it is smooth, remove the basin and add butter to it. Pour in one go over the top of the cake, and turn the cake to spread the icing about (try to avoid using a knife).

When the chocolate is set, pipe 8 blobs of cream round the cake and one in the middle. Cut the dates across the middle, throw away the stones, and put them cut side down, dome side up, on the blobs of cream. The tenth bit will not be needed – eat it. Store cake in the refrigerator.

DATE AND WALNUT BROWNIES

A good cake for family tea, from America.

60 g (2 oz/¼ cup) butter
200 g (7 oz/scant cup) soft brown or demerara sugar
1 egg, beaten

100 g (3½ oz/scant cup) flour
1 level teaspoon baking powder
½ level teaspoon salt
60 g (2 oz/¼ cup) each stoned dates and walnuts
vanilla essence (optional)

Melt butter over a low heat. When liquid and tepid, mix in the sugar, stirring until it has dissolved. Remove pan from heat – it should only be warm – and mix in the egg. Sift together flour, baking powder and salt and mix that into the pan. At this stage you can add ¼–½ teaspoon vanilla essence, an ingredient the Americans seem to be rather keen on.

Cut the dates into pieces – not too small. If you use fresh dates, remove their skins: with other kinds there is no need to do this. Mix with the walnuts into the pan.

Line a square shallow baking pan, measuring 18–20 cm (7–8 in), with Bakewell paper. Brush it over with melted butter. Put in the cake mixture. Bake at gas 4, 180°C (350°F) for about 25 minutes. Check after 20, but if the skewer comes out sticky leave another 5 minutes, or until it comes out clean. Baking time for brownies can be tricky. If you leave them longer than is absolutely necessary, they can be very hard on the outside.

Remove from the oven, cut into squares straightaway and leave in the tin to cool. The pleasure of this kind of cake is the crunchy top and the soft inside.

DATE NOUGAT

Thanks to a vegetarian father, I early became a devotee of Mapleton's fruit slice, made from crushed dates and other dried fruit, sandwiched between sheets of rice paper with coconut cream in the middle. Here is a homely version, which I like better as it has plenty of nuts to cut the sweetness. The quantities are large as it keeps indefinitely.

The recipe is more of a system than a law. Use boxed or fresh dates, and vary the nuts as you like. I avoid peanuts as they seem quite unsuitable for sweet dishes, but Brazils, cashew, macadamia or hazelnuts can be used just as well as pine kernels or pistachios.

500 g (1 lb/2½ cups) dates
300 g (10 oz/2½ cups) walnuts
200 g (6 oz/1½ cups) mixed pistachios and pine kernels

250 g (8 oz/1 cup) sugar
juice of a very large lemon
walnut or other nut oil
icing sugar
foil and rice paper

Stone dates and chop them finely. Chop the nuts. Dissolve sugar in a heavy pan with 4½ tablespoons (good ⅓ cup) water, bring to the boil when clear and boil without stirring to a golden brown caramel. Take immediately from the heat, add lemon juice, then the dates and nuts.

Cook over a moderate heat for about 10 minutes, stirring with a wooden spoon.

With oiled hands, roll the tepid mixture into sausages and leave to harden. Then roll them in icing sugar. Store in a cool dry larder, no need to refrigerate.

DATE SLICES

A crumbly sandwich cake that is popular at family teas. It can be served warm with cream or custard sauce as a pudding, too.

CAKE
125 g (4 oz/½ cup) lightly salted butter
125 g (4 oz/good cup) rolled oats
125 g (4 oz/1 cup) self-raising flour
60 g (2 oz/¼ cup) soft brown sugar

FILLING
250 g (8 oz/1¼ cups) stoned fresh dates, chopped
2 tablespoons (3 tbs) honey
1 tablespoon water

Mix the crumble first by rubbing the butter into the oats, flour and sugar. Rub a square 18-cm (7-inch) tin over with a butter paper and put in half the crumble mixture. Press it down, not too heavily, but so that it begins to look coherent.

Heat the filling ingredients together, and stir until they collapse into a thick paste. Spread over the crumble in the tin, and cover with the remaining mixture. Even out and press down very lightly, as the top should not be as firm as the bottom layer.

Bake for 30 minutes at gas 4, 180°C (350°F). If eating hot, cut into slices and serve immediately. Otherwise cool in the tin before slicing into oblong fingers.

COMMERCIAL DATES AND OIL

I understand from the catalogue of unsprayed dried fruit supplied by one firm in London, that:

Commercial dates are customarily dusted with Malathion or a similar toxic insecticide to keep down the date beetle.
Commercial dates when picked are customarily washed with detergent.
Because they are then wet, commercial dates must customarily be treated with a mould inhibitor.
Because of all this, commercial dates must customarily be coated with glycerine or something similar to make them look sticky again.

Since this seems to be the horror page, I will now quote a piece about the process of turning crude palm, soya and ground oils in to the kind we buy. You will understand why, in this book, butter, lard and olive oil are the fats preferred:

> First, the oils have to be degummed and neutralized. Phosphoric acid is injected into the oil and mixed under pressure to precipitate the gums.
>
> Then it is mixed with caustic soda which forms a soap containing gums and colour which can be separated easily from the oil.
>
> Next stage is to wash the oil, dry it, bleach with fuller's earth and filter it. At this point it's a fully refined oil but the original taste and smell still remain, making it unacceptable for consumption.
>
> The final stage, therefore, is deodorization to ensure a bland, odourless oil that won't tinge the flavour of what's cooked in it.

BOOKS IN CHINA

In the past, books were printed from wooden blocks carved from date and pear wood, and were sometimes referred to as 'dates and pears'.

DEWBERRY, see BLACKBERRY

FEIJOA or PINEAPPLE-GUAVA

What I know about feijoas or pineapple-guavas comes from books and friends, and I doubt if any of us will have a chance – I mean in an English garden – of picking up a basketful of ripe, reddish-green, pineapple-scented feijoas fallen from a 20-feet high feijoa bush.

Feijoa sellowiana comes originally from Paraguay and Uruguay. It needs sub-tropical warmth, so it grows well enough in orchards in California and Australia (also well enough in gardens on the French Riviera, where it is now a frequent escape, having been introduced there in 1890, by Edouard André on his return from South America). Feijoa bushes or small feijoa trees manage to grow – or survive – in some warm English gardens, particularly in several gardens in Cornwall but, like olive trees, they do not often set fruit, except occasionally after a long dry hot altogether exceptional English summer.

For berries, these feijoas are quite large, two inches or more long. Remove the skin, and you can eat them raw, or cut them into salads, if they are really ripe, slightly soft and sweet smelling; or do with them most of the things which can be done with the real guava (no relation, by the way), making feijoa jam, feijoa marmalade, feijoa jelly. Occasionally you may come across the fruit at the greengrocery by Covent Garden tube station in London. And I have a hope that one day they may be imported from New Zealand, since feijoa and Cape gooseberry (physalis) garnishes were served with a venison and sweetbread pâté at a centenary lunch in London, to celebrate the arrival of the first refrigerated cargo of meat from New Zealand in 1882 on *SS Dunedin*.

FIG

Every time I am in London and pass, or visit, the National Gallery and see the fig trees so nicely growing against its walls, I feel a resentment at our climate and our greengrocery trade. Not to live in a country where figs grow or can readily be bought in the season, seems to me a deprivation. Although I could not quite say that figs are my favourite fruit, they are the fruit I most long for, that I have never had enough of. Times I have eaten them remain clear. One year we happened still to be at Trôo when a friend had a basketful from a neighbour's tree, and he gave us a dozen of them. The other fig tree in the village should do well, growing against the heat of the rock: but it also grows by one of the steep public paths down to the river and the children pick the tiny fruit as they go to school.

The fig tree is one of those mythical ideas that one grows up with in the north. I heard about it first at school in County Durham – 'When her branch is yet tender and putteth forth leaves, ye know that summer is nigh' – because it was the first tree mentioned in the Bible. Adam and Eve pulled fig leaves to make skirts and hide their nakedness. That kind of role has clung to it, with patrons of art worrying about male nudity. All right for them of course, but bad for the public – you and me – and shocking for the ladies. This has always puzzled me. We give birth and clean up the messes of infancy and illness: are we not strong enough to look at marble nudity? Were we not strong enough in those even messier days? Apparently not, fig leaves were slapped on, round the galleries of the world, with elegant curves and exaggerations of size.

With that Genesis story, the fig acquired an aura of lovemaking, as a direct symbol of female sexuality. This I came to understand in a most embarrassing way. We were having lunch one day at a *pensione* in Florence – I was a student learning Italian. A basket of fruit was handed round, and there were figs in it. I had never tasted them, but knew the word for fig tree, *fico*. I also knew that, in general, fruits were the feminine form of the tree name, *melo* apple tree, *mela* apple. '*Una figa, per favore.*' The dining-room shook with laughter. A kind neighbour said, 'Cut open the fig, and you will see why they stick to the masculine.'

Well, yes. There inside is the brilliant deep mass of flowers, which later, in dried figs of certain varieties, has turned to seeds; there's a secrecy and flare of red warmth inside the discreet skin, as in Gide's poem:

> Now sing of the fig, Simiane,
> Because its loves are hidden.
>
> I sing the fig, said she,
> Whose beautiful loves are hidden,
> Its flowering is folded away
> Closed room where marriages are made:
> No perfume tells the tale outside.

The fig, with the olive tree, the vine and wheat, provided the four basic foods of Mediterranean eating until modern times. The uninterrupted enjoyment of them was every man's desire and dream. With them, he could live. All the tales and myths around these four foods reflect their central place.

In view of the deep Greek and Roman appreciation of figs, I was surprised to learn that the tree was not a native either of Italy or Greece but that it was most likely domesticated in Asia Minor. The best figs came from there in Roman times, and Smyrna figs have a reputation still. One night in France, the door banged open and an unrecognizable bearded young man fell across the threshold under the weight of his pack and the rain. As he asked for our daughter and at the same time brought a heavy necklace of dried figs out of the pack behind his head, we knew who it was. He had been in Turkey. They were Smyrna figs, bought outside Izmir, the right place, five days before, from the man who had grown and dried them.

Smyrna figs have travelled farther than that. Cuttings arrived in California in 1880, and the resulting fig, ambered and succulent, is known as the Calimyrna. The famous purple-black Mission Fig was taken there in the eighteenth century by Franciscan friars from Mexico where they had been given it by Spanish missionaries. A fig we all know from cans is the Kadota, a little translucent green fig in too-sweet golden syrup. I wish they produced it without added sugar, as they do so many different fruits these days. The name is difficult to understand – could it come from its plump resemblance to a Greek pot known as a *kados*?

My feeling towards dried figs is ambivalent – they remind me of the syrup of figs I once spat out on the nursery carpet and all over a blue dressing-gown. It must be the smell. Yet when I do have to eat one, whether dry or reconstituted, it always surprises and delights me. One member of the family eats a dried fig every night before he goes to bed.

That is a ritual I have come to admire, since tackling dried figs for this book.

HOW TO CHOOSE AND PREPARE FIGS

Most figs are thin-skinned and need no special preparation, unless it is just to rinse them and remove the stalk. Choose unbruised fruit, lifting each one out of the nice little paper they are often wrapped in. A good way of showing them off, is to arrange them on fig or vine leaves on a white dish with a low pedestal. Leave them in the sun for a while before serving them; warm figs are the best of all.

As far as the sweet course goes, avoid presenting them in ways which blur their shape, or which mean you have to chop them up. Shape is a good part of the fig's delight: any cooking should be gentle enough to maintain it undamaged. It should not be lost in a fruit salad but presented on its own.

The fig's allies, for sweet dishes, are nuts (excluding peanuts), candied peel of citrus fruits, fortified wines and orange liqueurs, cream and coconut cream. Caramelized figs – lightly caramelized – are lovely, or figs served with *crème brulée*, or figs scattered with praline.

I do not think it matters whether figs are black or white – or purple or brown or greenish or green-gold. Figs vary from region to region in taste, though they may look much the same.

Dried figs can be soaked in boiling water or steamed – the latter is particularly successful if you want to fatten them back into shape and stuff them with nuts.

I confess I could not bring myself to try Barnsley pudding or any of the other suet and dried fig puddings. Fig tapioca and fig ginger pudding have been avoided. You may not agree, but I concluded that dried figs need grand treatment to set off their antiquity and richness.

HAM WITH FIGS

In Assisi – very hot – on 9 July 1874, Ruskin wrote, 'Catherine brought me up as a great treat yesterday at dinner, ham dressed with as much garlic as could be stewed into it, and a plate of raw figs, telling me I was to eat them together!'

The usual hors d'oeuvre in Italy these days is figs with Parma ham – *prosciutto crudo di Parma* – which has been cured but not stewed. When figs are not around, melon is served instead. To me, it is the perfect summer lunch, beauty and freshness, sweetness and a salty bite.

Figs also go well, though not so brilliantly, with good cooked ham, the kind that is sliced from the bone as you wait. This is closer to Ruskin's meal, though noticeably lacking in garlic.

Be wary when buying ham. Read the article in the 1979 July issue of *Which?* if ham is regularly on your shopping list. Some canned hams have an added 38 per cent and 42 per cent of water. Other apparently real hams – real until you spot the line where the two halves of the mould met – are made of meat that has been taken apart 'during processing – pieces may even be *tumbled* and *massaged* (to help the curing solution spread evenly) – and then stuck back together so that the meat looks as if it's come straight from the pig's leg'. To complete the illusion, a bone will be stuck into the thin end. Those square and rectangular blocks of shoulder or ham may consist of one piece pressed into shape. More likely it will be a lot of small bits stuck together 'with gelatin or some other protein to act as a "glue"'.

Nothing shows up poor ham like perfect figs. Each must be worthy of the other.

DUCK IN PORT WINE AND FIGS

This recipe is based on one of Alice B. Toklas' in her *Cookbook*. Use two young duckling for 6 people. If you have to use 1 large bird, reduce the temperature to gas 5 190°C (375°F) and allow up to 2 hours.

24 figs
dry port wine
2 duckling
about 250 ml (8 fl oz/1 cup) veal stock

Put the figs side by side, close together in a closely fitting dish Pour on enough port wine to come about half way up their sides. Cover with plastic film and leave for 36 hours, turning the figs twice, so that every part is marinaded.

Put the duckling on a low trivet in a roasting pan. Set the oven to gas 8, 230°C (450°F), and when it is heated put in the birds. After 15 minutes pour off the fat, and baste with half of the wine. Put back for another 15 minutes, then baste with the rest of the wine. Leave for a further 15 minutes, then put the figs round the duckling, and baste with rather less than half of the veal stock. Baste once more after 15 minutes, with the remaining stock. After 1¼ hours the duckling should be cooked.

Remove them to a dish, and put the figs round them. Pour fat from the roasting juices, and boil it down hard to a concentrated sauce. Serve in a separate jug. Bread and a plain salad to follow – not a dish to be fussed with.

ESCOFFIER'S CARLTON FIGS
FIGUES À LA CARLTON

The culmination of the great Ritz-Escoffier partnership was the opening of the Carlton Hotel in London, in 1899. Escoffier was fifty-three, a small, strong, gentle man who ruled the splendid kitchens with discipline and order. He remained inventive and energetic. One new idea was the à la carte menu, another the adding of a raspberry purée to the dessert of peaches and vanilla ice cream that he had invented for Dame Nellie Melba (p. 302) six years earlier, at the Savoy.

If the figs are of perfect ripeness, just peel them and cut them in half downwards. Put them in a dish and stand the dish on crushed ice. Prepare some sauce Melba (p. 398) and mix it with twice as much whipped cream. Pour this sauce over the figs and cover them.

If the figs are not at their height, peel and poach them in syrup flavoured with a vanilla pod. Drain them, and finish them as above.

ESCOFFIER'S FIGS WITH CREAM
FIGUES À LA CRÈME

Another recipe for perfect figs. Peel, halve and arrange the figs in a dish. Put the dish on crushed ice. Pour over a few tablespoons of either Curaçao or Cointreau or maraschino or noyau. Before serving, mask with sweetened vanilla-flavoured whipped cream.

Note I do not think the cream is necessary and, if you are a grower of figs, you might prefer to follow the habit of one proud gardener who was unwilling to slice them in half, and introduced the alcohol – orange liqueur, Marsala, kirsch – by means of an eye dropper, through the top of the fig.

CARAMEL FIGS

Here is a simple way of baking figs that are ripe and soft, whole but on the verge of splitting. Use figs of a kind that do not need peeling.

First find an ovenproof dish that can be set on the table afterwards. A plain brown terrine is ideal, better than a gratin dish, as the higher sides give the figs support and prevent the juices burning. It should just hold the figs, stalk end up, close together.

Remove the figs, dip them one by one in water and roll them in caster sugar from the vanilla jar, so that they are quite snow-covered. Put them straight into the dish, as you finish each one.

Bake them in a hot oven, gas 7–8, 220–230°C (425–450°F) for about 20 minutes, but check after 12 minutes. They are done when the sugar has disappeared and turned into a rich brown juice at the bottom ot the dish.

Cool, then chill thoroughly. Serve them with cream if you like, and sugar or lemon thins or almond biscuits.

FRESH FIG TART

TARTE AUX FIGUES FRAÎCHES

A French way of presenting fresh figs which is most attractive to look at, as well as good to eat.

Make and bake blind a case of creamed pastry, recipe p. 456, of about 20 cm (8 in) diameter. It should be completely cooked, but barely coloured.

Rinse and dry $\frac{1}{2}$ kg (1 lb) of figs. If the skins are tough, peel them away. Halve the figs and put them on a plate. Pour over 3–4 tablespoons ($\frac{1}{4}$–$\frac{1}{3}$ cup) of an orange liqueur and cover them with plastic film. Leave in the refrigerator for several hours to macerate.

Heat gently and then sieve 3 tablespoons of apricot jam into a small basin. Mix in the liquid from the figs. Put the fig halves on the pastry case, as decoratively as you can. Spoon over them the apricot and orange mixture. Put back into the refrigerator to chill.

SPANISH FIG ICE CREAM

My sister's family have a passion for dried figs, and were delighted to be given this ice cream in Tours one day, at a small Spanish restaurant in the old town. I listened with amazement to their rapturous and most detailed explanations. Tried it reluctantly, and found myself converted. The brandy brings out the richness of the figs, and the praline crispness distracts your mind from the idea that dried figs are tough.

For six people, make the soft ice cream with eggs, on p. 473. It should be frozen to a firm softness, not a slosh. Divide it between six tall wine flutes or ice cream glasses, and put back into the freezer.

Have ready nine dried figs which have been chopped to small nuggets and soaked overnight in 6 tablespoons ($\frac{1}{2}$ cup) of brandy. Prepare some praline, p. 477, with 60 g (2 oz/½ cup) almonds, and crush it coarsely. Whisk 250 ml (8 fl oz/1 cup) whipping cream until very thick but just pourable.

To serve the ice cream, take the glasses from the freezer, divide the fig mixture between them and spoon some of the thick cream on top. It should trickle down slightly into the figs. Scatter the top with praline.

Note With the ice cream, the figs and brandy and the praline, it is not necessary to sweeten the cream as well.

POOR MAN'S NOUGAT

NOUGAT DU PAUVRE

In Sicily, I am told, they halve fresh figs, cutting them down through the centre but not quite through. Just enough of a cut to take a whole shelled walnut, both halves. The fig is pressed back together and left to dry round the walnut.

In Provence, they make a version of this poor man's nougat that we can copy. You slice a dried fig down from the top in the same way, and force a piece of walnut into the cavity. The fig is pressed back into shape. In Provence, too, at Montélimar, they make the rich white soft nougat that everyone knows about. That requires sugar, egg white, a variety of nuts and a lot of skill, certainly not a poor man's nougat. On Christmas Eve to finish the 'Gros Soupa' and to occupy the time before going to Midnight Mass, thirteen desserts are put on the table, signifying Christ and his disciples. There is no absolute list, but Montélimar nougat, poor man's nougat and black nougat (a dark almond brittle) are likely to be three of them. Others, might be:

a whole candied melon
sweet spinach tart
a collection of dates and prunes stuffed with nuts and almond paste, p. 277
calissons d'Aix (oval iced almond biscuits)
almonds and raisins
oreillettes (fritters)
tangerines
pompe à l'huile (brioche with glacé fruit)
wrapped sweets with jokes and mottos inside
quince and other fruit pastes

MME VERDIER'S BLACK FIG JAM

A recipe from the red village of Collonges, in the Corrèze near Brive, sent to me by a friend.

Prepare the figs by removing the stems and any skin which comes off easily. Weigh them and leave them overnight in a bowl with half their weight in sugar.

Next day, put the whole thing into a preserving pan and bring to the boil. As soon as it boils, remove the fruit with a skimmer or perforated spoon. Leave the syrup in the pan to go on boiling to the pearl stage, 106°C (222°F), when the sugar forms small pearl-like balls. Put back the figs and cook gently, while stirring, for about 15 minutes. Be careful that the fruit does not catch and burn. Pot in the usual way.

GOLDEN BERRY, see PHYSALIS

GOOSEBERRY

When Constable painted and drew elderbushes in flower, creamy panicles tilted against green, I wonder if he had ever tasted their muscat sweetness with gooseberries? Lyric painting we have been good at, lyric cooking has been rare with us. Yet with gooseberries in the spring of the year, with green easy-growing gooseberries, we had this bright thought of turning them to luxury with a few heads of wild elderflower which gives a muscat flavour.

Do not be so carried away with the poetry of it all that you drive out and cut the first elderflowers you see by the roadside. They will be grey from dust and exhaust fumes. You must look along some lane or in a field. Then the flowerheads will only need a light shaking to get rid of wild life and a brief rinse under the tap when you get home.

Seeing that gooseberries come early in the year, before strawberries, it is surprising that they are popular only in this country. People in the Midlands raised superb gooseberries, had gooseberry competitions at the beginning of the last century, made the vulgar mistake of breeding for size which put gooseberries down a notch or two in the social register. The Scotch were also 'great in gooseberries . . . When there was not a tree nor a shrub to be found in the Shetland islands and the Orkneys, there were gooseberry bushes in abundance; and it was an old joke against the Shetlanders, that when they read their Bibles and tried to picture to themselves Adam hiding among the trees of the garden, they could only call up in vision a naked man cowering under a grosart [i.e. gooseberry] bush. The gooseberries of Scotland are the perfection of their race, and for flavour and variety far beyond those of the south – just as English gooseberries are better than those of the Continent. On the Continent they are little prized, and not very well known. The French have no name for them, distinct from that of red currants.'*

Neither have we. Gooseberry has nothing to do with goose, apart from setting it off so well. Gooseberry is a culinary garbling, assimilated to goose, of old names groser, grosier, grozer, which are linked to the

**Kettner's Book of the Table* E. S. Dallas, 1877.

French *groseille*, red currant. All these names, French, English, Scottish, go back to the Frankish *krûsil*, crisp berry.

At least when the French remember the gooseberry, they call it *groseille à maquereau*, the mackerel currant, in appreciation of its goodness with oily food. There is not much evidence, though, that the two are married in French affections. Even manuals of *nouvelle cuisine*, in which fruit is put to new savoury uses, neglect the gooseberry. Its greatest French admirer, Monsieur Audot, who gave it a whole section in his popular book, *La Cuisinière de la Campagne et de la Ville*, of 1818, describes gooseberry sauce for mackerel as English.

Monsieur Audot began by saying: 'The English make great use of gooseberries for cakes puddings and desserts. We urge French ladies to try them and we can promise that they will find in this fruit a very great resource for filling dessert dishes and for adding to puddings in the season when fruit is short. Gooseberries can be used as soon as they are formed and all the time that they are green: their acidity makes their flavour.'

Then come recipes for a gooseberry tart, a boiled suet pudding, gooseberries set in jelly, the cake and pudding given later in this chapter, and a straightforward gooseberry pie flavoured with cinnamon, orange-flower water or grated nutmeg. Elsewhere in the book are the English sauce for mackerel I mentioned, a German sauce for chicken and veal, and a compote in which gooseberries are blanched in water, then cooked in syrup.

A satisfactory collection, which seems to have had no success at all with the French, in spite of the book's sixty-five editions. The only bushes I know in our part of Loir et Cher survive in the hedge of a cottage garden. Unless we pay them a visit, the fruit rots. Of course, their prickliness is against them. In Wiltshire, we have taken to standard gooseberry bushes, grown like standard roses, a style that originated in Germany according to the nurseryman who supplied them. The tall main stem brings the fruit up to convenient picking height, and by cutting back the branches we find the fruit grows thickly and to hand.

So far I have spoken of the sharp green gooseberry, the gooseberry for pies and fools, but there are many more gooseberries – white, yellow and red as well as green, which are grown for dessert, to be eaten raw like strawberries. These gooseberries are larger and sweet, they catch the eye – especially the eye of children, for they are 'the fruit, *par excellence*, of ambulant consumption'. That was Edward Bunyard's description in his *Anatomy of Dessert*, and he also commended one particular variety, the Lancer, for its well-flavoured, late summer fruit: 'As August gooseberries are in such demand when schools break up, lateness is a virtue which can hardly be overrated. No garden yet grew more gooseberries than the young people are able to deal with.'

TO CHOOSE AND PREPARE GOOSEBERRIES

As far as cooking is concerned, the first small green gooseberries have the best flavour. Although red ones may look mellow and tempting, they are often less good to cook or to eat raw than the green. If you have a choice when buying gooseberries, ask the greengrocer if you can taste them to see which you prefer. Unless they are being specially displayed, a few fine fruit in special punnets, he is unlikely to refuse.

For any recipes in which gooseberries will eventually be sieved, there is no point in topping and tailing them. Just rinse them and pick them over. For other dishes, top and tail them with a small sharp knife or a pair of scissors – whichever you find easier and quicker.

The Whitsun gooseberry pie is made like any other fruit pie – layers of fruit and sugar, mounded up in the centre above the rim of the dish, shortcrust pastry to cover the fruit brushed with egg white and scattered with sugar. A hot oven until the pastry is set, then a moderate temperature to cook the fruit. Plenty of cream.

Boiled gooseberry pudding is another favourite. Turn to the suet crust recipe at the back of the book, and fill the cavity with gooseberries, demerara sugar and butter. Boil or steam.

Raised gooseberry pies from Oldbury and other gooseberry recipes that I put into *Good Things* and *English Food*, I am not repeating here.

A number of currant and cherry recipes can be adapted to gooseberries.

GOOSEBERRY SAUCE FOR SMOKED MACKEREL

Smoked mackerel can be disappointing. Often the whole fish are hot-smoked until they are soft and unpleasant. The fillets may look less picturesque, but often they have suffered less in the curing. Chill them until they can be sliced like smoked salmon, parallel or on a slight slope to the skin. Serve them with brown bread and butter, perhaps a little cress or lettuce shreds, and this sharp sauce:

Stew about 200 g (6–7 oz/1¼ cups) gooseberries – no need to top and tail them – with a knob of butter and a tablespoon of sugar. Cover until the juices run, then remove the lid and raise the heat so that any wateriness evaporates rapidly. Sieve – processing or liquidizing makes them too smooth – and add to 150 ml (5 fl oz/⅔ cup) whipping cream whisked until very thick. Taste, add a little more sugar, but keep the flavour tart. Stir in a large teaspoonful of a good brand of creamed horseradish, to give a hint of earthiness. Fresh grated horseradish can be used, but it should come from a root recently dug from the ground. It may seem heretical, but I think that with this sort of sauce, a respectable commercial horseradish cream works better.

Add a pinch or two of salt to liven the flavour.

18TH-CENTURY GOOSEBERRY SAUCE

A sauce for sweetbreads, chicken, veal and poached fish.

Make 300 ml (½pt/1¼ cups) velouté sauce with a stock appropriate to the meat or fish and any juices from the cooking of it. Prepare a gooseberry purée as in the recipe above. Mix into the velouté. Beat 2 egg yolks and 100 ml (3½ fl oz/good ½ cup) whipping cream. Fold into the sauce and thicken over a low heat. Salt, pepper, a pinch of sugar. Enough for 8 people.

GOOSEBERRY STUFFING

For oily fish and goose. With the fish, it is a good plan to fillet them first and then fold them together again into shape round a long wedge of stuffing. If the fish are on the small side, make two of them into a sandwich – skin side out – with a layer of stuffing through the middle. In either case, bake the fish afterwards.

To fillet small fish, cut off the head, slit along the belly and clean the inside. Spread out the cut edges so that the fish are placed back upwards. Press along the back bone. You will feel it give. Then turn the fish over, pick out the bone and remove any whiskers left behind with a pair of tweezers. Season the fish inside.

FOR THE STUFFING

1 small to medium chopped onion
60 g (2 oz/¼ cup) butter
1 small clove garlic, chopped
125 g (4 oz) sharp green gooseberries
sugar
100 g (3½ oz/2 cups) white breadcrumbs
¼ teaspoon chopped marjoram
about 3 sprigs tarragon, chopped
salt, pepper, lemon juice

Cook the onion in butter over a low heat until melted and golden. Add garlic and gooseberries, which have been topped and tailed. When the fruit begins to collapse, mash it into a rough purée and add enough sugar to make the fruit palatable but not sweet. Mix in the breadcrumbs and herbs, off the heat. Season to taste, adding lemon juice only if extra sharpness is needed.

For a goose, the stuffing should be made in double quantities. You can bind it with egg if you like, but I prefer the crumbliness of the recipe above.

OSSI BUCHI WITH GOOSEBERRIES

Gooseberries were once so common with veal that it occurred to me to try them with *ossi buchi* instead of tomatoes. Why not? Their sharp spring freshness sets the veal off well. I hope any Italian readers will refrain from caustic letters, and at least try my version if they are housekeeping in this country.

Ask the butcher for slices of veal cut across the leg. They should be at least 2.5 cm (1 inch) thick, one for each person. Take particular care of the marrow by keeping the veal bones upright; it's the *bonne bouche* of the dish. In northern Italy, the tiny spoons that are used to extract it are called *esattori delle tasse*, tax collectors. Incidentally, if you have an amenable butcher, ask him to cut your slices from the hind shank because you get more marrow.

For 6 slices of veal, enough for 6 people, you will also need:

seasoned flour
butter, preferably clarified
1 medium onion, chopped
1 medium carrot, cut into neat little sticks
1 small clove garlic, finely chopped
300 ml (10 fl oz/$1\frac{1}{4}$ cups) dry white wine
200 ml (6–7 fl oz/$\frac{3}{4}$ cup) veal or beef stock
250 g (8 oz) green gooseberries, topped and tailed

GREMOLATA
2 heaped tablespoons (3 tbs) chopped parsley
grated zest of a lemon

Choose a sauté pan into which the veal slices fit closely in a single layer. Turn the slices in the flour, then sauté them in butter until golden brown on both sides. Remove from the pan, and put in the onion and carrot. When they are lightly coloured, return the veal to the pan, and add garlic and wine. Raise the heat, boiling hard until the wine is reduced by half. Add the stock and just over half the gooseberries; lower the heat, cover and leave to simmer for 45 minutes. Turn the veal over, put on the lid again and leave another 20 minutes. Taste the sauce and add seasoning, plus a little sugar to bring out the flavour of the gooseberries without making the sauce sweet. If the general tone is sharp enough, do not add the remaining gooseberries: if it seems too mild, put them in.

Leave until the veal is tender – another 20–30 minutes. If the sauce is on the copious side, do not cover the pan but leave it to reduce gradually.

Mix the parsley and lemon for the *gremolata*. Scatter over the top of the *ossi buchi* which can be served in their cooking pan if it looks reasonably respectable. Otherwise transfer them to a suitable serving dish. Put a bowl of grated parmesan on the table as well to go with the accompanying *risotto alla milanese*, which can be made while the veal is cooking:

60 g (2 oz/¼ cup) butter
2 shallots or a medium onion, chopped
about 1½ litres (2½ pt/generous 6 cups) veal or beef stock
500 g (1 lb) Italian arborio *rice*
generous pinch saffron filaments
extra lump butter
grated parmesan
salt, pepper

Melt the butter and cook the shallots or onion in it until transparent. Meanwhile bring the stock to simmering point in a separate pan.

Stir the rice into the shallot or onion and butter, and keep stirring until it is completely coated. Allow up to 2 minutes.

Pour in a ladle of stock, stir and as the stock disappears add some more – a slightly larger quantity. Go on doing this until the rice is creamy, but not gummy and wet.

Dissolve the saffron in a bowl with a ladle of hot stock from the pan. Leave it to infuse, while the rice cooks.

You will notice that the rice develops a characteristic flat look with occasional holes in it. When the rice is almost tender, but with a slight resistance in the centre of the grains, pour in the saffron stock and finish the cooking. Stir in the extra lump of butter, parmesan to taste and salt and pepper.

GOOSEBERRY MERINGUE TART

As a meringue top for a fruit tart, I prefer this recipe of Sheila Hutchins' to the usual egg white and sugar mixture. It cooks to a dry-looking crunchy hollow top, delicately browned and light to eat.

For a 20–23-cm (8–9-inch) tart tin, lined with sweet shortcrust pastry, baked blind and lightly coloured, you will need:

about ½ kg (1 lb) gooseberries, topped and tailed
60 g (2 oz/½ cup) butter
2 tablespoons (¼ cup) light brown or demerara sugar
2 large egg whites
1 large egg yolk, beaten
6 tablespoons (¾ cup) caster sugar
1 tablespoon (¼ cup) flour

You need enough gooseberries to make a close single layer on the pastry. The exact quantity will depend on the size of the gooseberries.

Melt the butter in a heavy pan of 20–23-cm (8–9-inch) diameter, stir in the brown or demerara sugar and when you have a liquid caramel, put in a layer of gooseberries, plus a few extra. Cover and cook briefly until the skins change colour. Remove the pan from the heat before any gooseberries burst. Cool. Put into the pastry case, with extra sugar if the gooseberries seem to need it. In the early season they can be very sharp and green-tasting.

Whisk the egg whites stiffly. Fold in the yolk, then the sugar and flour sifted together. Pile on to the tart, right to the pastry rim.

Bake at gas 1, 140°C (275°F) for 40 minutes, or until the meringue is cooked and nicely coloured. Serve hot, warm or cold.

GOOSEBERRY FOOL

I used to think – see *Good Things* – that the word fool came from the French *fouler*, to crush. Seemed logical, as to make a good gooseberry fool the berries should be crushed rather than sieved. But I was wrong. It's a word that goes with trifle and whim-wham (trifle without the custard) – names of delightful nonsensical bits of folly, *jeux d'esprit* outside the serious range of the cookery repertoire. The kind of thing that women are said to favour, but that men eat more of.

In *Kettner's Book of the Table* (1877), the author, E. S. Dallas quotes a passage from an old recipe book which connects the pudding with Northamptonshire:

> The good people of Northamptonshire maintain that all our best London cooks, in making gooseberry fool, are themselves little better than fools. There is no way, they insist, equal to their own, which is as follows: After topping and tailing – that is, taking off clean the two ends of the gooseberries – scald them sufficiently with a very little water till all the fruit breaks. Too much water will spoil them. The water must not be thrown away, being so rich with the finest part of the fruit, that if left to stand till cold it will turn to jelly. When the gooseberries are cold, mash them all together. Passing them through a sieve or colander spoils them. The fine natural flavour which resides in the skin no art can replace. The skins must therefore remain unseparated in the general mash. Sweeten with fine powdered sugar, but add no nutmeg or other spice. Mix in at the last moment some rich cream, and it is ready. The young folks of Northamptonshire, after eating as much as they possibly can of this gooseberry fool, are said frequently to roll down a hill and begin eating again.

In spite of E. S. Dallas's strictures, I do sometimes add a little muscat de Frontignan to the gooseberry mash, or cook a couple of elderflower heads with the gooseberries.

BAKED GOOSEBERRY PUDDING

Sounds homely by comparison with Whitsuntide gooseberry pie, but try it. The buttery, sharp-sweet flavour is good.

Melt 2 generous tablespoons ($\frac{1}{4}$ cup) of butter and 4 tablespoons ($\frac{1}{2}$ cup) demerara sugar in the bottom of a soufflé or pyrex dish. Put in a tight-fitting layer of topped and tailed gooseberries.

Over this spread *either* the pound cake mixture on p. 100, livened with 2 tablespoons (3 tbs) gin, kirsch or vodka. *Or* the variation given for baked cranberry pudding on p. 143.

Bake at gas 4, 180 °C (350°F) for 50 to 65 minutes, according to the depth of the mixture in the dish. The top should be a nice colour, and the cake should not stick to a warm skewer pushed in at an angle. Sprinkle with sugar, serve with custard sauce, cream or vanilla ice cream.

MONSIEUR AUDOT'S GOOSEBERRY CAKE (1818)

CAKE DE GROSEILLES VERTES

A recipe typical of the puddings and cakes that we still like and make at home. It is also very close to the French *quatre-quarts*, i.e. four-quarters or pound cake. Indeed, if you want to eat the cake cold at tea-time or on a picnic, I would advise you to follow the pound cake recipe and increase the quantity of egg to two whole eggs. A level teaspoon of baking powder sifted with the self-raising flour will also help to make it lighter.

125 g (4 oz/½ cup) butter
165 g (5½ oz/scant 1½ cups) self-raising flour
1 large egg yolk
125 g (4 oz/½ cup) caster sugar
1 tablespoon dry white wine
1½ teaspoons orange-flower or rose-water
½ tablespoon aniseed
⅓ nutmeg, grated
85 g (scant 3 oz) gooseberries, topped and tailed

Melt the butter and, while it is still warm, not hot, stir in the flour, egg, sugar, wine and flavourings. Beat well (this can be done electrically). Fold in the gooseberries, and turn into a buttered and sugared tin or baking dish. Alternatively, spread half the mixture in the prepared tin or dish, then all the gooseberries, then the rest of the mixture.

Bake at gas 4, 180°C (350°F) until the top is golden brown and a skewer pushed into the centre at an angle comes out clean. From 45 minutes, depending on the depth of the cake.

Sprinkle the top with sugar and serve with cream, or pouring custard if you prefer.

GOOSEBERRY AND ELDERFLOWER JELLIES

While you are making gooseberry and elderflower jelly to store in jars for the winter, it is a good idea to set some of the liquid aside to make a muscat jelly for dinner.

For the gooseberry and elderflower jellies, fill a preserving pan two-thirds full of green gooseberries. Barely cover with water. Simmer until the gooseberries have all burst, giving a stir from time to time. Drain through a jelly or stockinette bag, without squeezing extra juice through (the debris can be used to make gooseberry cheese, or gooseberry fool). If you are going to make a muscat jelly (below), measure the liquid, setting aside 725 ml (1 pt 4 oz/$3\frac{2}{3}$ cups).

Put the rest with warmed sugar back into the preserving pan, allowing 600 g sugar to every litre of juice ($\frac{3}{4}$ lb to every pint/$1\frac{1}{2}$ cups sugar to every $2\frac{1}{2}$ cups juice). Tie half a dozen elderflowers into a stockinette or muslin bag and suspend into the liquid from the handle of the pan. Bring to the boil and boil vigorously until setting point is reached. Taste occasionally and remove the elderflowers when the muscat flavour is to your liking: remember that flavours weaken as they cool. Pot in sterilized jars in the usual way. Store in a dark dry place.

MUSCAT JELLY To make the muscat jelly, warm the liquid you set aside with 3 heads of elderflower and sweeten it to your liking. Cover the pan and leave to infuse on the lowest possible heat, until the liquid is permeated with the muscat. Remove some of the liquid to a bowl and sprinkle on a packet ($\frac{1}{2}$ oz) gelatine. Stir until it is thoroughly dissolved, then add to the rest of the warm liquid. Pour through a strainer into individual glasses or a glass bowl. Cool and chill in the refrigerator until set.

Note If you have no access to elderflowers, you can always add muscat de Frontignan directly to the muscat jelly.

GREENGAGE, see PLUM

GRAPE

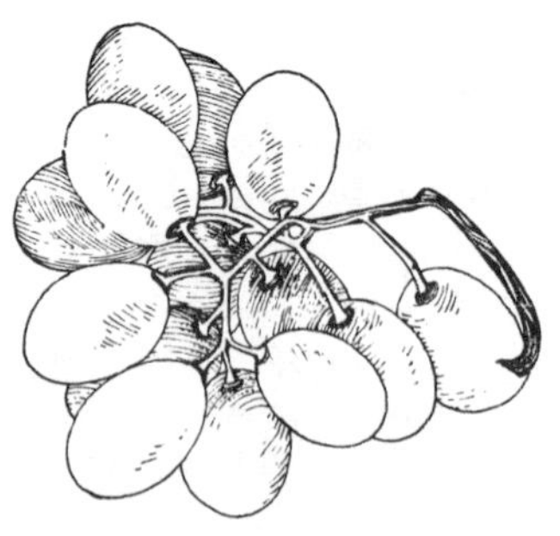

Grapes began on Mount Ararat, though rather before Noah, the vine being one of the earliest of cultivated plants and the Caucasus the earliest place known of its cultivation. I cannot imagine that anyone walking through a vineyard of ripe grapes could resist eating a handful, but their main purpose has always been for wine. Grapes were also dried and in the Middle Ages became a major export from the Mediterranean, Greece above all, to northern Europe, Britain in particular. It seems we could not have enough of their wrinkled sweetness, we craved sweetness especially in December and the following dark months when hoarded fruit was running out, often rotting in store. Domestic preservation of fruit with sugar, jams and so on, was limited before the era of cheap sugar began with the Caribbean plantations.

I daresay families in wine-growing districts had house vines, as they do today, but wine was the proper destiny of the grape.

With the European discovery and settlement of America, people tried to plant vines from back home. Mostly they failed on account of phylloxera aphid. Native American vines had developed with this bug which was to devastate European vineyards in the middle of the nineteenth century. Success came to American wine-makers after Thomas Jefferson undertook the business of growing American vines seriously and on some scale at his estate of Monticello. Now the whole successful wine trade depends on hybrids of European and American vines, as well as on cultivars of American vines. In Europe we have overcome phylloxera by grafting our vines on to American rootstocks which are resistant to it. I have read that people who knew about wine before and after phylloxera infestation in Europe, said that the earlier wine tasted better.

For people who think good food and cooking are important, it is helpful to know a little of the different vines and the wine they produce. At the lowest, it helps you to make sensible substitutions; at the best, it helps you to understand why a good dish tastes as it does, and why it tastes even better if you drink the right wine with it. Quite often in this book, wine is an ingredient, and quite often the precise kind is not specified. This is not because I think the choice of wine doesn't matter: it is because I think that the exact kind can be left to the judgment of the cook. It is also because where I learnt to cook – in France – people may have plenty of wine, but they have access to far fewer kinds than we do in this country – they use what they have, and make adjustments with other ingredients. And if they happen to be temporarily without wine, or the husband has gone off with the key of the *cave*, they will use the appropriate wine or cider vinegar and a couple of lumps of sugar, or in the season they will use grapes directly.

Famous wine-making varieties of grape will often appear on the label of middling wines (the best ones are attributed to their château or commune). Riesling for instance, or Cabernet (the claret grape), or Pineau white and black which are the Burgundian and Champagne varieties. The white Sauvignon is a favourite of ours because it is used for our easy local white wines, and in the grander Sancerre, Quincy and Pouilly-fumé. Down the river Loire, Vouvray, Savennières and Côteaux du Layon depend on the Chenin Blanc. Gamay is the Beaujolais grape; Grenache the main red wine grape in Rioja. The Muscat grape gives scented sweetness to Frontignan and Beaumes-de-Venise dessert wines (and in its hot house forms, it produces some of the best eating and cooking grapes). The finest dessert wines of all, from Sauternes, are mainly made from the Semillon.

This does not mean, alas, that Chilean Cabernet is claret in mufti. The patch of ground where grapes grow, the climate, the skill and aims of the growers, the blending of it with other grapes or not, are all things that influence the end product. But at least when Carême says 'put in a bottle of claret', you can judge whether you might not do better with a Chilean Cabernet than a Rioja. And if you have to use the Rioja, be on the look out for adjustments in other ingredients which might improve the final result.

A good thing that has happened in recent years is the return of sweet dessert wines to serve with the dessert. A glass of Beaumes-de-Venise or Quarts de Chaume for instance (Robin Yapp of Mere in Wiltshire, who supplies such wines, says that Quarts de Chaume drunk at a supper picnic with apricots and peaches has turned 'the second act at Glyndebourne into the most expensive siesta in the world' – which is not what I suggest, but quote to show what peace ensues from such combinations).

Sweet wines are often served as an apéritif in summertime in France. They may be suggested, too, with smoked salmon or with foie gras in grand restaurants. Such dishes may not come your way or mine, or not very often, but they may give you the idea of using a little sweet wine when making fish or poultry liver dishes of our own, instead of using a dry white wine or a fortified wine such as Madeira.

With such dishes, you may also need a garnish of dessert grapes, and the ones you would like to use are not always in the shops. In the provinces, we are not well served in the matter of grapes. We cannot rely on a choice between a fine grape and a crammed stemful of Almerian grapes with tough skins. They vary from small seedless grapes which often seem disappointing to eat, if handy to use, to large sizes which make skinning and de-pipping easier. Some are black and some are white. But how many have the wonderful perfumed fragrance that we think of when the word grape comes to mind? Most grapes outside the best shops in the largest towns are bought without enthusiasm and eaten without surprise. I look out for the muscat grape of the autumn, and now – amazingly – of February and March when they come in from South America, as preserving something of what I always hope grapes will be.

If you want to make the kind of dish in which game or pigeons are buried in grapes, the slight sharp grapes from a vineyard or those you might grow yourself are ideal. Their vigour is just right with that kind of meat. I learnt to cook such dishes in our undistinguished wine district to the north of Touraine where the only modest grandeur is produced at Jasnières near La Chartre, and where a bunch or two of grapes is gladly handed over for culinary experiment. I daresay the same would be true at the vineyards in this country.

The best and most famous eating grapes are grown in Belgian hot houses. They are worth eating and – up to a point – worth paying for. It depends whether you want them for the family vitamin supply or whether you are going to appreciate them one at a time. This would certainly be the only way to eat some grapes I saw last year, on sale for £13 the kilo, early hot house grapes, enormous, each one as perfect as if it had been carved in jade, the bunches tied with a smart red label. No good asking for 60 grams from these works of art, so I cannot report on them for you, or tell you if they were really the experience of a lifetime's eating. Belgium aims to be the Switzerland of fine food production; in other words, a country which by skill and high quality produces goods that command respect worldwide, that people will pay for. The superb grapes come from Hoeilaart near Brussels. It is not just glass and care that make the grapes special, but the special nature of the soil, the *goût de terroir*, that affects table fruit as much as it affects wine grapes.

At a more mundane level, you will come across recipes stipulating black or white grapes. This is only a matter of appearance, when grapes are to be used decoratively in their skins. There's a story about a grand doctor called to attend a grand invalid in the country. He made his visit, and was then driven back to the station for the return journey to Harley

Street. Just as he was getting into the train, a breathless servant rushed up to him to ask if the invalid could eat grapes. The doctor paused reflectively, his foot on the step, 'White grapes, yes! Black grapes, no!' Some recipes are written like that, but once grapes are peeled – which really is necessary unless skins are to be strained out or a few black grapes are needed for garnishing – there is no visual difference and little difference in taste between the two, assuming that both are of the type supplied by nine out of ten greengrocers. Underneath every black skin lies a white grape. It may have pinkish-purple flushes, but in the main it is white and will turn whiter in the cooking. It is skins that give colour.

Dried grapes come from several parts of the world these days, but in naming them we cling to their Mediterranean associations. Sultanas, in French *raisins de Smyrne*, came, and the best still come, from around Izmir, like the best dried figs. Currants, in French *raisins de Corinthe*, are dried in Greece south of Corinth. And raisins, in French *raisins secs* or just *raisins* which can be confusing as *raisins* also means grapes, are best from Malaga and should be dried muscatel varieties. The finest Malaga raisins are dried on the vine, with the leaves cut back, and the stalk nicked. They are the aristocrats of dried dessert fruit.

HOW TO CHOOSE AND PREPARE GRAPES AND DRIED FRUIT

The tricky thing about grape transportation is that odd grapes fall easily from the bunches. For this reason, the best kinds come cradled in tissue paper, and the medium kinds in plastic wrapping. The cheapest grapes are pushed into the box, without any attention being paid to the fact that they are gown in bunches. Some people claim that you can improve a middling bunch of grapes by hanging it up in the sun – fine for summer and early autumn, but there's not much you can do later on into the winter. So taste the grapes before you buy them. Except with the grand hot house kind, there are usually grapes fallen from the bunches which the greengrocer will not mind you trying.

Grapes for eating should be washed as soon as you get home, and placed on a cloth to dry thoroughly, before they are put on the table. A pair of grape scissors is not a pompous affectation: it means that people can help themselves to what they want without tearing the beauty of the bunch and dislodging other grapes.

Grapes for cooking need to be skinned and pipped (unless their debris will be strained out of the sauce). Pour boiling water over the grapes, just as you do with tomatoes, peaches, apricots or plums, and pour it off as soon as you can – try a grape to see if the skin comes away easily: the riper the grape the less time it needs in the water. You can try hooking out the pips with a new hairgrip, but easiest by far is to halve the grapes and dislodge the pips with the point of a knife.

If you grow grapes, use the sharp unripe bunches for pressing out your own verjuice. It can be used instead of vinegar and lemon juice, and to make up mustard. In the past, it was fermented for a couple of days, and a little salt added. It might also be boiled down to concentrate it. With a juice extractor, this would be worth trying.

DRIED GRAPES Do not necessarily go for the most beautiful and plump. Dried fruit disguised in plastic bags, heavily printed over, is difficult to judge even if it has storage attractions. The same applies, of course, to boxed fruit. If you have any choice in the matter, and can find a small speciality shop or Greek delicatessen, go for the more wizened fruit that they are selling.

Raisins are supposedly dried muscatel varieties – but some are better than others, and usually come from Spain. Remember that seed*ed* raisins come from better tasting varieties than seed*less* raisins. I always look out for Greek currants, from sentiment perhaps as there really is little difference between them and Australian currants. The thought that they have been spread out on sheets in the sun in southern Greece, around Kalamata for instance, every late summer for the last 2000 years gives me an affection for them that other considerations cannot shake. These are the currants that a fifteenth-century northern merchant watched being crammed into a ship at Zante for shipment to Margate, and so to London. Every corner was stuffed with currants: the sweet smell and the stickiness of the decks and the insects are beyond imagination.

About sultanas, Tom Stobart had much good advice in his *Cook's Encyclopaedia*. Turkish and Greek sultanas have the true taste, which is lacking in the seedless sultanas produced in California and Australia. Some countries allow sultanas to be coated with mineral oil to prevent them drying out and to give them a shine. This partly disappears in washing, but try to buy duller, dryer sultanas instead. To wash sultanas – and other dried grapes and dried fruit – pour on boiling water, leave for a moment or two, then lift them out with a strainer spoon or skimmer, going carefully so as not to disturb any gritty, sandy sediment.

GRAPES AND CHEESE

Large black grapes in particular make a good contrast to soft white cheeses, both for taste and for looks. The cheese can be anything from a soft curd or yoghurt cheese, beaten up with a little cream, to Brie, the double and triple cream cheeses of Normandy, and the thicker mild cheeses such as soft white Pyrenean cheese. In the autumn and winter, such cheeses with black grapes and walnut bread (or fresh walnuts and wholemeal, rye or granary bread) are a good simple fresh way of finishing a meal.

If you are making open sandwiches, butter the bread thinly, right to the edges. Spread them generously with one of the cheeses mentioned, and put halved, pipped grapes on top, dome side up. Use unskinned dark grapes for the sake of the contrast; or white or skinned grapes with toasted almonds or hazelnuts.

CAULIFLOWER SALAD WITH GRAPES AND WALNUTS

Theodora Fitzgibbon found this elegant salad in a coverless eighteenth-century cookery book, and published it in her *Art of British Cooking*. She has no idea who wrote the book. I always have a hope of coming across another copy, as it must contain other good recipes if this one is any guide.

1 cauliflower
small bunch white grapes, preferably seedless
8 whole walnuts
6 tablespoons (½ cup) olive oil
2 tablespoons (3 tbs) white wine vinegar
salt, pepper
chopped parsley

Boil the cauliflower, being careful not to overcook it. Skin the grapes meanwhile, and remove seeds if necessary. Pour boiling water over the walnuts and peel away their fine skin, too.

Make a dressing of olive oil, vinegar and seasonings and put it into a bowl. Separate the flowerlets from the cauliflower head – they should still be a little crisp – and put them into the dressing. Leave to cool, turning them over carefully once.

Half an hour before the meal, taste the cauliflower and see if more dressing or seasoning is required. Arrange it in a bowl, with the grapes and walnuts on top. Sprinkle with a little parsley and serve.

GRAPE SALAD

Among the beautifully printed cookery books of the Thirties, two of my favourites are by Ruth Lowinsky, *Lovely Food* (1931) and *More Lovely Food* (1935). Mrs Lowinsky was writing at a time when domestic refrigerators were beginning to come in from America. An extraordinary number of her dishes are iced, chilled or frappé. The two books are set out as menus. This grape salad, to be served with roast chicken, comes from a 'Dinner to impress your publisher and make him offer ridiculous sums for the privilege of printing your next book. Try to keep off controversial subjects such as politics, sex or religion, but let him see that he can still shock you. Attend to the wine.'

For the rest of the menu, see the rhubarb spring soup on p. 406. And have confidence. I felt nervous of the marshmallows, which make me think of the worst of American cornflake packet cookery, but they are an excellent idea. The salad is creamy but light, not at all strange to the taste but full of agreeable little surprises. Enough for 10 people.

½ kg (1 lb) white grapes
12 white marshmallows
2–3 (scant ¼ cup) tablespoons brandy
2 medium apples, e.g. Blenheim orange
celery (see *recipe)*
60 g (2 oz/½ cup) chopped walnuts
1 lettuce

DRESSING
2 large tablespoons (½ cup) flour
4 yolks
2 tablespoons (¼ cup) sugar
dash of wine vinegar
60 g (2 oz/¼ cup) butter
300 ml (10 fl oz/1¼ cups) whipping cream

Skin, halve and remove the seeds from the grapes. Dip and turn the marshmallows in the brandy, then cut them up. Weigh the apples and cut up an equal weight of celery: leave the apples for the moment. Mix the marshmallows, celery and walnuts in the salad bowl or wide shallow dish, lined with crisp lettuce leaves.

Next make the dressing. Mix the flour with a little cold water to a smooth paste in a saucepan. Add 300 ml ($\frac{1}{2}$ pt/1$\frac{1}{4}$ cups) boiling water and stir over the heat to thicken the liquid smoothly. Beat the yolks with the sugar and vinegar. Stir this into the hot flour base with the butter, and cook until the dressing thickens: do not boil vigorously, but keep around simmering point. Cool, and just before serving the salad, fold in the whipped cream. Pour on to the grapes.

Now halve, core and cut up the apples, leaving the skin on if you like. Mix into the salad carefully, then put the grapes and dressing over the top to cover the whole thing. Serve cool rather than chilled.

The chicken should have been roasted and allowed to cool down a little – at least, this is what I feel. Mrs Lowinsky did not commit herself.

OXTAIL WITH GRAPES
QUEUE DE BOEUF À LA VIGNERONNE

These days in France, especially since the rise of the *nouvelle cuisine* style, there are a number of meat and fish recipes about that include grapes. Inevitably, they are called after the vineyard owner's wife. Whereas, some recipes may have a genuinely country origin, I suspect that most others belong to the pastoral tradition, the highly artificial tradition of country food as seen from the town – or from the smart restaurant. Did anyone before Monsieur Malley, the *saucier* at the Paris Ritz, think of putting grapes into savoury dishes? Was there any precedent for his sole Véronique? I can certainly claim no vineyard source for this recipe: we ate oxtail with grapes at the Hole in the Wall at Bath recently, it tasted good but I decided to put the ingredients together in a different way.

Cook 2–3 oxtails in the manner described for oxtail with prunes (p. 365), using white wine instead of red. Put the pieces to cool in one bowl, the strained liquor in another (the carrots are not required). Next day assemble the following ingredients:

300–375 g (10–12 oz) large muscat grapes
30 g (1 oz/1 cup) fresh breadcrumbs
1 heaped teaspoon thyme
2 heaped tablespoons (3 tbs) chopped parsley
grated zest of a lemon
melted butter
4–6 tablespoons ($\frac{1}{3}$–$\frac{1}{2}$ cup) muscat or other fortified wine

juice of a lemon
90 g (3 oz/scant ½ cup) unsalted butter, cubed
cayenne pepper, black pepper, salt

Remove the cake of fat from the jellied oxtail liquor. Put the liquor into a wide shallow pan with the oxtail pieces, cover and heat through. Allow half an hour at least, keeping the liquid at simmering point and turning the pieces over occasionally.

Meanwhile peel, halve and pip the grapes. Set half aside. Add the rest with any juice to the oxtail pan.

Mix together breadcrumbs, herbs and lemon zest.

When the oxtail is properly heated, remove from the liquor and arrange the pieces in a single layer in a heatproof serving dish, cut sides up. Brush the oxtail tops with melted butter, and press on spoonfuls of the crumb mixture to make an even topping. Sprinkle a little melted butter over the crumbs. Place under a hot grill or in a hot oven to brown the tops.

Add the wine to the liquor and grapes which should be boiled down vigorously to a strong sauce. Strain into a smaller shallow pan, pressing through as much liquor from the grapes as you like. If you sieve them all through, apart from odd fibres, you will have rather a grainy sauce: I prefer just to press on the grapes so as to get the juices through without much of their substance. Taste the sauce, add a little lemon juice, and bring it to the boil. Off the heat whisk in the butter, and add seasoning to taste. Lower the heat on the stove, put the uncooked grapes into the sauce and warm the whole thing through gently. It should not come to boiling point, and the butter should not be allowed to oil. Vigorous stirring at the first sign of oiliness usually corrects this, and you should keep raising the pan from the stove. A tablespoon or two of very cold water will also bring the sauce back. Pour round the oxtail, being careful not to disturb or moisten the crumbs, and serve. Bread is all that is required, with a salad to follow.

Note Sauces enriched and thickened with butter are not as difficult to make as they are to describe. The thing is not to overheat them, and not to have either the main item or its serving dish or the plates very hot. Hot, yes; very hot, no.

GUINEAFOWL GEORGIAN STYLE

Fazan po kavkazki, from the Caucasus, should be made with pheasant and fresh walnuts, i.e. in October and November when both are in season. I first gave the recipe in the *Observer*, then in *Good Things*, and have now adapted it to summer use with guineafowl because we like it too much to eat it only in the autumn.

As far as I can discover, the first Westerner to record a recipe was Monsieur A. Petit. He published his *Gastronomie en Russie* in 1860 to help colleagues who arrived in grand Moscow and Leningrad kitchens without a word of Russian or any knowledge of Russian specialities. The diaspora of French chefs to Russia seems to have started with Carême and continued until the First World War. One of the last grand chefs to do this was Edouard Nignon. The marriage of French style and Russian ingredients can be seen for instance in *Bef Stroganoff* which was devised by a French chef in the kitchens of one of the households of this powerful Russian family.

2 guineafowl
250 g (8 oz/2 cups) shelled walnuts
milk
$\frac{3}{4}$ kg ($1\frac{1}{2}$ lb) grapes, preferably muscat, peeled, pipped
150 ml (5 fl oz/$\frac{2}{3}$ cup) strong green, i.e. gunpowder tea
150 ml (5 fl oz/$\frac{2}{3}$ cup) sweet muscat wine
3 large oranges or 4 blood oranges
salt, pepper
125 g (4 oz/$\frac{1}{2}$ cup) butter

Put the birds together into a deep casserole. Put the walnuts into a small saucepan and cover generously with milk. Bring to the boil, simmer 1 minute, then leave until cool. Strain off the milk (use it up in bread-making or soup) and peel the thin skin from the walnuts: the milk refreshes their consistency and flavour. Liquidize or chop two-thirds of the grapes, and put into the casserole with the walnuts, tea, wine, grated rind and juice of one orange ($1\frac{1}{2}$ blood oranges).

Slice remaining oranges thinly and place the slices overlapping each other down the breasts of the guineafowl. Season and pour over one-third of the butter, melted.

Cover closely and cook at gas 8, 230°C (450°F) for 15 minutes. Lower the heat to gas 4–5, 180–190°C (350–375°F). Leave 20 minutes, then check to see how the birds are doing. If they are still pink, put them back.

Transfer the orange slices from the breasts to a hot serving dish, arranging them round the edge. Carve or quarter the birds and put them in the middle, surrounding them with the walnuts that you remove from the pan with a slotted spoon.

Strain the juice into a wide shallow pan and boil down hard to make a concentrated sauce (if too much juice has evaporated during cooking, which will depend on the fit of the lid, add a little poultry stock before boiling it down: or leave the sauce as it is, if the flavour is good). Heat through the remaining grapes in the sauce. Off the heat, whisk in the last of the butter in little bits. Check seasoning and pour over the birds. Serve with rice, brown rice for preference.

Enough for eight.

Note In a version of the Russian dish, called 'Eccentric Pheasant', from T. Earle Welby's *Away Dull Cookery*, side dishes are served – salted almonds, orange salad and peeled grapes. This you could do instead of putting orange slices on the birds, and the peeled grapes into the sauce. The quantities would need to be increased.

CHICKEN STUFFED WITH GRAPES
POULET AUX RAISINS

Use as good a chicken as you can find for this recipe. One of the corn-fed chicken imported from France, for instance, unless you are lucky enough to have a private source of supply. The flavours are so delicate that they would seem insipid if you used a frozen bird. The recipe can be adapted to guineafowl.

1 roasting chicken
$\frac{1}{2}$ litre ($\frac{3}{4}$ pt/2 cups) stock made from the giblets minus liver
150–175 ml (5–6 fl oz/good $\frac{2}{3}$ cup) dry white wine
150–175 ml (5–6 fl oz/good $\frac{2}{3}$ cup) double cream
a large knob of butter
salt, pepper
12–18 large muscat grapes, peeled, halved, seeded

STUFFING
90 g (3 oz/$\frac{3}{4}$ cup) chopped onion
90 g (3 oz/scant $\frac{1}{2}$ cup) butter
1 clove garlic, crushed
$\frac{1}{2}$ teaspoon dried tarragon
50 g ($1\frac{1}{2}$ oz/$\frac{3}{4}$ cup) fresh white breadcrumbs
200 g (7 oz) large muscat grapes, peeled, halved, seeded
salt, pepper
1 large egg, beaten

Make the stuffing first. Cook the onion gently in half the butter until soft and golden. Add the garlic and tarragon and the rest of the butter and stir round for a minute or two. Tip into a basin, add crumbs, grapes, seasoning, and finally the egg. Stuff the bird.

Roast the bird in your usual style. When it is cooked, put it on to a serving dish and keep it warm. Pour off the fat from the pan, leaving the meaty juices. Pour in the stock and wine, and boil down fiercely to make a small amount of concentrated liquid, about 150 ml ($\frac{1}{4}$ pt/$\frac{2}{3}$ cup). Stir in the cream and bring to the boil. Strain into a pan, bring back to the boil, remove from the heat and whisk in the butter. Season to taste and put in the grapes. Stir over the heat for a minute or two, but do not allow the sauce to boil or to come near boiling point. Serve with the chicken. Green salad to follow.

Vegetables are a mistake with this kind of dish, apart from a very few small potatoes, better still some plain rice.

DUCK IN THE STYLE OF THE VINEYARD OWNER'S WIFE

CANARD À LA VIGNERONNE

Of all the meats enhanced by fruit, duck is the best. Its rich meat demands sweetness, and a slight acidity. The recipe below can also be used for oxtail: brown the pieces first, then go straight to the braising stage, and add stock or wine, or a mixture of both, to make sure that everything is well covered and will remain moist during the long cooking time.

$2\frac{1}{2}$ kg (5 lb) dressed duck
250 g (8 oz) each chopped carrot and onion
1 tablespoon (2 tbs) sugar
1 kg (2 lb) grapes, the larger the better
juice of a lemon
salt, pepper
strong duck giblet stock or meat glaze
60 g (2 oz/$\frac{1}{4}$ cup) butter, cut in dice

Prick the duck with a fork in several places. Then roast in a hot oven, gas 7, 220°C (425°F), for 45 minutes. As soon as the fat begins to run into the roasting pan, remove two tablespoonfuls (2 tbs) to a deep enamel casserole and brown the carrot and onion lightly. Stir in the sugar and continue cooking until the vegetables caramelize slightly – but don't let them burn.

Take the duck from the oven and place it on to the vegetables. Turn the oven down to gas 4, 180°C (350°F). Tuck two-thirds of the grapes round the duck, squashing them down a little. Add lemon, salt and plenty of pepper. Pour in 150 ml (¼ pt/½ cup) stock or a couple of tablespoons (3 tbs) of meat glaze. Cover with the lid or a piece of foil, and put the casserole into the oven. Cook for ¾–1¼ hours longer, according to how well done you like your duck.

Peel and halve and pip the rest of the grapes. When the duck is ready, transfer it to a warm serving dish. Strain the cooking juices into a small saucepan, pressing through a little of the pulp to give the juices body, but not enough to make them too grainy. Taste and reduce by hard boiling if necessary. Skim well when the sauce is boiling. Off the heat, whisk in the butter and add the grape halves. Heat through without coming near boiling point. Pour the sauce round the duck.

Serve with a few small new potatoes, no other vegetable.

DUCK LIVER WITH GRAPES
FOIE DE CANARD AUX RAISINS

The ideal for this classic dish of grand French cookery is *foie gras de canard*, fattened duck's liver. As *foie gras* of any kind is almost impossible to find in the shops in its fresh state, we have to substitute ordinary duck livers – turkey livers can be used as well, so can chicken livers although they are on the small side.

½ kg (1 lb) duck livers
salt, black pepper, pinch cayenne
36–40 large muscat grapes
6 slices brioche (p. 463), or good white bread
butter
4 tablespoons (⅓ cup) Frontignan or other muscat wine
125 ml (4 fl oz/scant ½ cup) duck or poultry stock

Pick over the duck's livers, cutting away any greenish patches or the green gall bladders. Season them well and set aside. Peel, halve and de-pip the grapes.

Cut the bread into large rounds or ovals, and fry them until golden brown in butter. Put on to a warm serving dish, and keep them warm in the oven or over a pan of boiling water.

Clean out the pan with kitchen paper, put in some fresh butter and when it is very hot, cook the duck livers, fast at first to seal them, then more slowly. Allow 3 minutes a side, then check them. They should be pink in the middle. Distribute them among the pieces of fried bread.

Deglaze the pan with the wine and stock, scraping it all together. Strain into a clean pan, put in the grapes to heat up, correct the seasoning and stir in a large knob of butter off the heat. Pour round and over the liver. Serve immediately.

GRAPE JELLY

Jelly, wobbly complex jelly in traffic-light colours, was the centrepiece of children's parties years ago. It tasted disgusting, particularly when slices of banana were embedded in the green layers. I never thought until a few years ago that jellies made with fruit juice and gelatine, rather than a rubbery slab, could be a treat. This grape jelly is simple and refreshing, ideal for the last course of an elaborate meal.

1 kg (2 lb) grapes of good flavour, muscat for instance
about 125 g (4 oz/½ cup) caster sugar
1 packet (½ oz/2 envelopes) gelatine
juice of a lemon
liqueur glass Grand Marnier or Cointreau, or juice of an orange
150 ml (5 fl oz/⅔ cup) whipping cream

Set aside a dozen or so of the largest grapes, having peeled, halved and pipped them.

Put the rest into a pan with 125 g (4 oz/½ cup) sugar. Simmer until the grapes collapse, leaving them on the heat as briefly as possible. Sieve them into a measuring jug. If necessary make up the quantity with water, to reach ½ litre (¾ pt/good 2 cups). Dissolve the gelatine in 4 tablespoons (⅓ cup) hot water. Stir into the warm grape juice. Add lemon and liqueur or orange juice. Taste and add more sugar but do not make the mixture too sweet. Leave in a cool place until almost set. Beat with a rotary whisk to a creamy froth. Pour into six glasses and leave in the fridge until firm. Decorate with the grape halves.

Whisk the cream until it is thick but still just pourable. Spoon on to the jellies just before serving, but do not cover the domed tops of the grapes which should show up like little islands.

Crisp almond biscuits or sugar or lemon thins (p. 461) should be served with this light dessert.

GRAPE BAVARIAN CREAM
BAVAROISE AUX RAISINS

There's no evidence, I think, that this delicious type of cream is a native of Bavaria. Perhaps some French cook working there thought that the light rich texture of the dish he was making belonged in spirit to the baroqueries around him. Anyway, it's in French cooking that the bavaroise first appears. In Carême's *Cuisinier Parisien*, of 1828, he devotes three chapters to *fromage bavarois.*

The basic recipe can be flavoured with green walnuts, pistachios or hazelnuts; with mint, violets, pinks and orange blossom; with strawberries, raspberries, apricots. I wonder if the Prince Regent, one of Carême's employers, was ever served with the pale pink *fromage bavarois au parfait amour*? Mrs Fitzherbert must have eaten it with a wry smile. There are no ironies, though, to this simple version flavoured with grapes.

generous ½ kg (1 lb) black or white grapes
125 g (4 oz/½ cup) caster sugar
1 packet (½ oz/2 envelopes) gelatine
juice of half a lemon
5 tablespoons (6 tbs) muscat dessert wine (optional)
250 g (8 fl oz/scant cup) double or whipping cream
extra sugar

Peel, halve and pip a quarter of the grapes. Put the rest in a pan with the sugar. Simmer briefly, until they soften and begin to burst. Push through a sieve, into a measuring jug. There should be about 220–250 ml (7–8 fl oz/1 cup). Dissolve the gelatine in 4 tablespoons (⅓ cup) of hot water and add to the warm grape purée. Stir in lemon juice and wine, if used. Add more sugar if the grapes were on the tart side. You should now have about 375 g (12 fl oz/1½ cups) – if not, add a little water, or water and lemon juice. Leave in the refrigerator until thick, with an egg white consistency, but not set.

Whisk the cream with a tablespoon 2 tbs of sugar until thick and light. Fold it into the grape jelly. Setting aside a few of the remaining grape halves for decoration, stir the rest carefully into the bavaroise. Turn into an elegant dish, or into glasses or custard cups. Decorate with the remaining grapes when set. Serve with almond biscuits or *langues de chat.*

THE FOUR MENDICANTS

LES QUATRE MENDIANTS

The first four orders of mendicant friars established themselves in twelfth-century France, during the reign of Louis the Fat. They were the Dominicans or Black Friars, the Carmelites or White Friars, the Franciscans or Grey Friars, and the Austin or Augustinian Friars. These gentlemen became famous for their begging techniques, the way they sucked the labouring classes dry. Anything however poor was good enough for their wallets.

When it came to Lent, these reverend holy men kept such foods as black raisins, white almonds, grey figs and hazelnuts for spinning out their meals, and helping them to get over the taste of salmon, trout and so on, which, to mortify their flesh, they were obliged to eat on fasting days.

It's for this reason, so they say, that a wit seeing these dried fruits and nuts together on a plate one day, cried out: 'Look! the four begging friars – *les quatre mendiants!*'

A nice explanation that I came across in *Historiographie de la Table*, by C. Verdure, published in 1833.

THE PRICE OF A BUNCH OF GRAPES

In 1175, before he was king, Richard Coeur de Lion assembled the notables of his duchy of Guyenne and said, 'Whoever takes a bunch of grapes from somebody else's vines will be fined five sous or lose an ear.' Which tells us what the ear of a Gascon was worth at the time: it's gone up since then.

C. Verdure, *Historiographie de la Table,* 1833.

GRAPEFRUIT, POMELO, UGLI, CITRON etc

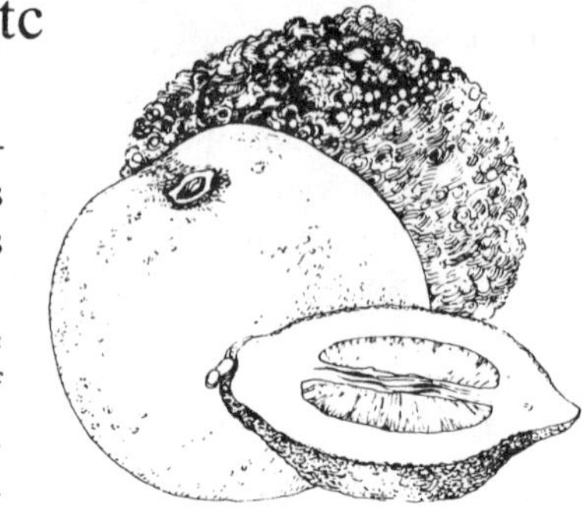

The oldest of the sharp citrus fruits – oldest as far as we are concerned – is the citron which, in this country, is known in the form of candied peel. This gives no idea of the strange elongated fruit and the noble size of it, the craggy look. I saw citrons in Crete once in November, when a friend was in the process of making candied peel (her recipe appears at the end of the section): it has a grand presence. The Romans knew it through Arab merchants and called it the Assyrian citron and the Median apple. Apart from its peel, it is useless.

Knowledge of the pomelo came to us in the seventeeth century. Again, it has a thick skin and handsome, large appearance. The segments of fruit inside have an attractive dryish look; as you bite into them, they are almost squeaky. I like this very much, especially as the fruit is at the same time sweet as well as sharp. Like the Dutch and the French, we also called the pomelo a pamplemousse. It is a native of Java and Malaysia, and early acquired a Dutch name – *pomp-* from the beginning of the name for pumpkin, -lemousse from *limoes* meaning a citron. In other words, it was a pumpkin citron or large citron. Lately, Israel has been sending us pomelos and they are magnificent. To confuse things further, we also called this fruit the Shaddock after the captain of an East Indiaman who left seeds in Barbados on his way home to England. If you ever read *Paul et Virginie* at school, and used a crib, you may have been puzzled by the translation of *eglise des Pamplemousses* as 'church of the shaddock grove'. Surely pamplemousse means grapefruit?

Certainly it does nowadays. But when *Paul et Virginie* became a best-seller all over Europe, in 1788, the grapefruit had not made an appearance anywhere. It was recognized as a species by 1830, and thought either to be a hybrid between the pomelo and orange, or a sport of the pomelo. As a commercial fruit, it was launched in the 1880s, first in Florida. In 1885, the English still needed to be told that the grapefruit had nothing to do with the grape: it had got its name from its habit of growing in clusters. One French writer on food plants recalled his first acquaintance with grapefruit, I imagine at the beginning of this century. He visited an old friend, who was a priest, on a day when he had received some grapefruit from a penitent who had gone to Guadeloupe. Neither knew how to eat a grapefruit; they cut it in four quarters as if it were an orange, and took a big mouthful . . . Not long afterwards, he learnt how to halve the grapefruit across, ease round the segments and sugar it – 'add port, sherry or kirsch, if you are a sybarite'. I imagine he died before pink grapefruit arrived in Europe, with their extra sweetness.

Another fruit we have been seeing more and more in recent years is the ugli. It is too new – in this country – to have made the first supplement of the great Oxford Dictionary, but perhaps it will be in the last volume of the second supplement, from O–Z. I suspect that the name is just an intriguing way of spelling 'ugly', which the fruit is, it cannot be denied. The colour with its strange mixture of green and orange, the baggy shape, are unpoetical, if endearing. The flesh tastes sweeter than grapefruit, sharper than a tangerine, an elegant flavour. It is a hybrid between the two, a tangelo (so is the mineola, which is smaller and more orangey). Ugli fruit peel candied is as good as the pomelo's, the citron's or the grapefruit's. Never waste it, for it's not a cheap fruit.

These fruits are best eaten raw – citron apart, which is kept for preserving. Or so I feel. The salads and jelly show them off best, though for a change one of the meat and grapefruit (or ugli, or mixture of pomelo and grapefruit) recipes in the section following may be pleasant.

HOW TO CHOOSE AND PREPARE GRAPEFRUIT ETC

Although you should avoid obviously damaged fruit, fruit with soft patches, I have sometimes found that by looking out for dryer specimens – or by keeping young ones for at least a week – I have ended up with mellower tasting fruit. This is difficult to test properly, of course, and may be just a coincidence. Pink grapefruit are quite a lot sweeter than yellow ones – in a salad, you might think of mixing the two with, for instance, avocado pear, or Sharon persimmon and watercress. Many people like soft cheeses of various kinds with grapefruit, *fromage frais, quark*, curd cheese, full fat soft cheese and so on. These can be mixed with glacé fruit and nuts: or with herbs. Grapefruit, soft cheese and high quality salted cashews are agreeable, though I confess it is one of the dishes I find easy not to eat, as I do not greatly like these mixtures.

All these fruit can be used in making marmalade with other citrus or sharp fruit (*see* the recipe for pineapple and grapefruit marmalade on p. 350). All the skins can be candied.

Everyone knows how to prepare grapefruit by cutting it in half, and easing the flesh with a curved knife. Less generally known is a way to prepare the whole inner fruit for skinning. Leave it, peeled, in the refrigerator for a day or two: the skin will become dehydrated and stiff, the flesh inside firm. It is then much easier to separate the two.

SWEET CORN AND GRAPEFRUIT SALAD

If you have fresh sweetcorn, cooked and scraped from the cob, this makes a good salad (some brands of frozen sweetcorn on the cob are well worth trying, which makes this a useful winter salad as well: tinned sweetcorn tends to be a bit sweet).

Weigh the sweetcorn, and add a roughly equal weight each of celery and of apple. Mix the juice of a grapefruit with a dessertspoon of cream, and use as a dressing for the salad. Depending on the sweetness of the apple and the sourness of the grapefruit, add a little sugar to the dressing. Chopped tarragon if you have any, can be scattered over the top, and mixed in lightly.

GRAPEFRUIT AND SHELLFISH SALAD

In hollowed-out grapefruit shells, put a salad made of either crabmeat or shelled prawns mixed with some of the skinned, diced segments, after first lining the shell with a lettuce leaf, so that it frills slightly over the edge. The rest of the grapefruit flesh can be used for another dish. If you like, also add some diced cucumber or tomato, with wedges of hard-boiled egg.

Put a spoonful of mayonnaise on top of each filling. Put a prawn in its shell or a neat piece of grapefruit on top. Serve chilled, with extra mayonnaise and brown bread and butter.

GRAPEFRUIT AND CHICORY SALAD

Slice two fine heads of chicory into slices about 2 cm (¾ inch) thick. Mix with the flesh from half a pink grapefruit, cut across in two. Squeeze the juice from the other half into a separate basin. Beat in 2 tablespoons (3 tbs) of sunflower oil, or a mixture of olive and ground nut oils, and a dash of sherry vinegar. Season, and pour over the salad.

Makes a refreshing first course for a winter lunch. Enough for four people.

GRAPEFRUIT AND SPROUT SALADS

GRAPEFRUIT AND CHINESE CABBAGE SALAD

Grapefruit is sometimes mixed with sliced raw members of the cabbage family. And if you are keen on the coleslaw idea, you will find it a useful combination in the winter when salad greens are difficult to buy or uninteresting.

My own feeling is that the finer and less cabbage-like the taste of the cabbage greens you choose, the more successful the salad. In other words, I find that young Brussels sprouts or Chinese cabbage (Chinese leaf) are the best, indeed the only bearable choice.

Be careful to remove the grapefruit segments from their skins. Allow a large grapefruit to about 175 (6 oz) of prepared sprouts or Chinese leaf. Dress with sunflower oil vinaigrette, if you find the taste of olive oil too strong with such pronounced flavours.

Peeled and pipped grapes, some diced boiled potatoes, a little chopped celery, toasted almonds or hazelnuts, can all be added.

BRAISED LAMB OR VEAL WITH GRAPEFRUIT

AGNEAU OU VEAU AUX PAMPLEMOUSSES

Brown a roasting joint of veal or lamb in a little butter. Put into a deep braising pot, with 18 small onions also browned in the butter, but lightly. Pour off surplus fat in the pan and deglaze with a glass of dry white vermouth (150–175 ml or 5–6 fl oz/good $\frac{2}{3}$ cup). Pour over the meat, plus the juice of two grapefruit. Cover and cook gently on top of the stove, or in a low oven at gas 3, 160°C (325°F), turning the meat occasionally. When tender, transfer to a warm plate with the onions. Skim fat from the pan juices.

Have ready a grapefruit which has been peeled, skinned and the segments removed and skinned. Heat them through in the pan juices. Put them round the meat. Taste and season the juices, adding a good pinch of cayenne pepper. Pour over the whole thing and serve with brown rice.

Note A coarsely shredded carrot may be put into the pot with the onion.

TURKEY OR CHICKEN WITH GRAPEFRUIT

This dish is best made with pink grapefruit. I approached it with caution as the mixture of poultry and grapefruit seemed an odd one, but it works well. The better the poultry, the better the result.

turkey or chicken joints for 6 people
salt, pepper, level teaspoon thyme
butter
3 tablespoons (scant ¼ cup) brandy
250 ml (8 fl oz/good ¾ cup) dry white wine
juice of a grapefruit
2 whole grapefruit

Season the joints with salt, pepper and the thyme. Leave for several hours or overnight.

Brown the joints in butter to a light golden brown, then flame with the brandy. Turn the pieces over in the flames. Now:

Either: transfer the meat to a buttered ovenproof dish, dot it with butter and pour in the wine and grapefruit juice. Bake for about an hour at gas 5, 190°C (375°F), turning occasionally and removing breast joints when cooked.

Or: add the wine and juice to the sauté pan, cover and leave until tender, turning the pieces over twice.

Meanwhile peel thin strips from one of the grapefruit. Cut them into shreds and simmer three or four minutes in water: set aside for garnishing. Strip remaining pith and peel from both grapefruit, and cut the wedges of flesh out, freeing them of thin white skin and pips. Do this over a plate to catch the juice, which should be added to the liquid in the poultry pan.

When the poultry is tender, arrange it on a warm serving dish. Heat through the grapefruit segments in the pan juices and place them round the joints. Taste the juices, boil down slightly and add seasoning. Beat in a couple of tablespoons (¼ cup) of butter, in little bits, and pour over the joints just before serving. Scatter the peel strips on top.

Cauliflower finished with butter and a chopped hot chilli goes well with this dish.

ROAST CHICKEN OR GUINEAFOWL WITH GRAPEFRUIT

Here is the roasting version of the recipe above, with one or two small alterations of flavour.

For six people, choose a 2 kg (4 lb) roasting chicken or two guineafowl. Stuff with the segments of two grapefruit, freed of all skin and pips. Place in a roasting pan. Soak a double-stockinette cloth or J-cloth in 150 g (6 oz/¾ cup) melted butter and lay it over the bird(s), making sure the tops of the legs are well protected. Roast in your usual way, at a fairly high temperature – gas 5, 190°C (375°F) is the one I use. Check the guineafowl after 45 minutes, the chicken after 1¼ hours.

From time to time, baste through the cloth with the juice of a grapefruit and 200 ml (7 fl oz/¾ cup) dry white wine.

Meanwhile boil 375 g (12 oz/1½ cups) long-grain rice – soak it first for a good hour if using Basmati rice, then boil it in twice its volume of water. Peel, quarter and fry 3 bananas. Arrange rice round the edge of a serving dish. Place the bananas on top.

Cut up the poultry or leave it whole. Put with its grapefruit in the middle of the rice. Skim the fat from the cooking juices, and taste. Boil them down if you like, be sure to correct the seasoning – a pinch of sugar is sometimes required – and strain some over the dish, serving the rest separately. The thing to avoid is a wet or sloppy dish: a small quantity of strong sauce, just enough to add zest and moisture, is what you should aim for.

GRAPEFRUIT JELLY

Put 375 ml (12 fl oz/good 1½ cups) grapefruit juice into a measuring jug. Add the juice of a large orange and 2 tablespoons (3 tbs) of lemon juice (if you used the juice of pink grapefruit, you may prefer to use less orange and more lemon). Add water to bring up to 600 ml (1 pt/2½ cups). Sweeten to taste. Dissolve a packet of gelatine (½ oz/2 envelopes) in 6 tablespoons (½ cup) of hot water. Allow it to cool to tepid, then blend in with the grapefruit liquor. Taste again, and adjust sugar, or orange or lemon juice as you think fit. Strain it.

Remove the sections from a pink grapefruit or a pomelo, leaving them free of all skin. Set some jelly in the base of six glasses, using the ice-making compartment of the fridge or the freezer. When it is just setting firm, but not rigid, place a segment of grapefruit or pomelo upright into each glass with jelly. Leave to set again – the jelly should come to the top of the fruit which will be leaning up against the side of the glass. Finally, add a top layer of jelly with what's left. The jelly should be served chilled.

If you have a fancy to set the jelly and grapefruit in a decorative mould, reduce the amount of liquid, making it up to 600 ml (1 pt/2½ cups) in all. My jelly above is wobblier, but much nicer to eat.

CANDIED GRAPEFRUIT AND CITRUS PEEL

I took to candying peel when we were buying a lot of pomelos, pink grapefruit, ugli fruit and other citrons for the purposes of this book (*see* p. 254): they were quite expensive and it seemed wrong that anything should be wasted. The results were better than I had expected. They had a much better flavour than the chopped bought kind, and were thicker and more luscious than the large pieces of peel bought in special boxes. Added to a pound cake, they seemed to flavour the whole mixture, but most of the peel was eaten as sweets. My advice would be to get into the habit of making it and store it away. Different peels can be prepared together, and I always include all the white pith which clings to the outer skin. With cooking in the sugar, it is transformed, becoming translucent and coloured and altogether delicious.

Take the peel of two fine grapefruit, removing it in neat sections shaped like a large curving bay leaf. Boil it in plenty of water until it is tender. Drain and boil it again in fresh water for 20 minutes. Do not be tempted to omit the second boiling – it really is necessary. Bring 300 g (10 oz/1¼ cups) sugar to the boil with 150 ml (¼ pt/⅔ cup) water, stirring at first so that the sugar is quite dissolved by the time the syrup boils. Put in the peel and simmer until the syrup has more or less disappeared; spread out on trays lined with greaseproof or Bakewell paper and leave to dry – this can take up to three days – on the rack of a solid fuel cooker or in a warm airing cupboard are both good places. Turn the pieces from time to time. Put into a paper bag with caster sugar and shake to coat the peel. Store in an airtight jar.

Once the peel has had its two boilings, or when it has cooled after candying, you can cut it with scissors into smaller pieces.

PANFORTE

The Christmas fruit-cake of Siena, white with icing sugar on top and rice paper round the sides. It's easy to make and very good, though you will miss the attractive flowery box in which it is sold at Italian groceries.

100 g (3½ oz/scant ¾ cup) hazelnuts, toasted, skinned, chopped
100 g (3½ oz/scant cup) almonds, blanched, toasted, chopped
100 g (3½ oz/good ½ cup) candied orange peel, chopped
100 g (3½ oz/good ½ cup) candied lemon peel, chopped
100 g (3½ oz/good ½ cup) candied fruit, chopped

3 teaspoons cinnamon
½ teaspoon mixed spice
100 g (3½ oz/good ¾ cup) plain flour, sifted
125 g (4 oz/⅓ cup) honey
125 g (4 oz/½ cup) sugar
rice paper
icing sugar

Mix nuts, peel, fruit, spices and flour thoroughly with your hands (best way to get an even mix). Bring honey and sugar to boiling point, then mix into the nuts etc. to make a sticky cohering mass.

Line two tart tins or rings of 18–20-cm (7–8-in) diameter with rice paper. Divide mixture between the two, pressing it down evenly but not too heavily.

Bake at gas 2, 150°C (300°F), for 30–35 minutes. Cool, then remove from the tins. If the rice paper base and edging are torn, add new ones, using dabs of egg white as glue. Sprinkle top with icing sugar.

To keep, wrap in foil. To serve, cut in small wedges. It is very much a medieval style of cake, like early gingerbread stuck together with honey and not expected to rise. I suppose that the modern descendant of these antique cakes is the impacted fruit bars that are sold in vegetarian shops.

DUKE OF CAMBRIDGE PUDDING

A nineteenth-century version of the old sweetmeat pudding, given a smart aristocratic title.

100–125 g (3–4 oz/⅔ cup) chopped candied citrus peel, eg lemon, orange and citron
200 g (7 oz) puff or shortcrust pastry
175 g (6 oz/¾ cup) lightly salted or unsalted butter
175 g (6 oz/¾ cup) sugar
4 egg yolks

Scatter the peel over the base of a 20–23 cm (8–9 in) tart tin lined preferably with puff pastry, though shortcrust will do perfectly well. Warm butter and sugar until dissolved together and beat into the yolks. Pour over the peel. Bake at gas 4, 180°C (350°F) for 35–40 minutes. Keep an eye on it after 25 minutes, and remove when it has a nice golden brown crust. Serve tepid or cold.

GUAVA

Guavas made an impression on this country during the last war, in the shape of tinned jam from South Africa. Delicious pink jam, though some people had so much of it that they can hardly bear to hear the fruit mentioned thirty five years later. Perhaps that is why canned guavas and canned guava jam have retreated to the shelves of specialized and very superior shops where they do not make much of a showing.

Guavas may submit to such processing with reasonable success, but the soul of the fruit has gone. If you ever go to Paris, make one excursion if you do nothing else. Basically, it costs nothing. Walk into the greengrocery department of Fauchon's, in a corner of the Place Madeleine, and sniff. Then sniff again, and relax into the rich and exotic air, and contemplate the fruit displayed with such skill. I think that that air, perfumed with guava and passion fruit and melon, is what the smell of the Garden of Eden must have been. And it is only a glass door away from the crash of taxis and cars swerving round the enormous roundabout of the church of the Madeleine. Nobody bothers you, unless you look hesitant. needing help. Good psychology, because even the most tight-fisted puritan could not emerge without some tiny delicacy, a white punnet of wood strawberries or one apricot-peach.

People have been trying to describe the guava's smell since the sixteenth century – 'the fruit guannaba unknown to, us, and somewhat like unto a quince', which is a brave try, giving some idea at least of the guava's pervasiveness. When I cook guavas, they fill the kitchen with their smell. It goes through the door, across the hall, up the stairs and right into my husband's study. The first time, he had to leave his work and come down to see what magic was going on.

Later travellers compared the guava to a fig – which gives a slight idea of the shape and size – and a peach. In a mid-seventeenth-century natural history of Brazil, the guava is described as being 'elegantly adorned with a crown, like the medlar, filled with many seeds and offering a smell as gentle and sweet as a strawberry's'.

Where the guava grows – and it has spread from America through the world's tropics – it grows with speed and abandon. Birds help most of all. I had never thought of guavas growing in Europe, but a friend in Crete happened to remark one day that the jam we were eating – delicately flavoured with root ginger – was made from guavas grown in a local garden. A South African had come to live in Crete a few years before, bringing a guava tree with him. Birds had spread its benefit to other gardens on the island.

In the north, we have tried our hand with guavas since the middle of the seventeenth century, according to J. C. Loudun's *Encyclopaedia of Gardening*: he commended one new species, *Psidium cattleiana*, a 'strawberry guava' with reddish-purple flesh that had come from China and been fruited by William Cattley, in 1820. Cattley was a distinguished patron of gardening, who has achieved a wider immortality than guavas would have brought him, an immortality he might not have relished. To honour his memory, a genus of orchids was named after him in 1846. So far, so good. Cattleyas became fashionable, so fashionable that they were worn by Paris courtesans, such as Odette, in Proust's novel, *A la recherche du temps perdu*. I suppose she might have been called *la dame aux cattleyas*, for she and her lover, Swann, used a phrase 'do a cattleya' to mean making love. On occasion he might appropriately have said –

> C'est trop peu que des fleurs: je veux t'offrir encore
> Goyave au court duvet que le safran colore.

Incidentally, guavas are good for you, having a higher vitamin C content than many citrus fruits.

CHOOSING AND PREPARING GUAVAS

The white and the strawberry guavas are said to be the best flavoured kinds, but I have never had the opportunity to make a proper comparison. As with all fruit, choose firm unblemished guavas.

Unless you are making a jelly, peel them first. They can be eaten raw, seeds and all, and have an unusual texture, gritty in a tender way, a honeyed roughness that is pleasant on the tongue. They taste good with cream cheese concoctions.

To cook guavas, go very carefully as they soon disintegrate in the syrup, shooting the seeds all through the liquid and clouding the juice with their soft pinkness. I suspect this is why guavas are so popular for creamy dishes, but if you go carefully they make a simple and attractive dessert on their own, or as part of a fruit salad. One way round the cooking problem is to put them into a cold light syrup, then bring them slowly up to boiling point. Remove from the heat and check their progress to complete tenderness, removing them from the liquid before woolliness sets in. This technique which I use with salted water for fish, works well with many soft fruits.

FILIPINO FISH AND GUAVA SALAD

This recipe, with slight alterations, comes from *Maria Y Orosa, Her Life and Work* (with 700 recipes) which was published in the Philippines in 1970. It was given in Alan Davidson's *Seafood of South-East Asia*, a book I would recommend for a number of reasons, but particularly for its range of fish and fruit dishes. We do not always take into account the sweetness of fish, and how well this is emphasized by fruit other than lemon. The intention of the recipe, in a guava-growing country, was frugality, extending expensive fish. For us, economy might mean cutting down on the guavas: there is no reason why the salad should not be served in small ramekins, which means a saving of fifteen guavas.

The fish used was a fusilier, which is a snapper. We do sometimes have a chance to buy snappers at the fishmongers, but bream or any good firm-flaked fish could be used instead. Weevers, for instance. Buy ½ kg (1 lb) and poach it in well-seasoned water. Cool and flake the flesh. You will also need:

25 ripe guavas
1 orange, peeled
3 bananas, peeled
125 ml (4 fl oz/½ cup) coconut cream, p. 61

Peel 10 guavas thinly, halve them and discard the seeds. Chop the rest into small pieces. Divide the orange into segments and remove the thin white skins and pips. Cut the banana into smallish pieces and mix all this fruit together, with the fish. Stir in the coconut cream gently. Cover the bowl with plastic film and chill thoroughly.

Cut lids from the remaining guavas. Scoop out the seeds to make 'shells'. Put into a plastic bag and fasten, then leave in the refrigerator to chill until needed. Just before serving, divide the salad between the shells and replace the lids.

CHILLED GUAVA SOUFFLÉ
SOUFFLÉ GLACÉ AUX GOYAVES

Peel and chop enough guavas – 300 g (10 oz) to give you 250 g (8 oz). Put with the thinly cut peel of half a lemon, all the lemon juice, the crushed seeds of a cardamom pod and 2 rounded tablespoons of caster sugar into a heavy pan. Simmer gently until the guava is soft, remove the lemon rind, then liquidize or process to a pulp. Then sieve.

Using three egg whites, 100 g (3½ oz/½ cup) sugar, a packet (½ oz/2 envelopes) gelatine and 300 ml (10 fl oz/1¼ cups) double cream, follow the parfait recipe on p. 476.

Put in extra lemon juice at the end if necessary to enhance the flavour.

Turn into a small, paper-collared soufflé dish and chill until firm. Decorate discreetly with chopped toasted hazelnuts.

GUAVA JELLY OR WATER ICE

Peel and slice 10–12 guavas into a processor or liquidizer. Add enough water to get them going, and whizz to a sloshy purée. Pour through a sieve to keep back the pips. Add caster sugar to taste – or icing sugar which dissolves very easily. Sharpen with lemon juice if necessary.

Set with gelatine, allowing a packet (2 envelopes) to a pint (2½ cups), and using 8 tablespoons (⅔ cup) of hot water to dissolve the powder. Avoid too firm a set. Serve in glasses.

FOR A WATER ICE Freeze in the usual way, returning to the processor beating electrically when the mixture is two-thirds frozen so that you can beat it smooth. Repeat this once more, and aim to serve the ice slightly soft. If you are using an ice-making machine, a *sorbetière* or electric churn for instance, freeze according to the instructions provided.

Note For a stronger flavour, you can boil the strained fruit and sugar to reduce it, but I prefer the freshness of the recipe above.

GUAVA CRISPS

A recipe based on one in the *Hawaii Cookbook* by Elizabeth Ahn Toupin, for a tropical flap jack, or tropical date slice. Macadamia nuts are tricky to find, but a good grocery or store selling American goods should have them. They have a light crunchy richness not unlike a fine Brazil nut, but not very like it either

PASTRY
150 g (5 oz/1¼ cups) self-raising flour
½ level teaspoon salt
125 g (4 oz/½ cup) butter
125 g (4 oz/½ cup) sugar
100 g (3 oz/scant cup) rolled oats

FILLING
60 g (2 oz/¼ cup) butter
100 g (3½ oz) guava jelly (see *following recipe)*
2 tablespoons lemon juice
2 level tablespoons sugar
¼ level teaspoon salt
1 large egg yolk, beaten
30 g (1 oz/¼ cup) chopped macadamia or Brazil nuts

Mix the first four pastry ingredients to a crumble, then mix in the oats. Spread half on an 18–20-cm (7–8-inch) square tin.

Next make the filling. Melt first five ingredients over a low heat. Whisk some of this into the yolk, then return it to the pan and stir over a low heat until the mixture thickens. Do not allow it to overheat – you are making a sort of guava curd, not a guava curdle. Remove from the heat, cool quickly by dipping the pan base into cold water, and stir in the nuts. Spread this over the crumble in the tin. Cover with the remaining crumble. Bake for about 25 minutes at gas 4, 180°C (350°F).

GUAVA JELLY (JAM)

Slice your guavas, without peeling them, and put them into a preserving pan. Cover with water and simmer until the fruit is very soft. Strain through a jelly bag, or doubled muslin or stockinette in the usual way.

Measure your juice back into the pan in cupfuls, and to every three cups add 2 cups of sugar, warmed first. Stir until the sugar is dissolved, then bring to the boil and boil hard until setting point is reached. Pot in the usual way, but in little jars, as this is something of a treat to be allowed out of the store-cupboard on special occasions.

GUAVA CHEESE OR PASTE

Turn to the recipe for cotignac in the quince section (p. 393), and use guavas.

INDIAN FIG, see PRICKLY PEAR
JAPANESE MEDLAR, see LOQUAT
KIWI, see CHINESE GOOSEBERRY
KUMQUAT, see ORANGE

LEMON and LIME

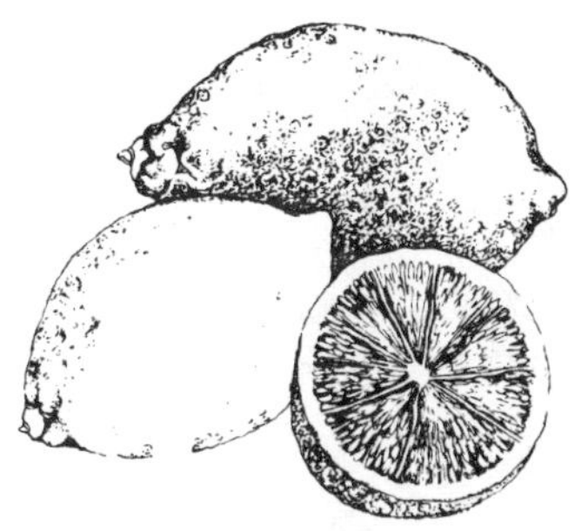

A pleasure northerners never forget is a first proof that lemons (and oranges) actually grow on trees – a first sight, somewhere round the Mediterranean, of actual lemons hanging from actual lemon trees, in a lemon orchard. I add to that a second pleasure, sitting out of doors at Limone, on Lake Garda, drinking in hot sunshine a first ice-cold *limone spremuta*, squeezed from lemons grown around the village, so improbably near the chill and the snow of the Alps.

Every romantic German sipping his *limone spremuta* at Limone and getting his watercolours ready surely – and quite rightly – remembers the best, if most hackneyed poem of this honeymoon with Italy and citrus fruits – Goethe's poem from *Wilhelm Meister*, which begins –

> Kennst du das Land, wo die Zitronen blühn,
> Im dunkeln Laub die Gold-Orangen glühn
> Ein sanfter Wind vom blauen Himmel weht.

> Do you know that country where the lemon-trees flower,
> And oranges of gold glow in the dark leaves,
> And a gentle breeze blows from blue heaven.

That cold *limone spremuta*, with sugar collecting at the bottom of the glass, will be so lemonish, so pleasurably acid that it may make your throat a little sore, for a while.

Lemons do not do well in the full tropics, limes do not do well round the Mediterranean. English housewives or French housewives who settle in West Africa, fall back on limes, and quickly discover that they are very much stronger. So be warned, when you try limes from the supermarket. I came across a recipe in a French cookery magazine, for a Senegalese dish *poulet yassa*. Ingredients listed a kilo or 2 lb of limes to go with one chicken. Obviously a mistake. Equally obviously the limes were intended to be important so I used four. The dish was uneatable, I would have thought by anyone's standards, or however hot the climate which might have made the sharpness agreeable. It was indeed a question of 'the piercing lime'. Two was the right number, the word kilo had slipped in from the line below. Later I came across the right version in *La cuisine sénégalaise* by Monique Biarnès. A modified version appears later in the section, which may also be made with the milder lemon. Many recipes are of course interchangeable, which is to say that the method works for both – though I repeat, always be on the alert to use less lime if it is a lemon recipe – but the taste will be different.

The similarity of the names of the two fruit is no accident. Both come from the Arabic word *limah*. No connection, though, between lime fruit and lime tree which should really be lime tree as in Linden.

Everyone knows the anti-scorbutic virtues of both fruit. Older people will recall the lemonade given to them in February and March – presumably when the first lemons of the new season came in – to 'cleanse the blood'. That was in the days of fewer vegetables and scarce winter fruit, when our diet was stuffier altogether. Sailors were given lime or lemon juice, after ten days of salt provisions, and as rum had been added to the juice to preserve it, the occasion was a happy one.

HOW TO CHOOSE AND PREPARE LEMONS AND LIMES

As these fruit are priced individually, pick out the largest and freshest you can find. Avoid dry wrinkled specimens. If you have no use for the thin outer rind, the zest, pour boiling water over the fruit and leave it to stand for 5 minutes before squeezing it: you will get more juice.

Generally speaking, though, the zest should never be wasted. Put it in a jar with sugar, if you cannot use it immediately (*see* p. 477). To remove the outer peel really thinly, there is nothing to beat a lemon zester. It removes the finest of shreds which will disappear into the cooked dish leaving only delicate traces and their flavour behind. If you need julienne shreds, you can try a canelle knife, but I find it quicker and better to peel off strips with a potato peeler, trim them and cut them with a small sharp knife. Soften the peel always in plain boiling water before cooking it in anything sweet, as sugar has a hardening effect.

There are many ways of squeezing juice from lemons, and many gadgets. I have a heavy juice press over forty years old, which is a delight if there is a lot of fruit to be squeezed. Most of the time, I use the traditional beechwood juicer or reamer, or my hands – squeezing the half with one hand through the fingers of the other, which are kept slightly apart to hold back the pips. The other kinds of small lemon squeezer are fiddly to wash up and difficult to keep absolutely clean.

The acidity of lemon or lime is required with most other fruit, once you begin to cook or use it with other ingredients, not so much as a flavouring but as a flavour enhancer, to emphasize the virtues of the main fruit. With some tropical fruit, it is essential: papaya, guava, avocado. It is also used to prevent discolouration – a tablespoon of juice added to the water into which apple slices are cut will keep them white. Cut up pears on to a plate containing lemon juice: turn the pieces over and they will not go brown. Moreover, their flavour will be much improved.

Light syrup simmered with thinly cut strips of peel from lemon or lime makes the ideal liquid for poaching many fruits. Extra flavour can be added by squeezing in the juice. Pay attention, though, and taste the syrup from time to time. Fish out the peel strips before they add any bitterness.

Both fruit go admirably with fish and light meats, such as chicken and veal. A wedge of one or the other is essential with fried food, to counteract the richness. Lemon or lime juice sharpens hollandaise and mayonnaise sauces, gives a lighter vinaigrette or a more fragrant *beurre blanc* when used instead of vinegar.

If you need lemon slices and lemon juice for a dish, cut the slices from the centre of the fruit, then use the tricky ends for squeezing. The same of course with lime (or oranges).

Do not discard the squeezed-out fruit. Rub it over copper pans and basins with a little coarse salt to shine them up, or whiten your elbows by sitting and reading with them resting each in a lemon half. I remember girls at school trying to bleach away their freckles with lemon juice, but it was not – happily – very successful.

GREEK LEMON AND EGG SAUCE
AVGOLEMONO

A lovely sauce that can be made either with fish or chicken stock, according to the dish it will accompany.

30 g (1 oz/2 tbs) butter
2 level tablespoons (½ cup) flour
300 ml (½ pt/1¼ cups) appropriate stock, heated
2 large egg yolks
juice of 1½ large lemons
salt, pepper
chopped fines herbes

Melt butter, stir in flour and cook for two minutes. Add stock gradually, whisking to make a smooth sauce. Cook gently, stirring occasionally, for at least 20 minutes. Beat the yolks with the lemon juice, and add a tablespoon of water salt and pepper.

Just before serving, stir a little of the hot sauce into the egg mixture, then tip it back into the pan. Stir over a low heat until the sauce is thick and smooth. Taste and add extra lemon juice if needed. Finally mix in the herbs, about a level tablespoon.

LEMON AND GIN SAUCE

A simple and elegant sauce for steamed puddings, and for the banana dumplings on p. 59: The gin can be omitted, of course, if children are sharing the meal, or the sauce could be divided in two, one part with gin added, one part without. The plainer the pudding the more sauce you will need. With a steamed suet pudding for six people, double the quantities. For six banana dumplings, the quantity below should be enough.

1 large lemon
3 heaped tablespoons (6 tbs) sugar
$\frac{1}{4}$ litre (8 fl oz/1 cup) water
2–3 tablespoons (3–4 tbs) gin

Remove the skin of the lemon in thin strips and put them into a pan with the sugar and water. Stir over a low heat until the sugar is dissolved, then cover the pan and leave to simmer gently for 20 minutes. Remove from the heat, add the lemon's juice and strain into a hot sauce boat or small jug. Stir in the gin and serve. If you want to keep the sauce hot for any length of time, stand it in a bain marie, and add the gin just before you put it on the table.

SHARP LIME SAUCE

A good sauce for substantial puddings, such as banana dumplings or a steamed pudding. It comes from the Caribbean.

Provide yourself with three limes. You will most likely find two enough, but it's worth having an extra one in reserve.

Remove thinly the peel of 2 limes and put into a pan with 200 ml (6–7 fl oz/$\frac{3}{4}$ cup) water. Cover and bring slowly to the boil, then transfer from the heat to a warm place and leave to infuse for about 15 minutes. Taste occasionally to make sure that the flavour does not become bitter.

Meanwhile, mix 250 g (8 oz/1 cup) sugar with a heaped teaspoon arrowroot in a basin. Bring lime peel water back to the boil and pour through a strainer on to the sugar, whisking as you do so to make a smooth amalgamation. Rinse out the pan, and put in the sauce. Simmer until thickened, stirring. Remove from the stove, add the juice of the two limes, or more to taste. A very little salt is sometimes a good idea. Reheat gently, or serve cold.

SEVICHE, CEVICHE OR CEBICHE

This Latin American dish shows how lime, and also lemon, can be used to 'cook' pieces of filleted white fish – in Peru they use corvina, which we do not have in this country but sea bass can be used instead, or any good fresh sole, brill, weever, John Dory.

Skin 1 kg (2 lb) of the fillets, and cut them into 1-cm (½-inch) strips or cubes. If you have a zester, shred fine threads of the peel of 2 limes over them, after putting them in a single layer in a close-fitting dish. Pour on enough lime juice to cover them – about 5–7 limes (boiling water poured over the fruit first, will make them juicier).

Cover and leave in the bottom of the refrigerator for 4 hours, turning the pieces once. Check the pieces to see if they are opaque – this means they are cooked. If not, leave another hour. Put into the dish 3 serrano chillis (cans can be bought from delicatassens or specialized shops), cut into slices with their seeds. If you cannot get them, use 2 hot fresh chillis, red ones, cut into strips and rinsed free of seeds. Add 2–3 large skinned seeded and sliced tomatoes, the good Marmande type or tomatoes grown out of doors. Leave for another hour, covered.

Put some lettuce leaves on a serving dish, one for each person, and arrange the drained *seviche* on them, with an equitable distribution of tomato and chilli. Add a few half-slices of purple-pink Spanish onion, or other sweet onion, and thin slices of cooked sweet potato or avocado pear. A little cooked sweet corn, if you have it fresh can be scattered on top, but don't overdo it.

SEVICHE FROM TAHITI

Cube skinned and filletted fish, sole or scallops or the white fish mentioned in the preceding recipe. Cover with lime juice and add a level teaspoon sea salt. After four hours, if the fish is 'cooked', drain off the lime. Mix the fish with a very finally chopped clove of garlic, and a heaped tablespoon of chopped spring onions. Pour on coconut cream (p. 61) as a dressing – not more than ¼ litre (8 fl oz/1 cup). Quantities for each ½ kg (1 lb) fish.

Slightly adapted from a recipe in *Hawaii Cookbook* by Elizabeth Ahn Toupin.

LEMON SANDWICHES

Cut fresh, soft-skinned lemons into very thin slices and sandwich them between thin wholemeal bread slices, thickly buttered. Serve with smoked salmon and marinaded fish.

YASSA OF CHICKEN (LAMB, FISH)
YASSA AU POULET (AU MOUTON, AU POISSON)

Yassa is a Senegalese system of cooking that can be applied to chicken, lamb or fish, with only slight variations of cooking time. It comes from the southern Atlantic corner of the country, the island of Casamanca. This version is less piercing than the original, as I have reduced the lime and onion by a half. I have to admit that the best chicken to use is the yellow maize-fed chicken which is sometimes imported from France. For six to eight people for this dish, buy two 1½ kg (3 lb) chickens, and cut them in half. Line up the following ingredients as well:

2 limes
salt, pepper
1 fresh hot chilli, red or green, seeded
ground nut or sunflower oil
2 large onions, halved, sliced

Put the chicken halves in a large ovenproof dish, in a single layer (or use two dishes).

From the centre of the limes cut 6 or 8 slices and lay them on top of the halves of chicken. Squeeze over the juice of the four remaining lime pieces. Season and chop the chilli, then scatter over the chicken. Pour over 4 tablespoons (⅓ cup) of oil. Cover with onion slices. Put cling film over the dish(es) and leave in a cool place overnight, the larder rather than the refrigerator.

Next day, remove the chicken, brushing it with a little more oil, to a grill rack. Fish out and set aside the lime slices. Distribute the onion neatly to make a bed in the dish(es). Grill the chicken to a golden brown, skin side up first. Then grill the other side. Put on top of the onion, baste with some of the juices in the dish(es) and put into a very hot oven, gas 8, 230°C (450°F) until tender – not more than 30 minutes. Baste occasionally with the juices.

Serve with rice, followed by a mixed salad, or one of the chopped

North African style salads (see Eastern salad in my *Vegetable Book*, or books of Arab cookery).

CHICKEN OR LAMB WITH LEMON AND OLIVES

A Moroccan dish. My version is taken from *Fez: Traditional Moroccan Cooking*, by Z. Guinaudeau, with slight alterations.

2 chicken (French maize-fed if possible) weighing about 1½ kg (3 lb); or boned shoulder of lamb (1½–2 kg/3–4 lb)
175 g (6 oz/1½ cups) sliced onion
1 clove garlic, chopped fine
bunch of parsley and coriander leaf, chopped
pinch saffron
pepper, salt
6 tablespoons (½ cup) olive oil
175 g (6 oz) black olives
juice of a lemon

Put the chicken or lamb into a deep pot, on top of the onion, garlic, herbs and saffron. Season the whole thing and pour in enough water to come halfway up the chicken. Use as close-fitting a pot as possible, to avoid adding a great deal of water.

Bring to the boil, then pour in the oil. Cover and simmer until the meat is tender, turning it from time to time. Take the meat from the pan and boil away the liquid until you are left with moist and oily onions, the water all gone. Put back the chicken or lamb to reheat.

Meanwhile stone and simmer the olives in water to reduce their saltiness. Change the water if necessary, to make them mild. Add to the pan when you put back the meat.

Put the chicken with its sauce, olives and onion mixture in the centre of a ring of rice. Pour lemon juice over the chicken, and garnish with slices of pickled lemon (*see* page 219).

Note This recipe can be done with lime juice and pickled limes. It is a milder version of the grilled chicken *yassa*, a more everyday dish.

LEMON TART
TARTE AUX CITRONS

The best lemon tart I know. The recipe comes from our young doctor at

Trôo, whose governess used to make it regularly. What a happy childhood! The three children devoted to each other, kindly if remote parents and a gastronomic governess. Never, it seems, did she give them prunes and rice pudding.

PASTRY
250 g (8 oz/2 cups) flour
½ teaspoon salt
60 g (2 oz/¼ cup) sugar
2 egg yolks, beaten
125 g (4 oz/½ cup) softened butter

FILLING
2 large eggs
100 g (3½ oz/scant ½ cup) caster sugar
150 g (5 oz/1¼ cups) ground almonds
100 ml (3½ fl oz/½ cup) whipping cream
6 lemons
200 g (6½ oz/good ¾ cup) granulated sugar

Mix the first four pastry ingredients to a sandy consistency, then beat in the butter to make a dough. Chill for an hour, then roll out and line a tart tin of 22-cm (8–9-in) diameter.

For the filling, beat together the first four ingredients. Then grate in the rind of four of the lemons, add the juice of two. This leaves you with two lemons untouched, and two without their outer skin. (If you want to use the processor, take off the skins of 4 lemons with a peeler and whizz them to a powder with the caster sugar, then add the eggs, almonds, cream and juice.)

Put the filling into the pastry. Bake at gas 4–5, 180–190°C (350–375°F) until nicely cooked and browned, about 25 minutes.

Meanwhile slice the 2 untouched lemons thinly, discarding the pips. Put the slices into a pan, cover with water and start simmering them gently. Remove the pith from the other two and slice them thinly as well. When the peel of the first lot of slices begins to soften, add the second lot of slices and complete the cooking. Drain the slices.

Make a syrup by dissolving the granulated sugar in 100 ml (3½ fl oz/½ cup) water in a shallow pan. When the liquid is clear, bring it to boiling point, then simmer gently for 2 or 3 minutes. Slip in the lemon slices and cook them for 15 minutes, or until candied.

Arrange the slices on the lemon tart, the ones with peel on the outside (but save one good slice for the centre, or more if you have them over).

The syrup is most likely to be very thick, but if not, boil it down. Brush it as a glaze over the lemon slices.

Serve hot, warm or cold, with cream if you like.

CHESTER PUDDING, 1868

Is this, I wonder, the original lemon meringue pie? I had always thought it was an American pudding because in the thick yellow cornflour version it seems to be a drug store staple. It can even be bought in ready-mix packages.

The *Oxford English Dictionary* took it back no further than the Edwardian era. This tied up with my husband's recollection of eating it as a small boy in the Cornish rectory where he grew up.

Then the other day, looking for another recipe, I turned to that thick double-columned compendium of mid-Victorian eating, *Warne's Model Cookery*, compiled by Mary Jewry, new edition 1868. Chester pudding caught my eye: pastry, lemon curd filling, meringue top. It was delicious, needing no more alteration than a little extra lemon juice.

We reflected that Chester being near Liverpool, the great port of emigration to America, might well have sent its pudding recipe across the Atlantic in some settler's baggage.

Mary Jewry's Chester pudding was made in a puff pastry case, but I think that a crisp *pâte sucrée* or plain shortcrust makes a better contrast. Line a 20–23-cm (8–9-inch) tin, prick the base and bake it blind until the pastry has set and is lightly coloured but not brown.

FILLING
60 g (2 oz/¼ cup) butter
grated rind of one and the juice of two lemons
125 g (4 oz/½ cup) granulated sugar
4 egg yolks
24 almonds, blanched and ground in a mill
natural almond essence
4 egg whites, stiffly beaten
150 g (5 oz/⅔ cup) caster sugar

Melt butter in a saucepan over a low heat. Remove from the stove and beat in lemon rind and juice, sugar, egg yolks and almonds. Put back and stir until the mixture is very hot, but well below boiling point. Add a drop or two of essence and put filling into the pastry case. Bake at gas 3, 160°C (325°F) until just set.

Into the egg white, beat 125 g (4 oz/½ cup) of the caster sugar, and continue beating until the mixture is shiny and thick (use an electric beater, if you like). Take the tart from the oven, let it cool a few minutes and then pile on the meringue. Sprinkle with the remaining sugar. Put back into the oven until nicely browned. This last process can usefully be speeded up by raising the oven temperature to gas 7, 220°C (425°F) when you put the pie back.

LEMON MERINGUE PIE

Of the various lemon meringue pies we have baked, this one from the *Dione Lucas Book of French Cooking* was the universal favourite. What sounded different, what first made us try it, was the cardamom seeds. In fact, they disappeared completely into the general lemon flavour. Disappointing. Next time I altered the recipe, so that some extra heat at an earlier stage helped the cardamom to breathe its sweetness through the curd. I also increased the quantity.

An interesting thing – nineteenth-century recipes for lemon curd sometimes include crushed Naples biscuit or ground almonds, to give it extra body.

PASTRY
250 g (8 oz/2 cups) flour
½ teaspoon salt
175 g (6 oz/¾ cup) unsalted butter
grated rind of a lemon
2 tablespoons (3 tbs) sugar
3 tablespoons (scant ¼ cup) each lemon juice and water
melted butter

CRUMB LAYER
2 heaped tablespoons (¼ cup) sugar
2 heaped tablespoons (1 cup) white breadcrumbs
½ teaspoon ground cardamom seeds

LEMON LAYER
rind and strained juice of 2 lemons
200 g (7 oz/scant cup) sugar
3 eggs, well beaten
250 g (8 oz/1 cup) unsalted butter

MERINGUE
6 egg whites
225 g (7 oz/scant cup) caster sugar
extra sugar

Make pastry in the usual way with the first six ingredients. Chill for 30 minutes, then roll out and line a 25-cm (10-in) tart tin with a removable base. Prick the base, brush with melted butter and bake at gas 5, 190°C (375°F) for 30 minutes. Mix sugar, crumbs and cardamom seeds and put immediately into the pastry when it comes from the oven.

Have the lemon curd ready – beat the first three ingredients in a pan. Stir over a low heat, adding the butter in bits. When the mixture is hot – keep it well below boiling point – pour it over the crumb-lined pastry. Return to the oven for 10 minutes, until the filling is just set but still a little shaky in the middle. Leave to cool.

Beat the egg whites until stiff, add the caster sugar gradually, beating again after each addition. Pipe or pile on to the tart; take it right to the pastry rim so that the lemon filling is covered entirely. Put the pie into the oven again at gas 5, 190°C (375°F) for 10–15 minutes, until the meringue is browned.

LEMON FILLING FOR TARTS

Mrs Beeton's filling, scaled down, is one of the best. Use the egg whites to make a meringue topping if you like, as in the two previous recipes.

Beat together the juice of 2 lemons and the grated rind of one, 125 g (4 oz/½ cup) caster sugar, 6 tablespoons (½ cup) double cream, 4 egg yolks, 60 g (2 oz/½ cup) ground almonds and 125 g (4 oz/½ cup) melted, lightly salted butter (e.g. Lurpak or Normandy).

KEY LIME PIES

On the low coral islands – the keys – of Florida, semi-wild limes grow on thick and thorny trees. Nobody bothers with them much by way of cultivation, and the fruit remains small and very acid. This key lime pie is a local speciality, made in many versions with condensed milk and lime juice. Not as revolting as it sounds. By adding eggs, you can head off the sweet-sharp excesses of the partnership. The second version comes from Evan Jones's *American Food*. 'It uses neither evaporated milk nor the gelatin common to most Florida lime pies: it is as fresh as any fruit ice cream.'

1) For a 23-cm (9-inch) shortcrust pastry case, baked blind, beat – electrically if possible – 5 egg yolks until they are thick. Slowly add a 397 g (14 oz/scant 2 cups) can of sweetened condensed milk, then the grated peel of 3 limes and 150 ml (5 fl oz/$\frac{2}{3}$ cup) lime juice. Beat 3 of the whites separately to soft peaks, then fold into the yolk mixture. Bake at gas 3, 160°C (325°F) until the filling is firm, 15–25 minutes. Serve cool, or chilled, with whipped cream.

2) Line a 23-cm (9-inch) tart tin or pie plate with Graham cracker (digestive biscuit) crust. *See* p. 342. Bake 10 minutes at gas 4, 180°C (350°F). Cool.

Beat (electrically, if possible) 5 egg yolks over simmering water until very thick. Gradually beat in 125 g (4 oz/$\frac{1}{2}$ cup) caster sugar, and go on beating until the mixture is pale and falling off the beater in threads. Remove basin from heat. Add 2 teaspoons grated lime zest and 150 ml (5 fl oz/$\frac{2}{3}$ cup) lime juice. Whisk 4 egg whites and a pinch of salt until thick and soft, add 60 g (2 oz/$\frac{1}{4}$ cup) sugar gradually, beating all the time. Fold a third into the lime custard, then the rest. Pile into the crust and bake for 15 minutes at gas 4, 180°C (350°F). Cool, then chill and freeze. Take from the freezer 10 minutes before required. Top with whipped cream and thin slices of lime, or fresh strawberries dipped in caster sugar.

LEMON OR LIME SOUFFLÉ

An American quick soufflé based on the principle that citrus juice turns condensed milk into a palatable cream.

With an electric beater, whisk 10 egg yolks until thick, add the juice of 5 lemons or limes, and the grated peel of 2. Fold in the 10 egg whites, beaten until soft and thick, then sweetened with 150 g (5 oz/good cup) icing sugar and beaten again. Blend the two mixtures. Put into a soufflé dish and bake at gas 8, 230°C (450°F) until risen and brown. About 30 minutes, but be prepared for a little longer. The soufflé falls quickly, but it tastes all right cold – like the filling of a key lime or a lemon pie without the crust.

LIME SOLIDS

I decided to try to adapt the idea of Elizabeth David's lemon solids to limes and found it successful. One caught the ginger flavour of the fruit over the lime-freshness.

3 limes
100 g (3–4 oz/about $\frac{1}{2}$ cup) sugar
4 large egg yolks, beaten
1$\frac{1}{2}$ packets ($\frac{3}{4}$ oz/3 envelopes) gelatine

Remove the peel from the limes as thinly as possible (with some limes this may mean all the peel). Put it into a pan. Roughly chop the flesh of the fruit and throw away pips. Put this into the pan too, with any juice. Pour on 600 ml (1 pt/2$\frac{1}{2}$ cups) cold water. Bring to just below boiling point, squashing down the fruit as you stir to extract the maximum flavour. Stir in the sugar. Put the lid on the pan, turn off the heat and leave to infuse for 5 or 10 minutes. Taste occasionally and immediately the flavour is strong but not bitter, strain on to the yolks, beating.

Discard the debris, rinse out the pan and pour back the mixture. Stir it over a moderate heat until the liquid turns opaque and begins to thicken.

In a measuring jug, put 150 ml ($\frac{1}{4}$ pt/$\frac{2}{3}$ cup) very hot water, and sprinkle on the gelatine. Stir it with a fork or small whisk, not too vigorously, until you have a clear liquid again. Pour in the hot lime/egg mixture. Continue to stir gently, to make sure the gelatine has melted. Fill little custard cups – there is enough for eight or nine – or glasses. Leave to set. Serve chilled. I like them quite alone, but some people prefer them with cream which is a good idea with so lively tasting a jelly.

LEMON ICE CREAM

zest and juice of 6 large lemons
300 g (10 oz/1$\frac{1}{4}$ cups) sugar
1 level teaspoon cornflour
4 large egg yolks
250 ml (8 fl oz/1 cup) whipping cream, whisked until stiff

Bring zest, juice, 200 ml (7 fl oz/scant cup) water and sugar slowly to the boil, stirring to dissolve the sugar. After boiling for a second or two, fish out the strips of zest (before they make the syrup bitter) and leave the syrup to boil for 4 minutes.

Put the cornflour and yolks into the bowl of an electric mixer. Mix until the yolks are thick, and the syrup has come to the end of its boiling time. Pour the syrup on to the egg yolks, still being beaten, and go on beating until the mixture is foamy and smooth. Cool and fold cream into the egg-lemon mixture. Freeze in the usual way. This mixture does not need beating half way through freezing.

TEA AND LIME SORBET
SORBET AU THÉ AU CITRON VERT

Sorbets have had an enormous revival with the new cooking in France, spoonfuls of different flavours on a plate with fresh fruit make up the favourite *Grand dessert* (p. 437). This particular ice is Anne Willan's version, from *The Observer French Cookery School*, of one of the double-flavoured sorbets that they favour. Recently I came across another version, made with orange pekoe tea, and given the punning name of *Grani thé*: the ice was embellished with 'sequins' made by freezing a teacupful of the mixture in an ice-tray, and then cutting out large fish scale pieces that are scattered over the ice when it is served. I can recommend orange pekoe tea for this recipe.

3 limes
10 high quality tea bags

SYRUP
1 litre (1¾ pt/scant 4½ cups) water
250 ml (8 oz/1 cup) sugar

Make the syrup first. Dissolve the water and sugar together over a low heat, then boil for 2–3 minutes.

Remove the zest thinly from the limes and chop into the smallest bits possible (if you use a zester and take it off in thin shreds, this task is much easier). Blanch by putting them into water, bringing it to the boil, and then tipping it into a sieve and cooling it under the cold tap. Simmer the blanched zest in 150 ml (¼ pt/⅔ cup) of the syrup for about 15 minutes, until tender. Keep an eye on it, stir occasionally, and be sure it does not boil away and burn.

Bring 300 ml (½ pt/1¼ cups) water to the boil and add the tea bags. Infuse for 5 minutes, then strain. Taste after 3 minutes in case the flavour is already pleasant. Too long an infusion will make the liquid bitter.

Squeeze the limes, and mix with syrup and tea, the zest and its syrup. Taste, adding extra sugar or lime juice if you think it needs it. Freeze as usual.

LEMON SYRUP CAKE

One of the best developments in recent cookery has been the revival of our old favourite pound cake (*see* p. 100). I suppose we have to thank the soft-fat merchants for the discovery that if you throw equal quantities of self-raising flour, sugar, eggs and soft fat, plus a level teaspoon of baking powder, into a processor or electric mixing bowl, you can avoid hours of work. What few cookery writers seem to have realized is that if you have the forethought to remove a packet of butter from the refrigerator in good time and cut it up into rough chunks, you can make a fast cake that is also excellent eating.

The simplicity and appeal of the method is not just its speed in a busy world. It's sensible that people do not wish to spend a day turning out some masterpiece of the pastrycook's art, knowing that with every mouthful their family and friends are likely to put on a kilo extra weight. If it only takes five minutes, the wickedness of it all is somehow lessened.

The various possibilities of embellishment, which is the enjoyable part, seem to be endless. The syrup idea is a particularly good one, and with orange or lemon syrup the fresh intensity of flavour takes you by surprise (and prevents you eating too much).

Turn to the recipe for orange syrup cake (p. 275), and substitute the rind and juice of a lemon. To the syrup, I also add a measure or two of gin. Use candied lemon or citron peel, rather than orange, though this is not always easy to find.

LIME SYRUP CAKE As above, but using two limes for a 175–250 g (6–8 oz) mixture. Lime syrup cake tastes good when used with papaya for an upside-down pudding.

PICKLED LEMONS OR LIMES
LAMOUN MAKBOUSS

This recipe is Claudia Roden's version, to be made when the fruit is at its cheapest.

Scrub and slice the fruit. Put in a single layer into a colander, or on a pierced strainer, or on a tilted plate. Sprinkle generously with salt. Leave to drain for at least 24 hours until they are soft and limp and have lost their bitterness. Put the slices in layers in a glass jar, sprinkling each layer with paprika – just a little. Pour on ground nut oil or sunflower oil to cover (olive oil is too strong in taste). Close the jar and leave for at least three weeks. The lemon oil can be used discreetly in one of the Yassa recipes (p. 210–11).

Serve with rice, and meat or fish dishes. Or as a pickle with cold meat or fish.

LEMON CURD

In Tom Stobart's *Cook's Encyclopaedia*, he writes: 'English lemon curd, for instance, made with freshly picked lemons in a Mediterranean country (where it is unknown) is a revelation to anyone in England who has had to depend on imported lemons from the local fruit shop.' I read this and wrote to a friend in the Var to see if she agreed. She used my recipe with less sugar, which improved it, but her family and English friends did not find it *very much more wonderful* than the home-made kind they liked so much in England. Her lemons had fresh green leaves on them, and were untreated, which made them a bonus – at least for the cook who had the pleasure of preparing them.

2 large lemons
90 g (3 oz/6 tbs) butter, cut in bits
175 g (6 oz/¾ cup) sugar
3 large eggs, beaten

Remove the thin rind of the lemons with a peeler, or – better still – a zester, and chop it finely. Put into the top of a double boiler with the butter and sugar, with the water simmering underneath. Stir until dissolved. Pour in the eggs through a strainer, stirring all the time until the mixture is thick. Do not let it boil or it will curdle. Pour into small pots and cover. Lemon curd will not keep beyond three months: it is best to store it in the refrigerator. For a milder curd, add another 30 g (1 oz/2 tbs) of butter and a fourth egg.

When making lemon curd tarts, bake the pastry blind, then put biscuit or cake crumbs, or coarsely ground almonds in the bottom, before spreading the curd on top.

LEMONADE

The late-winter drink of our childhood; it was intended to cleanse our blood – strange idea of the past, cleaning one's blood – and revive us with its freshness after stodgy winter food. Nowadays I make it in the summer in hot weather.

3 large lemons
sugar
soda water
lemon or cucumber slices, mint, borage etc

With a potato peeler remove the peel very thinly from the lemons. Put them into a 18-cm (7-in) heavy pan and cover with about 2½-cm (1-inch) of water. Cover with the lid and put on to a very low heat; the peel gradually flavours the water quite strongly. Do not let the water come quite to the boil, or it will taste bitter. Leave to cool, then strain the liquid into a jug. Add the juice squeezed from the lemons and sugar to taste. It needs to be quite sweet and strong, as it will be diluted.

Cover and chill in the refrigerator. Or add ice cubes if you are in a hurry. To serve, dilute with soda water, and float lemon or cucumber slices on top, with sprigs of mint or blue-flowered borage.

VARIATION: orangeade is made in the same way, using two or three large oranges and 1 large lemon and if possible, a final flavouring of orange-flower water.

LICHEE & LITCHI, see LYCHEE
LIME, see LEMON
LOGANBERRY, see RASPBERRY

LOQUAT or JAPANESE MEDLAR

The loquat (*Eriobotrya japonica*) is native to China and south Japan. The name was adapted from Cantonese *lu-kwyit*, meaning rush-orange. And orange-coloured it is, with a smooth waxy skin. Sizes vary. The original fruit is not much bigger than a gooseberry, but Japanese gardeners have worked on it and developed larger and larger varieties. Some I had from Spain last year were 5 cm long and 4 cm across. Greater size is important, because the beautifully fitting brown stones inside, more like curved segments than stones, take up a lot of the fruit.

The particular value of loquats is that, like gooseberries in the north, they come on to the market in the Mediterranean when last year's fruit is nearly finished, or looking tired, just before the strawberry season. To us, this does not matter very much these days, as we are such importers of fruit all the year round. We were down in the Hérault during May one year, in the hills behind Béziers, and loquats were very welcome. We managed to buy them on our expeditions to Béziers and Pézenas; they kept well and made a change from tough-skinned golden delicious apples.

To anyone who may be catering for a family in the Mediterranean in April or May, I would say, Look out for them. Their orange colour is creamy and soft, and although they are plum-shaped, they could not be mistaken for plums on account of their evident firmness. I suppose their elegantly lop-sided shape has a Japanese look – you could imagine them, three or four, some upright, some on their side, in a Japanese painting, with a few red characters down the right hand side. If they are labelled, the name will be some variation on the word medlar – *nèfle du Japon* in France, *nespola giapponese* in Italy. What they are called in Greece we did not discover, though we saw great orchards of them, near Phaistos, in flower.

In England they have been a rare plant of mainly hot-house cultivation ever since the first one was introduced at Kew in 1787. They were brought out into the open for their 'winter' in July and August, then taken back into a hot-house for their well-heated 'summer'. Sir Joseph Banks, who travelled round the world with Captain Cook in 1770–71, thought that loquats could be as good as the mango. He must have been unfortunate in the mangos he met with on his way. The loquat is not a fruit of the gods or the poets, even, I suspect, in its native lands. Princesses changed into peaches: mangos changed into princesses, but loquats have never been in this league. They are comely and refreshing, not glamorous.

HOW TO PREPARE LOQUATS

They can be eaten directly, just cut them across and remove the stones then the inner white skin if it seems unappetizingly thick. The outer skin can be eaten or not, as you please.

If you want to cook loquats, and this is quite a good idea as you can use lemon to give their flavour a little more zest, open and quarter them but do not peel them until after they are cooked. Poach them gently in a light syrup with strips of lemon in it, and the juice of a lemon. Then skin them and serve chilled. If you have an abundant supply, make jam or jelly, put them into tarts – always using lemon to bring out the flavour.

Loquats are sometimes used in China with chicken, as you might use lychees.

LYCHEE, LICHEE or LITCHI, RAMBUTAN and MANGOSTEEN

> Looking back at Ch'ang-an, an embroidered pile appears;
> A thousand gates among mountain peaks open each in turn.
> A single horseman in the red dust – and the young Consort laughs
> But no one knows if it is the lichees which come.

The young consort is the exquisite Yang Kuei-fei, the Imperial Concubine of Emperor Hsüan Tsung, the last great ruler of the T'ang dynasty. Their story was told by Chinese poets, and best of all by Po Chü-i, who was born eighteen years after the dynasty's collapse in AD 756, and the hanging of Yang Kuei-fei. Her passion for lychees made them a romantic symbol for the Chinese. To please her, the Emperor had them brought by a guard mounted on relays of fast horses, and told to make the journey from Ch'ang-an the capital to Canton, where the best lychees grew, in five days, there and back. The distance between the two places, measured straight across the map, is 600 miles.

Speed was essential. According to Po Chü-i, who was a *bon viveur* and drinker as well as a great poet, 'The lichee's color is altered in one day, its aroma in two, and that color, aroma and taste are all lost in four or five.' Perhaps lychees have been improved a great deal in the last twelve centuries, because the ones we import today are still a pleasure to eat. They are far more than five days journey from home.

I had hoped that these Imperial lychees would turn out to be the first example of fruit travelling by refrigerated transport, as ice was being used for this purpose by the T'ang dynasty, and snow. Professor Schafer has speculated on this extraordinary lychee courier in *The Golden Peaches of Samarkand* and concludes that some other cooling system must have been used. How could there have been ice in hot Canton?

The symbol remained. Perhaps because the reality was indeed so good. 'One of the keenest pleasures of my whole life was tasting the fresh fruit of the lichee in Canton: a fruit I shall never forget,' wrote Shen Fu in *Chapters from a Floating Life*. And a friend tells me that in southern China the lychee is a tall and noble tree, with hanging baubles of fruit like the fruits of our plane tree, but larger and more beautifully coloured. The thin shell, often an exquisite deep pink is easily cracked away from the fruit, so that it shows the greyish white 'grape' that gleams softly inside. The texture is chewy, with a melting resistance as you bite through to the brown stone. One of the things I should most like to do is to eat 'the fresh fruit of the lichee in Canton', straight from the tree, no delays.

There are Chinese salads of lychees, both with chicken and with other fruit. Fresh and dried – ask for Chinese nuts at an oriental supermarket – they are used to sweeten hot meat dishes. But the best thing is to eat them as they are.

The same thing applies to the rambutan (*Nephelium lappaceum*) of Malaysia, which is related to the lychee but looks like a small hairy animal, if you catch sight of it casually. The skin is covered with a shock of soft spines. The fruit inside tastes and feels like a lychee. An attractive way to serve it is to cut round the fruit and remove half of the shell, so that the edible part lies like an egg in a surrealist egg cup.

The third lychee-like fruit, the unrelated mangosteen I have only come across in small and inferior samples in this country. A shame. It is one of the best and most beautiful of fruits; one writer on tropical fruits has described its 'abundant juicy flesh, white as snow, with an exquisite taste and perfume of infinite delicacy'. It looks dull enough, a dark reddish-brown, apple-shaped fruit, but the scoop-shaped curves of the calyx might stop you passing by. When you slice it across, you will have a surprise. No jewel was ever more beautifully set than the mangosteen, with its pearly white fruit, in five fat segments, enclosed in a pith of a deep and lively pink. In a while, the pink darkens in contact with the air. You must break it away and eat the delicious fruit quickly. Again, like the rambutan, cut it round the centre, and remove the top part of the shell, so that the fruit humps above its cup and can be properly admired.

If the fruit is so wonderful, you may well ask why it is so rarely on sale. Why is it not as easy to come by as the mango? The trouble is that the tree grows slowly. It takes at least eight years before it bears fruit, and more likely ten. Only at fifteen years does it begin to reach its full splendour. No country seems to have decided to develop mangosteens as an export. Once the region around Saigon was famous for them – I imagine that is one of the reasons they are easier to find in France. There must be many people who came across them in the days of French occupation and now, back home, like to be reminded of the good things they ate there.

THREE KINGS SALAD

I make this salad after Christmas to go with cold ham, turkey and spiced beef. It was first made in desperation when the salad greenery had been finished and the shops were still closed. Now it is part of the festival for us.

4–5 slices Chinese cabbage
1 ripe avocado
half a lemon
2 Chinese gooseberries (kiwi)
1 dessertspoon cider vinegar
4 dessertspoons hazelnut or sunflower oil
salt, pepper
60 g (2 oz/scant ½ cup) toasted hazelnuts or almonds
4–5 lychees
1 purple Italian onion or sweet Spanish onion

Cut the cabbage about ½ centimetre (¼ inch) thick across the wide end. Rinse, drain and dry in a cloth. Put in the bottom of a salad bowl or deep dish.

Not too long before the meal, peel, stone and cube the avocado Squeeze the lemon over it to prevent discoloration, then arrange in the centre of the cabbage. Peel and slice the Chinese gooseberries, putting the slices round the avocado. Pour on the vinegar and oil. Scatter lightly with salt, more generously with black pepper. Put on the nuts, then peel and cut up the lychees discarding the stones and arrange them on top. Finally cut a few very thin onion rings and put them between the lychees. Be careful not to overdo the onion.

Serve cool rather than chilled.

LYCHEE SAUCE

A simplified version of the kind of sweet-sour sauce that the Chinese serve with duck, chicken or deep-fried pork. I like it much better than the more correct elaborations that are described in Chinese cookery books. It tastes fresh. If you cannot buy fresh lychees, use canned ones: drain and rinse off the syrup first.

300 ml (½ pt/1¼ cups) orange juice
150 ml (¼ pt/⅔ cup) duck or chicken stock

1 tablespoon soya sauce
1 level tablespoon (¼ cup) arrowroot or cornflour
12 lychees, peeled, halved and stoned

Bring juice – which must be fresh juice squeezed from oranges, not the concentrated kind – to the boil with stock and soya sauce. Mix the arrowroot with a little cold water or extra cold stock if you have some, and stir into the pan briskly. Continue to stir as the sauce thickens. If it is too thick for your taste, dilute with extra juice and stock.

Just before serving, heat through with the lychees, and any juice that fell as they were prepared.

COCONUT CREAM WITH LYCHEES

A good way of setting off tropical fruit is to serve it with coconut cream. Mango, papaya marinaded in lime juice, or as in this recipe lychees and Chinese gooseberry with a strawberry purée. It's a refreshing dessert for a winter party; frozen strawberry purée makes an excellent sauce.

150 ml (5 fl oz/⅔ cup) each single cream and soured cream
vanilla pod
125 g (4 oz/1⅓ cups) desiccated coconut
1 packet (½ oz/2 envelopes) gelatine
1 generous tablespoon coconut cream from a tin or block
sugar
lemon or lime juice
300 ml (10 fl oz/1¼ cups) whipping cream
300 ml (10 fl oz/1¼ cups) strawberry purée
icing sugar
about 18 lychees, peeled, stoned
2 Chinese gooseberries, peeled, sliced, halved

Put single and soured cream into a pan with vanilla and coconut. Stir in 200 ml (7 fl oz/generous ⅔ cup) water. Cover and bring very slowly to simmering point, stirring occasionally. Simmer for 10 minutes, then switch off the heat and leave to cool to tepid. Remove the vanilla pod, process or liquidize, and then put through a sieve, pushing down the coconut. You will end up with a generous 300 ml (½ pt/1¼ cups) of liquid. Retain the coconut debris.

Dissolve the gelatine in 6 tablespoons (½ cup) hot water, and add to the liquid. Then mix in the coconut cream, with sugar and lemon or lime juice to taste – the juice is used to emphasize the flavour, not to be recognizable.

When the mixture is set to a thick egg white consistency, whip the cream and fold it in, with a good tablespoon of desiccated coconut from the sieve. Pour into a lightly oiled deep mould, and chill until set.

Turn out onto a dish. Sieve any pips from the strawberry purée and add icing sugar as needed. Pour some round the coconut cream, to make a pink sea (put the rest in a small jug). On the strawberry sauce, arrange lychees and the Chinese gooseberry half-moon slices.

MANGO

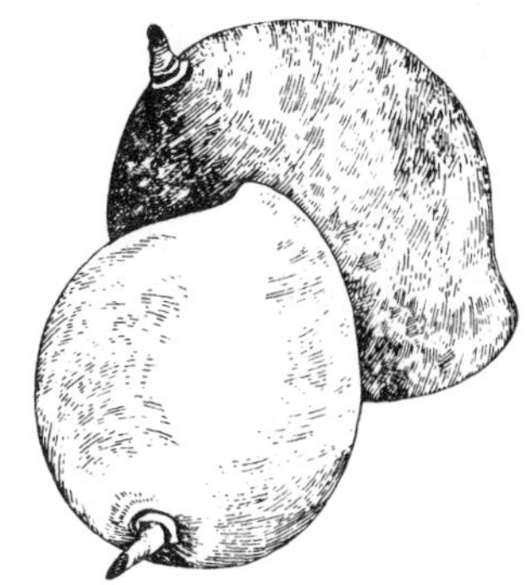

Long before mangos were grown world-wide in the tropics, everyone who had eaten them in India and written about them, agreed that mangos were the world's most delicious fruit. Dr John Fryer, who went out to Persia and India in the 1670s for the East India Company wrote that the apples of the Hesperides were nothing but fables to a ripe mango – 'for Taste, the Nectarine Peach and Apricot fall short'. We can appreciate that, now they have reached our supermarkets. They may be Brazilian mangos or South African mangos, but golden, green and red outside, and golden inside, they still have a sense about them of Indian luxury.

The mango is much esteemed as well as enjoyed by the Hindus. And no wonder for it once enclosed the daughter of the sun in its golden comfort. To escape from a wicked sorceress who was after her, she jumped into a lake and became a golden lotus. The king of the land fell in love with the golden lotus. The sorceress burnt it to ashes. From the ashes grew a tree, the tree flowered and the king fell in love with this second flower. The flower became a fruit, a glorious mango, and the king fell in love with the mango. When the mango was ripe, it fell to the ground and split. Out stepped the daughter of the sun in her beauty, and the king recognized her as his wife whom he had lost long ago.

Leconte de Lisle, familiar with the mangos that had naturalized themselves in his childhood island of Réunion, wrote an Indian poem about beautiful Leila dozing and dreaming of her lover in the royal orchard of reddening mulberries. There are mango trees there as well and in the silence – no sound of wings or running water – the Indian 'mango-bird', the golden oriole, slyly and secretively sucks the juice, 'like golden blood', from the ripe mangos.

The golden-yellow and red-splashed flesh of the summer mangos we get make them the most conspicuous and blazing of fruits. The flesh is an even deeper orange than you might expect, and of a slightly acid sweetness that convinces you that this is the best fruit in the world. There are recipes which include mangos, but with these beautiful objects it is best to leave them as they are. They make the finest of desserts.

The only reason for doing anything at all, apart from removing the stone when you halve them, is to make one mango stretch round more than two people. My advice would be to do as little as you possibly can – add syrup and make an ice cream, mix them with whipped cream to make a fool, mash the pulp and serve it in little tartlet cases baked blind. The flavours of whatever you use should be simple and direct, butter and almonds in the pastry, the best Jersey cream, no fussing with spices or alcohol.

Unripe green mangos have a fine acidity of their own, with a hint of the flavours to come. They are used in curries, to make fresh chutneys, and chopped to a pulp as a marinade for lamb to be barbecued.

CHOOSING AND PREPARING MANGOS

There are many mango varieties. Not surprising when you consider how people have loved this paradisal fruit. The original problem seems to have been a particularly fibrous flesh, so development has been towards succulence. I feel we are lucky to have mangos at all in this country and rejoice that they seem to become more and more popular. By the time this book is published, there may be more kinds of mango on sale, but at the moment there are three.

Most beautiful is the large summer mango, orange and red-skinned, a spectacular sunset glory of a fruit. Inside the flesh is bright orange. So is the flesh of the dark green-skinned mango like the favourite Alfonso variety. To eat either kind raw – the best way – buy it soft, or else keep it in a warm kitchen until it gives in your hand. In this, it is like avocados, except that the whole fruit should be soft, not just the neck end. If they are ripe too soon, wrap them in cling film and put them in the vegetable drawer of the refrigerator. Should you want to make your own mango chutney, pick out either of these mangos while they are still firm: you can then peel them quite easily with a potato peeler and slice them down and across from the stone.

The third mango is to be found at Indian and West Indian shops rather than the supermarkets or greengrocer's. It is smaller, about the size of an avocado, and green all through. Darkish green skin, paler green flesh. The taste has a slightly sharp refreshing tang, with a hint of the sweetness of the ripe orange fruit. For a number of Indian dishes, for fresh chutneys, this is the fruit to buy. They also can be peeled and sliced quite easily.

The great problem with mangos is parting the soft ripe flesh from the flat stone. Anyone wrestling with their first mango will see the point of that enterprising greengrocer's slogan, 'Share a mango in the bath with your loved one!' Mangos are the tightest clingstones of them all, the limpets of the vegetable world except that the stone they fix themselves to is internal.

Publicity leaflets make it all seem easy. 'Slice round the middle, or round the edge like an avocado. Twist the two halves in opposite directions, until they come apart, and remove the stone.'

Elated with anticipation, knowing, you make a cut. You give a delicate twist – nothing happens beyond an internal lurch. A stronger twist, a couple of curses, then more strong twists. You are now covered with juice to above the wrist, and the mango looks battered. Salvage what you can with any implement to hand, and turn the pulp – exactly the right word in this instance – into a mango fool or water ice. At least you have the pleasure of licking your hands and arms before washing them clean in plenty of water.

Another system which works slightly better is to stand the mango upright, narrow side towards you. The idea is to isolate the stone by making downward cuts on either side of it. This way you will find that two neat halves fall to left and right, leaving you with a central slice that contains the stone. Peel off the skin from this slice and scrape what pulp you can into the neat halves. Then suck the stone, chewing what you can from the fibrous threads that beard it.

I sometimes treat a mango like an oyster. First the cut all round. Then I slip a very sharp little knife into the fruit and feel my way over the upper surface of the stone. The blade needs to be wiggled as if there was a recalcitrant oyster inside. Having removed the top half, I put the knife in under the exposed stone and work it away from the fruit.

If you have in general a firm attitude towards family and friends, you can always adopt the escape system. 'It is best to leave guests to tackle the fruit their own way. Provide spoons and a small dessert knife and fork. Provide finger bowls also, or a moist towel and an extra napkin.' (Of course you need to take into account the time you will spend next day, cleaning table and furniture and carpet of a mess that you have not seen since the children were learning to feed themselves.)

Much as I admire the writers of those sentences – Leslie Johns and Violet Stevenson in the *Complete Book of Fruit* – I do think that their solution has a tinge of moral cowardice. Or perhaps of desperation, after preparing mangos for a party of twenty people.

You may ask, Is it worth it?

Yes, it is.

MANGO PRAWNS

'I was a long way from the sea when I heard about this one,' wrote Alan Davidson in the *Seafood of South-East Asia*, 'precisely, at Ban Houei Sai, the enchanting village which stands on soil studded with sapphires in the Golden Triangle area of Laos. I was on the trail of *Pangasianodon gigas*, the giant catfish of the Mekong. Peter Law, a narcotics expert from Hong Kong, was following other trails of his own; but he is also a gastronome and was moved by some turn in the conversation to impart this recipe to me . . . may be used for crab instead of prawns. Whichever you use, the amount of crustacean and the amount of mango should be about equal. The quantities given are right for 4 people.'

Use the large cooked Mediterranean prawns that a number of fishmongers have taken to selling, otherwise go for crab.

8 prawns
4 mangos

DRESSING
300 ml (10 fl oz/1¼ cups) mayonnaise or thick cream
2 tablespoons (3 tbs) freshly grated horseradish
a squeeze of lemon or lime juice
1 teaspoonful sugar
a little freshly ground pepper
a little single cream or creamy milk to thin dressing, if necessary

Peel the prawns, and cut up into chunks. Prepare the mangos so that you are left with the empty skins and the flesh cubed.

Mix prawns and mango. Mix the dressing ingredients, then stir it gradually into the prawn and mango. Stop when the mixture is nicely bound and not in the least liquid. Divide between the empty mango skins, and arrange on a dish. Garnish with strips of sweet red or green pepper and mint leaves. Serve chilled.

INDIAN MANGO AND CREAM
AAM MALAI

Malai is milk boiled down to a cream, and is a most necessary ingredient in Indian cookery. It can be faked up with milk powder and evaporated milk, but the results do not compare with the real thing. When it is painstakingly made, stirred more or less continually while it slowly reduces to an ivory liquid, it is finely delicious. Even made in a rough and ready domestic way – the occasional stir as you flash by on a hundred other chores – it is good: it is also the basis of *kulfi*, Indian ice cream, and the lovely rice pudding known as *keer*.

3 large mangos
caster sugar
lemon juice
1¼ litres (2 pt/5 cups) Channel Island milk

Peel, stone and slice the mangos. Put them into six little bowls, sprinkle with a little sugar and a squeeze of lemon juice to bring out their flavour. If you consider their flavour cannot be improved, dispense with the extra ingredients.

Pour the milk into a wide slope-sided pan. Set over the stove and bring to boiling point. Now lower the temperature as much as you can: with gas, use a metal mat to diffuse the heat as much as possible (the ideal stove for this is a solid fuel or oil-fired cooker). Stir to break up the skin from time to time, and scrape down the sides and any crust that forms on the base. The milk must not burn, but the bottom crust can become a pale brown without ruining the *malai* You will notice lumps forming towards the end, but do not worry about them. Just stir again.

When the milk has reduced to 375 ml (12 fl oz/1½ cups), more or less, tip it into a liquidizer or processor, and whizz as smooth as possible. Pour through a strainer into a jug, and chill. Stir up before serving with the mango.

MANGO PIE

A delightful way of serving mango, from Elisabeth Lambert Ortiz. It comes in her *Best of Caribbean Cooking*, and is attributed to the English-speaking islands. The idea is a simple one, but it works elegantly. If you cannot find a single mango weighing ¾ kg (1½ lb), buy two which weigh rather more – say a good kilo.

PASTRY
175 g (6 oz/1½ cups) plain flour
1 tablespoon (2 tbs) sugar
¼ teaspoon salt
90 g (3 oz/6 tbs) unsalted butter, chilled and cut into small bits
30 g (1 oz/2 tbs) lard, chilled, and cut into small bits

FILLING

1 large ripe mango, weighing about $\frac{3}{4}$ kilo (1$\frac{1}{2}$ lb)
3 tablespoons (scant $\frac{1}{4}$ cup) lime juice
125 g (4 oz/$\frac{1}{2}$ cup) sugar
1 level tablespoon mixed with 3 tablespoons (scant $\frac{1}{4}$ cup) water arrowroot

Sift dry ingredients for pastry. Rub in the fats until the mixture is crumbly. Use 3 tablespoons (scant $\frac{1}{4}$ cup) cold water to mix to a fairly stiff dough. Chill for at least 30 minutes. Roll out and line a 23-cm (9-inch) pie plate, brushed with melted butter. Crimp the edges and prick the base of the pie. Bake at gas 4, 180°C (350°F) for 25 minutes, or until golden brown. Check occasionally, and if the pastry balloons up during the first 10 minutes, prick the base again. Cool.

To make the filling, peel the mango carefully over a plate, so that no juices are wasted. Cut two good wide slices down each side, to the central seed, and remove them to a plate. Cut the slices into long strips and sprinkle them with some of the lime juice. Remove the rest of the pulp from the seed and measure it. There should be about $\frac{1}{4}$ litre (8 fl oz/1 cup). Put the pulp into a pan with the rest of the lime juice, sugar and 3 tablespoons (scant $\frac{1}{4}$ cup) water. Simmer until soft, about 10 minutes, then sieve or liquidize and return to the pan. Stir up the arrowroot and add to the mango pulp. Cook until thickened, stirring all the time. Cool slightly. If the consistency is too gluey, dilute with extra water.

Arrange the mango slices in the cold pastry shell in an overlapping pattern. Use the shorter pieces to fill in the sides. Spoon the purée over the strips, to make an even coating. Chill thoroughly. Serve plain, or with custard, whipped cream or vanilla ice cream.

AMBAKALAYA

A sweet green mango relish from *101 Parsi Recipes* by Jeroo Mehta, to serve with lentil dishes, especially the lamb and lentil dish called *dhansak*.

Peel and slice 1 kilo (2 lb) soft ripe green mango into six pieces each. Cook without water over a medium heat for 5 minutes. Boil 6 tablespoons ($\frac{1}{2}$ cup) water with $\frac{1}{2}$ kilo (1 lb/2 cups) jaggery or pale soft brown sugar, to make a syrup. Add with 5 cm (2 in) piece of cinnamon to the mangos. Cook for 15 minutes, uncovered, until the liquid is a medium thick syrup. Cool.

MANGOSTEEN, see LYCHEE

MEDLAR

From its unmistakable and curious formation, the medlar has long been the happy target of jokes in much of Europe. The first ancient – and descriptive – English name of openarse tells you why: 'gradually and politely it was superseded by the French-derived *medlar*'.

Away from the refinements of Parisian culture in the countryside, however, impoliteness flourished to match our own. 'In some provinces' – this is a French writer of 1833 – 'it is called the *cul de chien* (dog's arse) and some forty years ago if a peasant went to Les Halles in Paris and asked how much *culs de chien* cost, he provoked a flow of filthy language. The Comte d'Artois, now Charles X, loved to go there and purposely provoke abuse in this way. Who would believe it now? The language at Les Halles has much improved. Soon it will be no more than it should, in a market where seller and buyer politely debate their interests.' H'mm. Forty years later, when Zola published *Le Ventre de Paris*, a novel about Les Halles, language had not advanced much on the d'Artois days. The dog's-arse joke would have amused Zola, so would the filthy flights of fancy that followed the Duke's antics (he made a rotten king, incidentally, and I am sorry to think that the light delicious puff pastry Dartois may be called after him).

More decorously the medlar was appreciated for its medicinal quality. You took it *against* looseness, women ate it to prevent haemorrhages. In the Bibliothèque Nationale at Paris, I found a splendid paper, *La Nèfle dans la thérapeutique d'antan (The Medlar in Ancient Medical Practice)*, published in 1952. The author was the cultivated Dr Henri Leclerc who wrote much about fruit. He quotes Abraham Cowley and reflects that medlars, spread on straw, ripen by their corruption. Medlar jelly was recommended by doctors in 1907, and a chemical analysis of 1939 showed that the medlar was the perfect regulator of the stomach on account of its assembly of '*matières pectiques, gommeuses, sucrées et tanniques*'. There were several pages more of such information.

By way of a break at lunchtime, I went off to see what was on sale in the fruit department of Fauchon's; among the glowing exotic fruits were white cardboard punnets of the largest medlars I have ever seen. They were worth eating, exactly right, not too soft, though perhaps the taste lacked the pleasant sharpness of ours in Wiltshire.

They had come from Italy, I think, but were much too early for the St Martin's Eve game of Ventura, a family game that used to be played there on 10 November when the various new wines were sampled. On account of this game, medlars were particularly dear to children. After dinner everyone gathered round the fire. The father took a basket filled with medlars, two for each person there, plus two extra for any poor man who might come begging. Three small token coins were slipped into three of the medlars, without anyone seeing which they were. The father then called his youngest child over and said, 'Put your hand into the basket and take out two medlars for the poor.' The child chose them and put them on the table, then removed two for himself. The basket was handed round. Bad form to examine them before the basket was empty. Then anyone who found the coins was given a larger sum in proportion. If they were in the poor man's medlars, the first beggar who came along next day had the money. The medlars were eaten, and a few chestnuts, everyone tried each of the wines and went merrily to bed.

The medlar makes a charming tree in the garden. It grows and droops over to make a sheltered house for children to play in. In spring, the flowers are white, spreading cups. In autumn the leaves turn a deep yet brilliant red, and fall to show the greenish brown medlars displaying their ancient name. Pick them when they begin to turn soft and darker brown, and do not despise the windfalls. The best can be eaten, as they are. Turn the others into:

MEDLAR JELLY

Aim to assemble one-third firm medlars, medlars that are barely softening, and two-thirds bletted medlars. Cover them with water, and simmer steadily until the fruit is soft. Strain off the liquid, then boil with sugar, 500 g to 600 ml or a pound to a pint (2 cups sugar to 2½ cups water), until setting point is reached. Pot in small jars, enough for one meal. Serve with game, lamb, the meats that go well with tart fruit jellies.

MELON

Among fruit, the melon has been a particular favourite. No other fruit that I know of has had a long ecstatic poem devoted to it. Of famous melon lovers, I have chosen three, a scholar monk, a queen, and the rumbustious poet who wrote *The Melon*.

First the monk, the ninth-century scholar Walafrid Strabo, Walafrid Squint-eye, student then abbot of Reichenau, the island monastery near Constanz, on the Unterzee. He wrote learned poetry, edifying and courtly allegorical poetry, he was mixed up in politics and backed the wrong man. He had many friends and to one of them, Grimald, who had been his master, he dedicated his only poem that anyone outside medieval Latin academic circles could enjoy. It was about his garden and the things he grew in it. The correct name is *De cultura hortorum*, but it is usually known by an affectionate diminutive of a nickname, *Hortulus*, the little garden. Reichenau seems to have been hospitable to melons: 'When the iron blade strikes to its guts, the melon throws out gushing streams of juice and many seeds. The cheerful guest divides its bent back into many slices. He tastes this delightful succulence of gardens. Its pure taste charms his throat.' Then he goes on to say how tender it is (I imagine that much food in the past was excessively tough, tough roots, tough meat, tough bread with little bits of grit in the flour that eventually wore away your teeth), and how it cools the belly. It is strange suddenly to hear a human, sympathetic voice from over ten centuries ago.

The queen, Catherine de' Medici, queen of France, was less sympathetic. Her husband died leaving her with a collection of useless or delicate children (some were both). She ruled for them. A particular worry that hung over her was the thought that if none of the brood survived or fathered surviving children, the crown of France would go to the Protestant son of the Protestant Jeanne de Navarre. And indeed it did, to Henri IV, our Charles II's grandfather, who turned Catholic as he rightly assessed that Paris was worth a mass. What, I wonder, would Catherine have thought of that? Meanwhile, she sought consolation in astrology and food.

One day, she complained of feeling unwell, presumably from indigestion as she was inclined to stuff herself. Jeanne de Navarre, whose position gave her a certain freedom, tactlessly remarked that it was no wonder, seeing how many melons she had eaten. Catherine flared back: 'It's not the fruits of the garden but the Fruits of the Spirit that upset me!'

My third melon lover was an exuberant character with the delightful name of Saint-Amant, Holy Lover. His energetic affections encompassed a melon, a particularly fine melon grown in Anjou. I do not know if the first Charentais melon had been raised as early as the seventeenth century, in the green Marais to the south of Anjou in the Charentes, but its orange flesh and golden seeds inside the pale green skin fit his description. His melon is firm to the centre, with few seeds like grains of gold. It's better than his beloved apricot, than strawberries and cream, than the holy pear of Tours or the sweet green fig. Even the muscat grape that he loves is bitterness and muck compared to this divine melon. He loves the way a melon grows – 'O sweet grassy snake, crawling on a green bed'. It's Apollo's masterpiece. The brothels of Rouen (his home town) will be free of the pox . . . tobacco smokers will have white teeth . . . I will forget my love's favours before I forget you –

O fleur de tous les fruits! O ravissant MELON!

HOW TO CHOOSE MELONS

In general terms, good melons should seem heavy for their size. Weigh a couple that look the same, and take the heavier. A ripe melon will be slightly soft at the stalk end and some kinds will also have a delicious smell. Cut the melon in half, if you are going to divide it into sections. Or cut off a lid, if it is to act as a container. Scoop the seeds into a strainer set over a basin, so that none of the juice is wasted. Melon seeds are edible: dry them off and toast them in the oven. Wrap melon in cling film before chilling, or storing in the refrigerator since its smell tends to permeate everything else.

The kinds you are most likely to see in the shops and markets are:

OGEN: a green skin with orange lines, dividing it into sections, and green flesh. I had thought that this melon was first raised in kibbutz Ha-Ogen, but a fruit importer has told me that it belongs to Hungary. Certainly the commercial development started in Israel, at Ha-Ogen, and it was named there. It is now grown throughout the world.

GALIA: a netted skin, easy to spot, which turns from green to a brownish yellow when ripe, and green flesh. Related to the Ogen, and named after the daughter of the man who raised it. Larger than the Ogen.

HONEYDEW: looks like a pale creamy or green rugger ball. Turning yellowish as it ripens. Usually on the tasteless side, though sometimes one has a honeydew that is delicious. Its main advantage is cheapness.

LAVAN: a round Honeydew, with a bit more flavour.

CHARENTAIS: the best of all that are easily available in France, less available in England. Green skin and glorious fragrant orange flesh. Although developed in the Charentes, most seem to be grown on the opposite side of France around Cavaillon.

CANTALOUP: a melon to look out for; what one thinks of as a melon from seventeenth-century paintings, craggy, sectioned into wedges, a thoroughly real-looking fruit that will catch your eye among its smooth relations.

Water-melons have their own section, *see* p. 428.

MELON SOUP

A Russian soup so delicate that it makes an ideal first course for a summer party.

1 melon weighing generous 1 kg (2 lb)
150 g (5 oz/⅔ cup) sugar
250 ml (8 fl oz/1 cup) dry white wine
lemon juice
150–300 ml (5–10 fl oz/⅔–1¼ cups) soured cream or crème fraîche

Discard pips and scoop the flesh from the melon. Ideally you should have ¾ kg (1½ lb) weight, but do not worry if you have less or more. The other quantities can be adjusted to your taste – and this will depend more on the strength of the melon's flavour than on its bulk.

Dissolve 600 ml (1 pt/2½ cups) water and the sugar in a pan over a low heat, then simmer for four or five minutes.

Liquidize the melon with the wine, and gradually add the cooled syrup to it. Stop when it tastes right, having regard to the fact that you will be adding a faint sharpener of lemon juice, and at least 150 ml (5 fl oz/⅔ cup) cream. More if you like.

As you see, an easy recipe and one that always turns out well. Serve chilled, with tiny almond biscuits, or tiny meringues.

MELON SALAD

Choose a ripe, fragrant melon. Dice the flesh, discarding seeds and peel, and sprinkle it with lemon juice and pepper.

Serve with cold ham and poultry. Particularly good with smoked poultry, and with Bayonne or Parma ham, *lachschinken* or *kassler rippenspeer* from the delicatessen.

Or serve the melon salad around 500 g (1 lb) curd cheese, well-drained *fromage frais* or yoghurt cheese, or petit Suisse, or Gervais, that has been mixed with 4 or 5 tablespoons (5 or 6 tbs) of double cream, and seasoned with a little sugar. Sprinkle with chopped pistachio nuts.

THREE-MELON SALAD

Simple to make, delightful to eat – ideal for a summer party.

Mix together the different coloured flesh of three melons, in roughly equal quantities. Sprinkle with caster sugar, and leave for a few hours so that the juices mix with the sugar to form a light syrup.

Taste and flavour with lemon and orange juice to taste, especially lemon. Chill in a bowl covered with cling film; serve in the hollowed-out shell of a water-melon. On a bed of crushed ice, if you like.

Note If you keep an empty shell ready in the freezer (*see* p. 431), you do not need to include more than a slice of water-melon, seeded and chopped.

MELON, PEAR AND CUCUMBER SALAD

Equal quantities, in cubes, with a French dressing; or with a hazelnut oil and cider vinegar dressing, with chopped toasted hazelnuts. Put for one hour into the refrigerator. Serve with Melba toast or thin slices of bread baked until curved and not too brown.

If the salad turns very liquid, stir it up well and then drain it before putting into a fresh serving dish. Peel the cucumber or not as you feel inclined, or ridge the peel with a fork or canelle cutter.

MELON WITH WILD STRAWBERRIES

Buy small orange-fleshed melons, of the Cavaillon type. If you can find the really small ones, a bit larger than an orange, take off a lid and remove the seeds from each one. The larger size will have to be halved, one between two people.

Sugar wild strawberries, or the Alpine or *fraises de bois* that you can grow in the garden. Add a little alcohol if you like – port is the usual one, but a light orange liqueur can be used instead. Or a good wild strawberry brandy. There are some strawberry liqueurs about, but they are too sweet and heavy and a little false in taste.

Just before serving, distribute the strawberries between the melons or the melon halves. If you are short on the strawberries, remove a little of the melon flesh and dice it. Then mix it with the strawberries, before filling the melon shells.

MELON WITH RASPBERRIES

The recipe above using raspberries – yellow ones if possible. An orange liqueur or kirsch or maraschino, or an *eau de vie de framboise.*

MELON AND GINGER SALAD

A good winter salad for ham and duck, for cold game. You can also reduce the salt, and serve it with toasted hazelnut brioche as a first course or pudding in the summer.

The point to remember is that the stronger the flavour of the melon, the less preserved ginger you use.

For a Honeydew melon of 1 to 1½ kgs (2–3 lb), chop three knobs of preserved ginger and slice a fourth thinly. Toast 60 g (2 oz/a good ½ cup) hazelnuts in the oven at gas 2, 150°C (300°F) – this takes about 10 minutes – and rub off the skins. Chop the nuts coarsely. Whisk 150 ml (5 fl oz/ ⅔ cup) whipping cream with a teaspoon of lemon juice until it is thick but not stiff. Add the chopped ginger and most of the nuts. Season with a pinch of salt and some cayenne pepper, or with a little ginger syrup for a sweeter salad.

Cut the melon into 8 slices, discard the seeds and slice the flesh into cubes. Put them into a wide bowl.

Leave until serving time, then pour over the cream dressing and decorate with the ginger slices and remaining nuts.

MELON TART
TARTE AU MELON

Use an orange-fleshed, well-flavoured melon for this delicate tart, which can be served warm or cold.

1 melon weighing $\frac{3}{4}$ kg (1$\frac{1}{2}$ lb)
1 large egg yolk
60 g (2 oz/$\frac{1}{4}$ cup) sugar
30 g (1 oz/$\frac{1}{4}$ cup) ground almonds
grated peel and juice of $\frac{1}{2}$ lemon
1 large egg white, stiffly beaten
30 g (1 oz/$\frac{1}{4}$ cup) blanched, split almonds
sweet shortcrust pastry, chilled

SYRUP
150 g (5 oz/$\frac{2}{3}$ cup) sugar
tablespoon kirsch (optional)

Slice melon, discarding peel and seeds. Keep the slices as even as possible.

Mix yolk, sugar, almonds, grated peel and lemon juice. Fold in the egg white and the almonds.

Roll out the pastry and line a 25-cm (10-in) tart tin. Pour in the egg filling. Bake at gas 5, 190°C (375°F) for 25 minutes, or until the filling is well risen and slightly browned.

Meanwhile make the syrup by boiling together the sugar and 125 ml (4 fl oz/$\frac{1}{2}$ cup) water until reduced by half. When tepid add the kirsch.

Arrange the melon slices decoratively on the hot tart. Brush over with the syrup, to make a glaze.

MELON FOOL

Hollow out a good-sized Charentais melon, after slicing off a lid and removing the seeds. Mash the flesh – do not liquidize it, or it will become too sloppy – with 6 dessertspoons of vanilla sugar. Lightly whisk 150 ml (5 fl oz/$\frac{2}{3}$ cup) whipping cream until it is very thick, but not stiff. Fold in melon mash and add 1$\frac{1}{2}$ tablespoons (2 tbs) kirsch.

Taste and adjust flavours. Pile back into the shell, replace the lid. Serve chilled, on a dish of green bay leaves.

MELON ICE CREAM

GLACE AU MELON DE L'ÎLE ST JACQUES

For a cream melon ice, I recommend Elizabeth David's recipe. For those who do not know it, I give a summary below. It tastes so fragrant and magical, that she called it after the French Caribbean island 'conjured out of the ocean, only to be once more submerged, by Patrick Leigh Fermor in *The Violins of St Jacques*'.

1 large Charentais, or medium Cantaloupe melon
125–175 g (4–6 oz/$\frac{1}{2}$–$\frac{3}{4}$ cup) sugar
4 egg yolks
1 glass kirsch
juice $\frac{1}{2}$ lemon
300 ml (10 fl oz/1$\frac{1}{4}$ cups) double cream, whipped

Slice top from melon. Scoop pulp carefully from the shell, discarding the pips. Cook the pulp with 125 g (4 oz/$\frac{1}{2}$ cup) sugar until soft – about 2 minutes. Beat yolks until thick, add to the tepid purée and cook very gently until mixture is like thin custard. Taste and add extra sugar if necessary.

When the mixture is cold, stir in kirsch and lemon. Fold in the cream and freeze.

Serve in the melon shell, replacing the top, and setting the melon on a bed of vine or bay leaves.

MELON WATER ICE

An especially good and simple recipe, so long as you use a melon of character.

Simmer 125 g (4 oz/$\frac{1}{2}$ cup) sugar with 250 ml (8 fl oz/1 cup) water for four minutes. Liquidize enough melon pulp to give you 300 ml (10 fl oz/1$\frac{1}{4}$ cups). Add cooled syrup gradually to taste. Sharpen a little with lemon juice. Freeze at lowest possible temperature, until sides are firm, the middle semi-liquid.

With an electric beater whisk an egg white until stiff. Stir up the ice and add spoonfuls to the egg white, still beating, until you have a mass of creamy-looking foam. The ice can be rapidly poured into chilled glasses and served at this point.

Or it can be put back into the freezer to set firm.

The first method is delightful for a summer meal out of doors, so long as you and your guests are relaxed and leisurely. The second method is for slightly trickier occasions, when you want to reduce last-minute work to a minimum.

SPICED MELON

Orange-fleshed melons make a fine and beautiful pickle to serve with ham or duck, but they are tricky. The pieces need to be just cooked. A few seconds too long and they go squashy, which is not nearly so nice. Remember that the hot syrup poured on to them in the jars continues to cook them very slightly.

With melon, some fresh ginger is a good idea – turn to the general recipe on p. 449 for details of the method. To provide 1½ kg (3 lb) prepared melon cubes, you will need three melons weighing something under 1 kg (2 lb) each. The melon juice can be poured into the pickling liquor, but it should not be included in the fruit weight – drain the cubes well before weighing them.

MELON RIND PICKLE

Any melon rind can be used to make this pickle – one kind alone, or a mixture. The part you want is the white layer between the outer rind and the inner flesh. In some varieties this layer is much thicker than in others, they are the best kinds to use.

Remove the hard outer rind as thinly as possible and discard; then cut the rest into roughly 2½ cm (1 in) pieces and weigh them. Dissolve 2 level tablespoons (3 tbs) of sea or kitchen salt in a litre (1¾ pt/scant 4½ cups) water. Pour over the rind and leave until next day. Drain off the brine, rinse the rind, then simmer it in fresh water until tender.

To every 1 kg (2 lb) rind, put in a pan:

1 kg (2 lb) sugar
½ litre (¾ pt/scant 2 cups) water
½ litre (¾ pt/scant 2 cups) vinegar
1 thinly sliced lemon
1 stick of cinnamon
1 teaspoon cloves
1 teaspoon allspice or a head of star anise

Bring slowly to the boil, stirring at first until the sugar dissolves. Then boil steadily for 5 minutes without stirring. Put in the drained rind, bring back to a quiet boil and leave until the rind looks transparent and candied.

Fish the rind from the syrup with a perforated spoon, and put it into warm sterilized bottling jars. Bring syrup back to the boil and pour over the rind, which should be covered. Close the jars. Store for at least a month before using.

Note Taste the syrup occasionally during the cooking, to see whether the spices should be removed. They and the lemon can be included in the final pickle, but will continue to flavour it: you may prefer to strain them out.

MELON AND GINGER JAM

Some years ago, one could buy melon jam in cans, from South Africa. It was delicious and well worth making when melons come down to a reasonable price. Ginger is a good flavouring with melon, or at least chopped preserved ginger is – I am not so keen on powdered ginger which is sometimes scattered over melon wedges. Use green-fleshed melons.

Weigh the melon after peeling it and removing the seeds. For each kilo (2 lb), you will need:

1 kg (2 lb/4 cups) granulated or preserving sugar
finely grated rind and juice of 4 lemons
150 g (5 oz/⅔ cup) drained preserved ginger in syrup

Slice or cut the melon into cubes. Layer it with the sugar into a large basin, cover and leave in a cool place overnight. Best not to put it into the refrigerator.

Tip the melon and what is now its juice into a preserving pan and add the finely grated rind and juice of the lemons and the ginger. Bring to the boil, lower the heat and simmer until the melon looks transparent – 30–45 minutes. Raise the heat and boil hard until setting point is reached. Pot in the usual way.

MELONS IN ITALY, THE LAND OF LANDS

A girl bare-footed brings, and tumbles
Down on the pavement, green-flesh melons.

Robert Browning.

MULBERRY

Four things I like about the mulberry.

If you can get hold of a branch from somebody else's tree, and put it into the ground, it will grow. Just as the willow tree does – but far more slowly. Give a mulberry tree for a wedding present, if they seem like a staying couple.

One of the last emperors of Constantinople called the southern extent of Greece the Morea, because it is the shape of a mulberry leaf and *moron* is the Greek for mulberry. Or so the story goes. I hope it is true.

The mulberry is a symbol of prudence. Not a leaf appears until all danger of frost is over. So reliably late, that prudent John Evelyn gave his readers this infallible rule for their orange trees and 'more tender curiosities' – 'observe the mulberry tree; when it begins to put forth and open the leaves (be it earlier or later) bring your oranges etc boldly out of the conservatory; 'tis your only season to transplant and remove them.'

The fourth thing, best of all, is the mulberry fruit itself, the dark long oval fruit that will stain your mouth and your clothes as efficiently as any bilberry, and more lavishly. This is why the mulberry is unsuitable for shops and markets. And because people do not see it, they do not think about it. Even those with mulberry trees in their gardens often ignore the fruit.

I remember years ago, one August, we suddenly needed to see the Constable painting at Sudeley Castle. We drove there fast, jumped out, put the protesting baby into her push chair and rushed to get our toe into the drawbridge before it was pulled up for the night. Emerging some time later, absorbed with the beauties of the place, we had a shock. Our car was sploshed over with purple. Mulberries dripped off the roof rack and lay squashed on the bonnet. Too late to go back and ask if we could pick some (nobody else seemed interested). We drove home sadly, and recall the lost opportunity every time we see the few mulberry leaves on our young tree.

Profit from our loss – whenever you go visiting castles, manor houses, old vicarages, keep an eye open for a mulberry. A tree of any age is a spectacular sight, it spreads in great beauty and is often supported on posts and an iron framework. You will often know when the season has arrived from dust-sheets spread below the trees to catch the fruit. Old mulberry trees can be covered with so much fruit over several weeks that they are the despair of their owners who will be pleased for you to take some away. But do go ready for the picking, in dark purple clothes.

The best way to eat mulberries is with cream, just as they are. Elizabeth David suggests that you put some of the leaves on a dish and then pile them up carefully, so that you end with a pyramid of fruit as beautiful as Chardin's pile of *fraises de bois*. For this, you need a steady hand and perfect fruit, not too ripe.

Good advice comes from two other mulberry lovers, Leslie Johns and Violet Stevenson, in their *Complete Book of Fruit*. They point out that as mulberries often have to be picked from the ground, they need washing. And washing crushes and spoils such tender fruit unless you are careful. They suggest putting a few into a colander and dipping it in and out of a bowl of water a time or two. Leave them to dry in the colander undisturbed. A slow process. I suggest you line up a few colanders, sieves, chip baskets, cream cheese draining plates and so on. Or you would do better to pick carefully in the first place, taking precious mulberries from the tree and putting the ones on the ground into a separate basket for washing and turning into syrups, jelly and so on.

If you think you would like to grow a mulberry tree yourself, you will also find plenty of growing advice in *The Complete Book of Fruit*. Mulberries can do well here. James I encouraged them, hoping for a silk trade which never came off. People went on planting them, though.

There are not many recipes in this section, as mulberries can be treated like blackberries, and do well in a number of blackcurrant dishes. For instance, they make a good sauce with lamb or pheasant. Mulberry jelly – pudding or preserve – is made in the usual ways. Elizabeth David recommends mulberries for summer pudding: cook them lightly with sugar until their juice flows, and be sure to use good bread. Mulberry sorbets and ice creams are always successful. I have never made a *crème de mûres* (*see* the recipe for *crème de cassis* in the black currant section), but it should work well.

A word of warning – in using French cookery books or translations of them, be on your guard. The word *mûre* can mean blackberry (i.e. *mûre de ronce*) or mulberry. It probably will not matter, as recipes are interchangeable as I have said, but there might be cases where it would be a good thing to understand which is meant. In general, older books and books of southern cookery indicate the use of mulberries. Books of this century, free food and self-sufficiency, or books of northern French cookery, indicate blackberries.

Another point, the silkworms' mulberry, the mulberry of the Lyonnais for instance, is the white mulberry: they produce better silk than if they eat the leaves of the black mulberry that we know so well. In the case of fruit, virtue is the other way round, black mulberries are the thing.

A MULBERRY AND ALMOND DISH

A recipe of Elizabeth David's suggested to her by a sixteenth-century Italian sauce or relish called *sapore de morone* which appeared in *Epulario*, first published in Venice in 1516. It was a tart sauce for meat – which meat is not mentioned – of mulberries, bread and almonds, spiced with cinnamon and nutmeg.

375–500 g (12–16 oz) mulberries
175 g (6 oz/¾ cup) sugar
45–60 g (1½–2 oz/about 1 cup) soft white breadcrumbs
45–60 g (1½–2 oz/about ½ cup) blanched and finely ground almonds

Cook the mulberries with the sugar, not too fast, until the juice flows. Mix the crumbs and almonds in a basin, then add the mulberries folding them in gently.

Divide between glasses or white custard cups and chill. Before serving, float a little cream on top.

MULBERRY PUDDING

When Louis Eustache Audot, author of the best-selling *La Cuisinière de la Campagne et de la Ville* (1818), came to England, he must have stayed in a comfortable middle-class house, a small manor perhaps with a good garden. I conclude this because he so loved our puddings, gooseberry pudding in particular (*see* that section). This *pudding de mûres* may seem an unsuitable dish for summertime, but I have known days in late July and August when the comfort of mulberry pudding would have been most welcome.

Line a basin with suet crust (500 g or 1 lb self-raising flour, 250 g or 8 oz/1¾ cups freshly chopped suet, mixed with water), setting aside enough for the lid. Fill the cavity with mulberries, interspersing them with sugar. Allow 175 g (6 oz/¾ cup) to each 500 g (1 lb) of mulberries.

Cover with the remaining suet crust, then with greaseproof and foil or cloth in your usual way. Boil or steam for three hours. Serve with an egg custard or cream. Put a bowl of granulated sugar on the table for extra sweetness if necessary, but mainly for that agreeable grittiness that sets off this kind of pudding so well.

MULBERRY COMPOTE

Pick out your best mulberries and put them into a bowl. Reduce the rest – or about 1 kg (2 lb) of the rest – to juice. This can be done by liquidizing or processing, and then straining. Or by simmering the fruit over a low heat until the juices run. Or by putting the fruit into a jug in a low oven, or a bain marie, until they float in their own juice.

Sweeten the juice to taste, having regard to the ripeness of the mulberries in the bowl. Stir well together and bring slowly to the boil. Simmer for 5 minutes, then pour boiling over the fruit. Leave to cool.

Another way is to make the syrup as above, and leave it to cool. Put the fine mulberries into it, bring to the boil, allow it one good bubble, and remove from the heat. Leave to cool with a lid on the pan. Avoid cooking the mulberries in any proper sense of the word – as Sir Harry Luke says in *The Tenth Muse*, surplus mulberries or any mulberries that are not quite ripe or that have a hard core are best turned into water ice for storage and later use.

MRS BEETON'S MULBERRY PRESERVE

Using damaged mulberries, reduce them to 600 ml (1 pt/2½ cups) of strained juice – *see* recipe above.

Put into a preserving pan with 1 kg 150 g (2½ lb/5 cups) sugar. Bring to the boil, stirring, and then skim. When moving vigorously, tip 1 kg (2 lb) mulberries into the pan: they should be ripe but not soft enough to mash. Take the pan from the stove and leave until the fruit is warmed through (taste one). Then put back on the heat and boil gently for about 15 minutes.

Pour carefully so as not to break the fruit into a large bowl, however if you are using a stainless steel preserving pan, you can leave the mulberries in it.

Next day, bring to the boil and simmer steadily for 15 minutes, or until setting point is reached. The point of this careful method is to keep the mulberries as intact as possible. You can avoid breaking the fruit 'by gentle stirring, and by simmering the fruit very slowly'.

Pot in the usual way.

DRIED MULBERRIES

At some groceries of the kind loosely called oriental, you will occasionally find dried white mulberries, soft and a fawnish brown, the reduced fruit of silkworm trees. They have a strange, modest flavour. I find myself wandering into the larder to eat a few in an absent-minded way, but they are best mixed with nuts – pistachios and pine kernels, almonds and hazelnuts lightly toasted but not too strong – which make a contrast of texture and taste.

A MULBERRY PARADISE

> It was in a mail coach such as this that Babel [the writer] and I were one day travelling from Sukhum to New Athos . . . We travelled with no undue haste, enjoying the heat and the ripe mulberries. They were spread on the road like a thick carpet.

Konstantin Paustovsky:
Southern Adventure

NECTARINE, see PEACH

ORANGE, TANGERINE, KUMQUAT

Stendhal defined the true south as the place where orange trees grow in the ground. As a child in Grenoble – which seems south enough to us in spite of the snow – he would watch the yearly ritual of rolling out the municipal tubs of orange trees in the spring, and rolling them back under cover in the autumn, and sigh for his imagined true south.

More northern poets have pictured the land 'where oranges in green night wink', and contemplated the magic way in which the bright fruit share the tree with the waxy flowers of the next crop. This odd conjunction, this way in which oranges are harvested in their own flower-scented air, gives the fruit a glory that talk of vitamin content can never spoil. To be in Cyprus or Greece at that time of year, late February, early March, has a special sweetness. Roads are bordered with the fruit tumbled from lorries. The space between suburban houses on a city road at night is suddenly fragrant with the scent of orange flowers. Causeyed roads take you level with the tops of citrus groves, undulating dark green blobbed with bright orange. And if you wander off the roads, to look for some small chapel or spring, you move more and more slowly in the warm, enclosed, aromatic air between orange groves.

Oranges are not native to the Mediterranean. They come, in all their varieties, from the bitter orange to the tiny kumquat, from China. Even the tangerine whose name suggests a North African origin comes from China. Its other name of mandarin is 'explained as an orange the colour of the robes of mandarins, or an orange eaten by mandarins, or an orange which looks like a mandarin'.*

Of all the developments of that delicious loose-skinned fruit – *Citrus nobilis* var. *deliciosa* – the Satsumas, Wilkins and so on – the one I look out for every autumn is the clementine. It has an extra edge to its sweetness, having been raised near Oran in Algeria *c.* 1900, by the priest

**A Dictionary of English Plant Names* Geoffrey Grigson (1974).

Père Clément, from the Mandarin and the bitter Seville orange. This sharp strain makes it the ideal tangerine for bottling with syrup and alcohol, or for using with spinach or as the base for *mandarines givrées*.

These Christmas oranges, small and silver-wrapped, seem to have a longer season every year as new varieties are raised. They came to us in the nineteenth century, and are now everybody's favourite, for their easy skin and their charming sweetness. It seems that we knew the tiniest of all oranges, the kumquat, in the seventeenth century, yet it is still a rarity. It is sold in little white cartons at the best greengroceries; you eat the whole thing, peel, pips and all. The name, which is sometimes spelt cumquat, means 'gold orange', from the Cantonese *kam kwat*.

The word orange itself shows an ancient delight. It goes back over 3000 years to Dravidian India, *narayam* meaning 'perfume within'. The Arabs took the word from the Persians as *narandj*. Italians softened it to *arancia*. In medieval France *arancia* slipped naturally into *orange*, because the town of Orange was a great centre for the fruit.

Oranges arrived in Italy towards the end of the Roman Empire, brought to Europe like many other things – spices, silk, sugar – by Arab traders. As John MacPhee points out in his book, *Oranges*, groves of orange, tangerine and lemon trees in a wide band from India to Spain now mark the conquering track of Muslin armies in the sixth and seventh centuries.

John MacPhee goes on to give a version of how oranges came to the North. A band of Norman pilgrims on their way back from Jerusalem, in the eleventh century, saved the Christian prince of Salerno from a ferocious Muslim attack. This gave the prince an idea. He sent an embassy with the pilgrims to the Duke of Normandy, 'accompanied by a mountainous gift of beautiful oranges, frankly tempting the Duke to conquer southern Italy – which he did, taking Sicily, too. The Norman conquest of Sicily turned into something of a scandal. Norman minds dissolved in the vapors of Muslim culture . . . The Normans went Muslim with such remarkable style that even Muslim poets were soon praising the new Norman Xanadus. Of one such place which included nine brooks and a small lake with an island covered with lemon and orange trees, the poet Abd ur-Rahman Ibn Mohammed Ibn Omar wrote:

> The oranges of the Island are like blazing fire
> Amongst the emerald boughs
> And the lemons are like the paleness of a lover
> Who has spent the night crying . . .

Oranges became a sign of opulence. The richer you were, the more dishes were flavoured with oranges at your table. Baths and make-up were scented with a distillation of orange flowers, which was also used to flavour food – and still is, especially in southern Europe. Oranges were taken as their symbol by the Medici (the five golden balls of the Medici coat of arms are oranges). Once you know this, you begin to perceive a silent assembly of orange trees taking place. In the backgrounds of Quattrocento painting, orange trees are tucked – inaccurately – into the nativity or behind the Mother and Child. The Aphrodite and Primavera landscapes have the dark foliage of orange trees contrasting with vivid light.

The first orange tree in France – it came from Pamplona in Spain – belonged to the grand Constable of France, the Duc de Bourbon. The king, François I, coveted that tree, kept his eye on it I would say. When the Constable fled to join François's enemy, Charles V, in northern Italy, the first thing the king did was to seize the orange tree. It descended to Louis XIV, who took it to Versailles, where it died not very long ago. I should add that François I married his eldest son to Catherine de' Medici. Evidently oranges were much to his liking.

The great king for oranges – as for other fruit – was Louis XIV. Henri IV may have built the first orangery, at the Tuileries, but Louis his grandson had one put up at Versailles so magnificent that it was the envy of all Europe. Of course, Louis had the benefit of one of the greatest of gardeners, de la Quintinie, who was stimulated and at times, I suspect, unnerved by his master's passion for garden produce. Strawberries nine months of the year, asparagus in January – that was the kind of thing he had to aim for. When the orangery went up, de la Quintinie found a way of holding back the flowers, so that when they did come, their scent was all the stronger and sweeter.

An orangery was to shelter the trees from frost. On a nasty winter day, you could take a saunter there to sniff the rich smell of orange blossom. As spring began to show, you had the delightful task of deciding, with your gardener, the day when the orange tubs could be wheeled out on to the terrace. The fruit of these northern trees has never come to much, in spite of all the de la Quintinies, Evelyns and Millers. But I doubt it matters: orange trees in their tubs, family heirlooms nowadays, are a cherished symbol.

All those precious early oranges were bitter. It was the aromatic skin and sharp juice that was treasured, the scent of the flowers. Oranges for eating, sweet oranges, came via Portugal, first from India, then – and this was the success – from China. The first China orange tree – given the botanical name *Citrus sinensis* – reached Lisbon in the 1630s, and flourished. In time it ousted the bitter *Citrus aurantium*, which is still grown in Spain but for the British marmalade market only. How long will this continue?

Thirty years after that first tree, sweet oranges from Portugal became a smart theatre refreshment in London, with pretty girls like Nell Gwynne to sell them. Pepys tried some orange juice warily, frightened it might upset his stomach. He liked the taste. So did other people, but such oranges remained very expensive, and in cookery books the word continued to mean bitter orange, an alternative to lemon, a flavouring, something to candy as a sweetmeat, but not a fruit to be eaten out of hand.

How can you tell the difference when reading older recipe books? Sometimes you will see that Portugal or China oranges are required and that is the sweet kind – but up until the early nineteenth century, it is safe to conclude that the word orange means what we call a Seville or marmalade orange. Of course, you can use a sweet orange in most cases for such recipes, but the flavour balance will be wrong, too sweet, or too insipid. A little lemon juice corrects the matter, though inevitably the extra aromatic air of the dish is lost. When you do come across recipes of this kind, mark them with a slip of paper and try them out as soon as the first bitter oranges arrive in the shops after Christmas. For baked orange creams, orange fools, orange ices, there is nothing to equal their juice, either on its own, or mixed with sweet orange juice. Never waste the peel. It can be candied – *see* p. 198 – or put into a litre jar with syrup and alcohol to make an orange ratafia for drinking, or for adding character to many dishes as a substitute for the branded liqueurs such as Cointreau, Grand Marnier, Triple Sec, Aurum and so on.

BUYING AND PREPARING ORANGES

Oranges should look healthy, with a soft shine to their skin. They should be unbruised and unwrinkled, feel heavy in the hand for size. In theory, it is easy to pick a good orange, but what about those nets of bargain oranges? The mesh often seems designed to be thick enough to prevent you getting a good look, and feeling the oranges properly all round. You have to weigh up the prices, and work out if one bad orange in the pack means it is no longer a bargain.

Technically, it is easy to spot a good orange, and we soon sort out our favourites. Mine are the navels, partly I admit on account of the charming shape, pregnant with a tiny orange. Blood oranges are another pleasure. So are ortaniques from Jamaica, an orange hybrid of delicious flavour. Jaffas have a splendid look of orange roundness; they almost convince me that oranges must have been the Golden Apples of the Hesperides (except that they did not arrive in Europe soon enough).

Having taken care of all these points, you may still be disappointed when you get inside your carefully chosen orange. It may be boring, or sharp, or even dry. There seems no way to guard against this. Oranges are after all growing things, with the variations of quality and character that this implies. Some years will be better than others – think of the despair of the citrus farmers in Cyprus and the Eastern Mediterranean at the rainy season of 1980 which spoiled their crops and prevented harvesting. Only dehydrated foods made from soya bean granules always taste the same. In 1981, in January, several of us came to the conclusion that our marmalade was the best we had ever produced. The oranges seemed to come to setting point more quickly, the marmalade tasted fresher, with exactly the right balance of jelly and peel. Yet we all followed our usual recipes, and all the recipes were different. It must have been the quality of the oranges.

Try not to waste the orange peel. It has all kinds of uses. By peel, I mean the thin bright orange outer layer, which can be removed in various ways. But first scrub the oranges. They may have been coated with something waxy to improve their shine; they probably have some shipper's name stamped in ink on them. Do not despise the occasional greenish orange – some manage to slip through. Many of the bright fruit you buy would be the same colour if they hadn't been treated with ethylene gas. Some French consumers are concerned about this, and in shops you will find boxes of citrus fruit labelled *non traité*, untreated. Treated or no, I have been using fresh orange peel for many years without any ill effects: and I reflect that some fruit exude ethylene gas themselves, to the extent that if you put, say, an apple in with pale oranges, they will gradually turn bright under its influence.

Peel can be removed in several ways, according to need. The simplest and quickest is by potato-peeler. You can take off the peel in a long long strip, and hang it up to dry as Mediterranean housewives do: when dry and leathery, it can be tied into a bouquet garni for beef and veal stews and for some fish dishes. You can also peel it downwards, which is quick if all you want it for is to infuse with syrup for orangeade or ices: this is the best way, too, if you need julienne strips, each piece can be squared off and cut with a small sharp knife.

Julienne strips can be taken off directly by a gouging tool called a canelle knife. It can also be used to score oranges, small oranges, decoratively before cooking them in water until tender, then finishing them in syrup. Easier to buy is the zester, which removes the peel in tiny shreds or longer wisps according to the way you handle it. The advantage of the wisps is that they can be used to decorate food without the need for boiling. This is because they are cut so finely from the zest; with a potato parer, sharp knife or canelle knife you inevitably get a thicker cut which means the strips must be cooked in boiling water until they are tender – 5 minutes is enough.

To peel an orange to the quick – in French, *peler à vif* – remove the outer orange rind in any of the ways described above, then take a small sharp knife and cut away the pith so that you also remove the skin enclosing the edible part of the orange. This means you are through to the juicy flesh, so hold the orange over a bowl while you work. This mode of preparation is required for Italian caramelized oranges, or for slicing oranges into a fruit salad.

If you wish to take perfect segments out of an orange, for a garnish perhaps, peel it and put it into the freezer or coldest part of the refrigerator until it is very firm (not frozen). This makes it easier to cut down between the segment skins and remove perfectly shaped pieces.

To get the most juice from an orange, it should be warmed. Again, use the airing cupboard or rack of a solid fuel stove overnight. A quicker way is to pour boiling water over it, and leave for 5 minutes.

When it comes to the squeezing, there are many implements on the market to make life easier. And if you are squeezing oranges in quantity, quite often, you could do with a juice press or electrical juicer (*see* p. 206).

Seville oranges apart, there is little reason for most of us to clutter up the freezer with bags of oranges. But if you do have some special purpose or reason for doing so, oranges do freeze well. So do prepared orange slices: sprinkle them with sugar for dry freezing, or cover them with syrup. In this form they are useful, if you suddenly want to make a tomato and orange salad or prepare duck with orange. Syrupy sweetness can be quickly rinsed away.

Frozen concentrated orange juice is a good standby if you like orange ice creams. As a drink it is unsubtle, only a grade or two above the orange squashes and 'orange drinks'. For drinking, especially first thing in the morning, fresh orange juice is best: later in the day, home-made orangeade, well chilled, can be as agreeable.

Orange juice and a little zest can also be added to tomato and carrot soups, as a final flavouring.

MALTESE MAYONNAISE

'Maltese' in classic cookery means something flavoured with blood oranges. This mayonnaise can be served with asparagus, with chicken salads, and with fish.

2 blood oranges
2 large egg yolks, or 3 small
300 ml ($\frac{1}{2}$ pt/1$\frac{1}{4}$ cups) light olive oil, or mixed olive and sunflower oils
salt, pepper

Grate the zest of the oranges into the basin in which you intend to make the mayonnaise. Squeeze out the juice into a jug, and set aside.

Put the yolks into the basin and beat vigorously. Add a teaspoon of orange juice, and beat again. Have the oil slightly warm, and start adding it to the yolks drop by drop, whisking all the time. As the mayonnaise starts to thicken, add it a little more quickly but be careful. If the mayonnaise gets too thick to beat easily, thin it down with more orange juice. Finally season with salt and plenty of pepper, and add a tablespoon of orange juice. You may need a little lemon juice as well: this depends on the flavour of the blood oranges.

If the mayonnaise curdles, break another egg yolk into a clean bowl, beat it, then beat in gradually the curdled mixture.

For making mayonnaise, I use a spoon-shaped whisk and a heavy pudding basin or mortar. Blender and processor I find unsatisfactory, but an electric beater works quite well as long as you can stand the protracted noise.

CUMBERLAND SAUCE

See the recipe in the Currant section, p. 94.

ORANGE CREAM SAUCE

A mild version of *sauce bigarade* that goes well with calabrese or purple sprouting broccoli, asparagus (supposing you have a supply of frozen Sevilles to use in May and June), seakale, and white fish.

3 Seville oranges
150 ml (5 fl oz/$\frac{2}{3}$ cup) dry white wine

2 large egg yolks
150 ml (5 floz/⅔ cup) whipping cream
1 level tablespoon butter
1 level tablespoon flour
salt, pepper, cayenne, sugar

Remove and shred peel from 2 of the oranges. Boil it in water to cover for 3–5 minutes, then drain. Put juice of the three oranges into a pudding basin or the top of a double boiler, over simmering water. Add wine and yolks. Bring cream to boiling point. Whisking vigorously, pour the cream on to the egg mixture. When the ingredients are amalgamated, stir the sauce with a wooden spoon until it is creamy – about 5 minutes.

Mash butter and flour to a paste. Add little bits to the sauce, stopping when it is nicely thickened. Keep stirring. Season, adding a good pinch of cayenne and a little sugar to bring out the flavour. Put in the orange strips to reheat.

THE CARDINAL'S ORANGE
ARANCIO CARDINALE

This excellent way of serving an orange with salt makes me think of those great Quattrocento banquets, given by churchmen and princes alike, in which oranges were used as an indication of luxury. Oranges, at that time sour oranges, accompanied everything.

The modern Cardinal's orange is a Roman speciality of today's restaurants. Score the peel down in sections, so that it can be bent out from the fruit like petals. The segments are then gently opened out from the centre, to make an inner 'flower'. Season with salt and a little olive oil.

Another way, after bending back the peel, is to remove the whole orange. Slice it across, season with salt and olive oil. Put the slices back into shape, and place in the centre of the peel again.

ORANGE AND OLIVE SALAD

Our Italian son-in-law pointed out to me how well salt brings out the flavour of citrus fruits, and that around Rome such salads as this are common. The olives to use are the small wrinkled Greek olives, or the tiny Niçois olives.

6 large oranges
4 tablespoons olive oil

1 tablespoon wine vinegar
salt, pepper
2 heads chicory (optional)
125 g (4 oz/¾ cup) black olives

Grate the peel from two oranges. Mix it with the oil, vinegar and a little salt. Peel the oranges, removing as much pith as possible, then slice them thinly and spread them out in a single layer on a large dish or tray. Pour the dressing over them. Grind on plenty of pepper, it should cover them like a very thin layer of sand. Leave for 2 hours in a cool place, or longer.

To arrange the salad, separate the leaves of the chicory and rinse them. When they are well drained, arrange them round the edge of a dish, like the rays of the sun. Lift the orange slices from their juice, and arrange them inside the chicory. Leave a space in the centre to put the olives. Sprinkle a little of the orange dressing and juices over the salad, but not too much. Serve lightly chilled.

ORANGE AND ONION SALAD

When the large Spanish onions come in, the fine sweet onions, make an orange salad as above, without the chicory and with fewer olives. Cut four or five slices from the centre of a large onion, separate out the rings and arrange over the salad.

SPINACH IN ORANGE CUPS

An attractive way of serving spinach with roast lamb, duck or veal, with boiled and glazed ham, or with a *fricandeau* of veal. If you use mandarins or satsumas, they look pretty and delicate with the turbot or brill *à l'orange* on p. 260. Most of the preparation can be done in advance, with only the reheating for the last moment.

It is difficult to be exact over quantities; allow a medium orange, or 2 mandarins or satsumas or clementines, for each person. When serving 6 to 8, buy 2 kg (4 lb) spinach or 1½ kg (3 lb) frozen leaf spinach. If there is too much, it can be kept for another meal or soup.

Wash and pick over the spinach, cutting away tough stalks. Either cook it in its own juices, or in a huge pan of boiling salted water (each system has its merits). Drain very well indeed in a colander, then chop and season with salt, pepper and nutmeg.

Slice the stalk end from the oranges. Scoop out the pulp with a pointed metal spoon, and put it into a sieve over a basin. Press it down to extract the juice. If you have the time, cut the orange cup edges into a pointed Van Dyck affect. Arrange them on an ovenproof dish or metal pan.

Not long before the meal, warm the orange cups in a low oven. Reheat the spinach with butter and cream, according to your tastes and the demands of the main item. Correct seasoning and when the creamed spinach is thick, flavour with orange juice. Keep the rest of the orange juice for the sauce or reduced liquor of the meat or fish.

Pile the spinach into the orange cups, and arrange them round the meat or fish on its serving plate.

VICHYSSOISE CARROTS WITH ORANGE

Allow one medium-sized carrot for each person, plus one or two extra to be on the safe side. Top and tail and scrape or peel them thinly according to their youthfulness. Cut them in half across, then cut each piece down several times, so that it falls into wedge-shaped batons.

Put the carrot pieces into a heavy pan with a lump of butter and a level tablespoon of sugar. Cover with water – Vichy water is correct, but really it makes no difference to the final flavour. Bring to the boil, take the lid from the pan and boil steadily until the carrot is cooked and the liquid reduced to a spoonful or two of buttery yellow juice. Watch the pan and test the carrots, so that you can slow the cooking if they are not tender. Squeeze in the juice of one or two large oranges (depending on the amount of carrot you cook). Turn the panful thoroughly and boil again until hardly any juice is left. Be careful that they do not burn or brown. Season the carrots with salt and pepper.

To serve, put the carrots in a dish, or round the meat etc, and scatter them with a little parsley and some chopped orange peel, cut thinly from the squeezed orange halves.

TURBOT OR BRILL WITH ORANGE
TURBOT OU BARBUE À L'ORANGE

Orange with fish was once as popular as lemon is nowadays. Sweet oranges did not come in until the 1660s and for quite a time after that 'an orange' in cookery books continued to mean a Seville type of bitter orange. It was the sweet orange that needed characterizing with such names as Portugal or China orange. The extra aromatic peel and flavour of the Seville type made it appreciated above the lemon, for sauces with meat and fish.

Although this dish seems French, it is one I have adapted from English recipes of the eighteenth century, and it can be used for other firm white fish. Quantities will serve four as a main course, six if meat is to follow, or if the rest of the meal is abundant.

1 kg (2 lb) turbot or brill
¼ litre (8 fl oz/1 cup) dry white wine
seeds of three cardamom pods (optional)
2 large oranges
1 Seville orange or half a lemon
250 ml (8 fl oz/1 cup) double cream
sugar, salt, pepper, cayenne
60 g (2 oz/¼ cup) butter, diced

Put fish into a sauté pan that fits it as closely as possible. Pour on the wine and add cardamom seeds. Add water to cover (if you have unsalted fish or light chicken stock to hand, that can be used instead of water). Remove the zest thinly from the oranges. Cut a neat bundle of julienne strips from the best orange strips: simmer in water for 3 minutes, then drain and keep for the garnish. Put the remaining peel into the pan, bring to simmering point and cover. Keep heat low so that the water barely moves. When the fish is still pink at the bone, remove it to a serving dish and keep warm in a very low oven.

Strain the liquor into a wide, slope-sided pan and boil it down hard to about 375 ml (12 fl oz/1½ cups). Add all the orange (and lemon) juices. Boil 2 minutes, then add the cream, and continue boiling to reduce the sauce to ½ litre (¾ pt/scant 2 cups) or a little more. Add seasoning to taste, then off the heat whisk the butter into the sauce. Pour over the fish, arrange the julienne strips on top and serve.

Laver reheated with orange or spinach are the obvious vegetables – chicory or new potatoes are also good.

SCALLOPS WITH ORANGE LAVER SAUCE

Laver is a prepared seaweed, or rather two seaweeds – *Porphyra leucosticta* and *Porphyra umbilicalis* – much sold in west country fishmongers and in Wales. You can buy it by post from the food department of James Howell Ltd, of Cardiff. It has a slightly spinach-like appearance, dark dark green with iodine coloured overtones, and does not taste at all 'fishy'. It's a standard accompaniment to Welsh lamb, when heated through with orange as in this recipe, and with salmon. Mixed with oatmeal and fried in cakes, it makes a good breakfast dish with bacon.

With scallops, it is a delight.

As an alternative to puff pastry shells, you could serve brown rice, or toast.

375 g (12 oz) puff pastry
500 g (1 lb) laver-bread
2 large oranges
18–24 large scallops
butter
175 ml (6 fl oz/¾ cup) dry white wine
salt, pepper, cayenne
250 ml (8 fl oz/1 cup) double or whipping cream
thin orange slices to decorate

Bring home from the fishmonger 6–8 deeply curved scallop shells. Scrub and dry them, remove any tiny barnacles, and oil them well on the *outside*. Roll out the puff pastry thinly, without stretching it, and cover the outside of the shells, bringing the pastry slightly round the hinge to tether it. Prick the pastry with a fork. Chill thoroughly and bake at gas 8, 230°C (450°F) until nicely coloured and crisp. Remove from the oven, cut the pastry neatly at the hinge (unless it has come away of its own accord) and ease from the shells. This can all be done in advance, the pastry stored in air-tight containers and reheated before serving.

In a small pan, mix the laver-bread with the grated rind and juice of one orange. It should be reheated when the scallops are ready. Remove the corals from the scallops, and trim away the tough bit of muscle. Fry the scallops quickly in butter until lightly browned, adding the corals towards the end (they should not be browned). Lower the heat and pour in the juice of the second orange and white wine. Cook the scallops, turning them, until they are barely cooked and remove them from the pan to cool. Season them.

Boil down the scallop juice to an essence and stir in the cream. Season with salt, pepper and cayenne. Reduce again to about 250 ml (8 fl oz/ 1 cup).

Slice the scallops across according to their size. They should be creamy and barely opaque, with a hint of transparency at the centre. Don't slice the corals.

For the final assembly, reheat pastry and laver-bread. And reheat the scallops and corals gently in the cream sauce.

If you are sure of your audience, put some laver sauce in the base of each pastry shell. If not, put it into a separate bowl. Put the white scallop meat into the shells, arrange the corals on top, and spoon over a little sauce. Decorate with quarter-slices of orange.

Note If you use frozen scallops for this dish, choose large ones and thaw them completely on a pierced plate, so that the liquid flows away. Pat them dry in kitchen paper.

Left-over laver-bread sauce can be mixed with porridge oats until just stiff enough to handle. Form into cakes, and fry in bacon fat with bacon rashers, to eat at breakfast or supper.

CAPON WITH APPLE AND ORANGE COMPOTE

CHAPON AUX POMMES

A fine recipe from the Polish section of Audot's *La Cuisinière de la Campagne et de la Ville*, which first appeared in 1818 with the intention of helping the housewife to achieve simplicity with good living. The last edition, the sixty-fifth, came out in 1887. To the main body of the book, arranged normally according to classes of food, soups, poultry, game etc, are added foreign sections: 'I have profited from my journeys outside France to give readers an idea of the taste of gastronomy in other countries, acquainting them with unknown dishes, and giving them the ability to remind foreign visitors of some of their national delights.' A charming thought. I observe that his foreign sections begin with the cookery of Provence and Languedoc. The English part consists of meat and game with an odd mince pie recipe, and then *Sandwichs* (sic). The author remarks that other English recipes, including 'la célèbre mock-turtle, les puddings, cakes . . . le célèbre roastbeef' are given elsewhere in the book. He was much struck by our use of gooseberries – six recipes to seduce the French housewife into taking advantage of this useful fruit that comes at a season when there is little other fruit around.

Audot's capon was wrapped in a large and solid piece of paper. We use a roastabag or foil.

1 large capon
4 thin rashers green streaky bacon
1 large chopped onion
2 chopped carrots
half a lemon, thinly sliced
salt, pepper, 3 cloves
175 g (6 oz/¾ cup) sugar lumps or granulated sugar (see *recipe)*
2 large oranges

10 sharp eating apples
kasha (recipe follows)

Wrap up the capon with the next four items, plus the seasoning including cloves. Cook in your usual way, or at gas 7, 220°C (425°F) for about 2 hours, until tender. Check after 1¾ hours. With smaller chickens, reduce the time accordingly.

Meanwhile, rub sugar lumps over the oranges to remove the zest. Or use a zester. Put zest, sugar and juice of the oranges into a wide pan. Peel, core and cut the apples into wedges, turning them in the orange juice as you go, to prevent discolouring. Simmer until apples are tender, turning the wedges so that the cooking is even.

Arrange the apple pieces round a warm serving dish and put the capon in the centre. You may decorate with a slice or two of lemon and the bacon cut up, if they have survived the cooking in reasonable enough shape for a garnish.

Strain and degrease the capon juices, and add extra seasoning as required. This makes the sauce, which can be poured over the kasha if you serve that as well.

KASHA: lightly cook two teacups of kasha – buckwheat groats from a health-food shop – in some oil in a pan, stirring it about. Pour in four teacups of water, add salt. Leave to cook until the kasha begins to swell and thicken. Cover the pan, remove from the heat and wrap the whole thing in thick towels until required. Stir in a good knob of butter and correct the seasoning.

No towels handy? Then finish off the kasha in a low oven until it is split and fluffy looking.

Enough for 8–10 people.

DUCK BIGARADE
CANARD/CANETON À L'ORANGE

A dish of January and February when the Seville oranges come in. At other times of the year, use 2 large sweet oranges and one lemon instead: the taste is not the same, although perfectly agreeable.

The business with the duck before you start cooking it sounds laborious. In fact, like bread-making, it is no more than a matter of organizing a few simple procedures into your normal routine. The end result is a wonderfully crisp skin with a good flavour. I now always roast duck in this way.

1 large or 2 small ducks
2 tablespoons (3 tbs) gin/vodka/whisky
1 heaped tablespoon honey
1 trimmed leek, or four spring onions

SAUCE
2 level tablespoons (3 tbs) butter
3 level tablespoons (4 tbs) flour
150 ml ($\frac{1}{4}$ pt/$\frac{2}{3}$ cup) dry white wine
$\frac{1}{2}$ litre ($\frac{3}{4}$ pt/scant 2 cups) duck giblet stock or other poultry stock
3 Seville oranges, or 2 large sweet oranges and 1 lemon
salt, pepper, sugar
extra knob of butter
orange liqueur (optional)

Rub the duck over with the spirit. Tie string round the thighs, leaving enough over to suspend the duck from a broom handle laid across two chairs, or from a convenient drawer knob. Put a large dish underneath to catch drips. Train the cold blast of a convector heater on to the duck for an hour. Alternatively, rig up a hair-dryer. If the weather is blowing a dry gale from Siberia, hang the duck outside in the wind, out of the reach of cats.

Put the honey and leek or onion into a self-basting roaster or other oval pan. Add enough water to give a 4–5-cm ($1\frac{1}{2}$–2-inch) depth. Bring to the boil. Using the string put the duck into the liquid and leave for 30 seconds. Turn and give the other side 30 seconds. Put back to blow dry again, for a further hour. The skin will look very smooth, with a faint sheen. This can be done a day in advance, but keep the duck suspended and swinging free.

Put the duck on a rack, breast side up, over a roasting pan. Put a centimetre of water ($\frac{1}{4}$ inch) into the pan. Roast the duck for an hour at gas 5, 190°C (375°F). Turn it over, lower the heat to gas 3, 160°C (325°F) and cook for 30 minutes. Finally turn the duck back to its original position, and, if the skin is not too dark, return the heat to gas 5, 190°C (375°F), for 20–30 minutes, until the duck is cooked. Should the skin be very brown, keep the heat at gas 3, 160°C (325°F) for the last part of cooking time.

If you are cooking two smaller ducks, reduce the various cooking times in proportion to their weight.

When you finally take the duck from the oven, you will find that the skin is puffed away slightly from the flesh. With a pair of scissors, cut down the centre, then remove each side in squares. Put round the hot serving dish. Carve the meat and put it in the centre. Pour some of the sauce over the meat only and serve the rest in a sauceboat. Watercress or chicory and a few potatoes can be served as well.

To make the sauce, heat the butter until it's nut brown. Stir in flour off the heat, cook for two minutes to make a pale brown roux. Add the wine and stock, whisking it into the pan to prevent lumps. Return the mixture to the boil, skim and leave to cook down to a good moderately thick consistency. Peel oranges (and lemon if used) thinly, and cut the pieces into matchstick strips. Boil 3–5 minutes until tender in enough water to cover them. Drain. Squeeze out the orange (and lemon) juice. When sauce is ready, pour in the juice, season to taste and put in the peel shreds to heat through. Off the heat, stir in the butter and liqueur to taste.

LAZY WILD DUCK WITH ORANGE

A splendid recipe from Julia Drysdale's *Game Cookery Book*, a book I treasure as I do not cook game so very often and need to remind myself what to do. About this recipe, she is apologetic. In fact, it is one of the best ways with wild duck that I know. It can be used for guineafowl, too.

The low oven temperature is to prevent the marmalade catching. I often start the cooking at gas 7–8, 220–230°C (425–450°F), and turn the heat down to moderate, gas 4, 180°C (350°F) after 15 minutes. If you like wild duck pink, 30–35 minutes is enough. If you prefer it well done, and follow Mrs Drysdale's temperature, allow 45–60 minutes. Baste often and keep checking. Here is the recipe:

Mash a large lump of butter with some marjoram and parsley and put it inside the duck, with half a sweet orange cut in three wedges. Spread the duck with butter, then with marmalade, generously like breakfast toast.

Squeeze the juice of 2 sweet oranges into the cooking pan, add a little stock and the duck. Cover loosely with foil and place in the oven at gas 4, 180°C (350°F). As you baste, check the liquid level and add more juice, water or stock if it is drying up. Should you be dealing with an older duck, put it in a casserole with a well-fitting lid, and cook it for longer at a lower temperature, gas 3, 160–170°C (325°F).

Ten minutes before the duck is ready, remove foil to brown the breast. Keep basting. When the bird is done, remove it to a serving dish. Put the pan of juices over the heat, and scrape in all the bits and pieces that may have stuck. Taste and see what extra seasoning – sugar, for instance, or orange liqueur, even a spot more marmalade – is needed. Strain into a clean pan. Pick out of the strainer a few of the best bits of orange peel from the marmalade, and add them to the strained juice. Heat up, reduce briefly, and off the heat whisk in a good knob of butter, in little bits, to freshen the taste.

Serve with game chips or straw potatoes, watercress, and some coarsely grated carrots stewed in their own juices with a bit of butter until cooked but slightly crisp.

ORANGE-FLOWER SALAD

SALADE D'ORANGE AUX FLEURS D'ORANGER

A fresh simple way of serving oranges, in which the delicate air of the flowers hangs over the taste of the fruit.

Allow a large orange for each person, plus two or three extra.

If you have a zester, remove the peel from two or three of the oranges, and set it aside. Otherwise remove the peel thinly, cut it into shreds and simmer them for a few minutes in water until they are tender: then drain.

Peel the oranges, removing as much pith as possible. Slice them, discarding the pips and the end slices as you go. Arrange the slices overlapping each other in concentric circles in a shallow dish.

Mix together the juice left from the slicing, and two tablespoons (3 tbs) of water plus 3 tablespoons (scant $\frac{1}{4}$ cup) of orange-flower water for every six oranges. Pour over the fruit. Sand it evenly with icing sugar. Cover and leave in a cool place – not the refrigerator – and baste the slices with their juice from time to time. Taste it and see if more sugar, or more orange-flower water is needed (but go carefully, orange-flower water is powerful). Leave for 2 hours at least, until next day if you like. Scatter with the peel before serving. Vanilla or orange-flavoured sugar thins (p. 461) can be served with the slices, or almond biscuits.

If you have the luck to live where oranges are grown, add a few orange flowers to the dish and tuck some of the dark leaves round the side.

CARAMEL ORANGES

ARANCI CARAMELLIZZATI

In Venice, and in many Italian restaurants elsewhere, you will find a bowl of caramel oranges among the desserts. They have become a cliché, so you choose something else very likely. But making and eating them at home is another matter.

Allow at least one seedless orange per person. Remove the peel thinly from half or two-thirds of the oranges, and cut the strips into julienne shreds.

Next, peel all the oranges to the quick – *see* the buying and preparing section – over a plate to catch the juice. Slice them across and put them back into shape, with a cocktail stick through the centre to hold them together. Lay them in a serving bowl, with the juice from the plate.

For 6–8 oranges, bring 300 g (10 oz/$1\frac{1}{4}$ cups) sugar slowly to the boil with 150 ml ($\frac{1}{4}$ pt/$\frac{2}{3}$ cup) water and a small cinnamon stick. Stir to dissolve the sugar before the mixture boils. Then leave it to cook to a rich golden-brown caramel.

Have ready 150 ml ($\frac{1}{4}$ pt/$\frac{2}{3}$ cup) very hot water. Wrap one hand in a cloth and remove the pan from the heat. With the other, pour in some water gradually – it will spit furiously – then stir with a wooden spoon, and continue until the water is used up. If the caramel is lumpy, do not worry. Just put it back over a low heat and stir until it is smooth. Pour most of the syrup, with the cinnamon stick, over the oranges, but keep back just enough to cook the shreds of peel in. They should end up as candied wisps. Distribute them over the oranges just before serving them.

Note If you cover the bowl with cling film and keep it in the refrigerator, the dish can be made a day in advance without losing freshness. Keep the peel back for serving, so that you can turn the oranges from time to time to keep them looking juicy.

ANDALUSIAN TART
TARTE ANDALOUSE

The combination of aromatic eating apples with orange is remarkably – surprisingly – successful. I tried the recipe with misgivings when I saw it in a French newspaper, but now it is a favourite of ours.

Start by making a sweet shortcrust pastry, in which you have included the finely grated zest of a large orange (the orange will be needed later on for the filling. Let it rest, then line a tart tin of 23 cm (9 in) and bake it blind for 10 minutes at gas 6, 200°C (400°F): it should be firm and creamy-coloured still, rather than brown.

500 g (1 lb) reinettes or Cox's or Blenheim Orange
sugar
2 large oranges, seedless if possible
3 tablespoons (about $\frac{1}{3}$ cup) apricot jam
juice of half a lemon

Peel, core and slice the apples. Boil peelings and cores in $\frac{1}{4}$ litre (8–9 fl oz/

1 cup) water to flavour it. Strain the water into a shallow pan – there should be about 125 ml (4 fl oz/$\frac{1}{2}$ cup), but precision is not important; all that is needed is enough liquid to prevent the apples from burning as they cook.

Put the apple slices into the pan and cook them gently at first, turning the pieces. As they begin to soften, raise the heat to evaporate the juice as much as possible. Remove from the stove when the apples are tender. Mash them well and sweeten to taste with about 75 g ($2\frac{1}{2}$ oz/good $\frac{1}{4}$ cup) sugar.

Meanwhile, peel the oranges to the quick with a sharp knife and slice them thinly, discarding any pips.

Spread the apple purée into the pastry case. Put the orange slices on top. Bake for about 15 minutes at gas 6, 200°C (400°F), giving the fruit flavours a chance to blend, and the pastry time to brown nicely.

Have ready a thick apricot glaze made by boiling together the jam and lemon juice, then sieving it. Brush over the tart when you take it from the oven. Serve hot, warm (which is best) or cold.

CRÊPES SUZETTE

Crêpes Suzette is one of the most famous of restaurant dishes. It is also one of the simplest – though you might not think so from the showy business of flaming the sauce (or from the bill).

Like other well-known dishes, it has acquired its myths of origin, but it seems as if the inventor may have been the chef Henri Charpentier, who was born in Nice in 1880. By his middle teens, he was working at the Café de Paris in Monaco. One day the Prince of Wales (later Edward VII) came in with several friends to have lunch. They were all men, except for a little girl who was called Suzette. Forewarned of the royal party's arrival, Charpentier had prepared a special new sauce for the pancakes: it was a grand version of the simple orange sauce made by his peasant foster-mother, who was a splendid cook in spite of her poverty.

As Henri Charpentier was nervously reheating the pancakes in the sauce beside the Prince's table, the whole thing caught fire. He tasted the result, found it delicious beyond belief and recklessly added some more of his mixture of liqueurs. It flamed again, and the pancakes were served.

The Prince, being in M. Charpentier's opinion 'the world's most perfect gentleman', ate his pancakes with a fork, then spooned up the sauce enthusiastically, and asked the name of the dish. 'Crêpes Princesse,' replied the astute young man.

The Prince recognized that the name was ruled by the feminine style normally used for recipes, and that the compliment was meant for him. He protested with mock ferocity that there was a lady present. She was alert and rose to her feet. Holding her skirt wide with her hands, she curtsied. 'Will you change Crêpes Princesse to Crêpes Suzette?' he asked.

Charpentier tells the story in his autobiography, and gives the recipe.

SAUCE SUZETTE
1 heaped tablespoon (2 tbs) vanilla sugar
2½ cm (1 inch) square orange peel
2½ cm (1 inch) square lemon peel
5 tablespoons (6 tbs) Maraschino
5 tablespoons (6 tbs) kirsch
5 tablespoons (6 tbs) Curaçao
125 g (4 oz/½ cup) unsalted butter, diced

CRÊPES CHARPENTIER
3 eggs, beaten
2 rounded tablespoons (6 tbs) flour
60 ml (2 fl oz/¼ cup) water
60 ml (2 fl oz/¼ cup) milk
pinch salt

Start a couple of days in advance by putting the sugar and peels for the sauce into a screwtop jar. Mix together the liqueurs.

Next make the pancakes by beating the ingredients together in the order given. Using a 20-cm (8-in) omelette pan, heated and buttered, cook 8 pancakes. Fold each one in half, then in half again. Arrange on a dish.

Now you are ready for the flaming part, which can be done discreetly in the kitchen unless you have a flair for culinary swank.

Melt the butter in a presentable pan, over the cooker or table burner. When it melts and bubbles, stir in half the liqueur mixture. Tilt the pan so that it flames. When the fire dies down, put in the pancakes and the flavoured sugar, minus the peel squares. Turn them over in the sauce to reheat. Add the remaining liqueur and set it alight again. Take flaming to the table, if you have been working in the kitchen unobserved.

Note Crêpes dentelles, p. 465, can be used instead of the Charpentier pancakes. For the liqueurs in the sauce recipe, you can substitute 9 tablespoons (scant ¾ cup) orange juice, 3 tablespoons (scant ¼ cup) each of lemon and brandy, but the sauce won't flame.

ST CLEMENT'S SOUFFLÉ GLACÉ

Turn to p. 287, and for the passion fruit juice, substitute juice and rind of one large lemon, one large orange and two tangerines. Use juice of two large oranges for the sauce.

ORANGE CUSTARDS

1 Seville orange
1 tablespoon brandy or orange liqueur
125 g (4 oz/½ cup) caster sugar
2 eggs and 2 egg yolks
½ litre (16 fl oz/2 cups) whipping cream
8–10 pieces candied orange peel

Simmer the rind of half the orange in water to cover for 15 minutes. or until fairly tender. Drain and liquidize or process to a smooth cream with alcohol, juice of the orange, sugar, eggs and yolks. Bring the cream to boiling point, then add gradually. Put in extra sugar, or a pinch of salt, if the flavour requires it.

Pour into 8 or 10 small soufflé dishes or custard cups or little cream pots (the kind with lids). Stand them in a roasting pan half full of very hot water, and put into the oven at gas 2–3, 150–160°C (300–325°F), for about 30 minutes until they are just set. At the lower temperature, they may need 35 minutes if the pots are deep rather than wide. Put a thin-bladed knife into one of the creams to test. If it comes out clean, or with a very slight creaminess, the custards are done (they will firm up as they cool, which you should bear in mind if you wish to serve them chilled). Take the pots or dishes from the water and, when cold, store overnight or several hours until chilled. Put a piece of candied orange peel in the centre of each one, and a tiny bay leaf if you have your own tree.

ORANGES CONFIT

Once when we stayed at the Hôtel de la Gare at Digoin, before Monsieur Billoux acquired his second star, we were served these semi-candied oranges with the chocolate mousse. They were proper oranges, but small ones, and you ate the whole thing.

They are not at all difficult to make, and can be kept in the refrigerator for several days. Choose oranges with tightly fitting skins. Knock off any hard bits at the stem end, and gouge lightly down the skin with a canelle knife at regular intervals to expose the pith beneath the colour. They should end up looking like a slashed puff sleeve on some Elizabethan court dress.

Put them into a pan, with a lid on top to keep them down, and cover with water, plenty of it. Simmer until tender, but remove from the water before the slashes burst (if one or two go, the damage can later be concealed by displaying the orange the other way up). This takes from 35 minutes.

Drain the oranges well, then simmer them in a light sugar syrup to cover until they look half-candied and syrupy – about 45 minutes. The sugar in the syrup hardens the fruit, so that any damage is unlikely to increase.

Cool in the syrup. Put into a dish, with the syrup, cover with cling film and chill until needed.

ORANGE AND CARDAMOM ICE CREAM

This lovely recipe of Josceline Dimbleby's is served with an orange-blossom and walnut syrup that goes well, too, with orange soufflé glacé and other ice creams. For this recipe, do not shudder at the idea of using concentrated orange juice. I found out years ago that it works well with ices and am grateful to the American book which first put me on to it. As a drink it may be unconvincing, but tamped down with other substances – used as a flavouring that is – it can be most useful.

1 can frozen concentrated orange juice
3 large eggs
½ teaspoon salt
175 g (6 oz/¾ cup) caster sugar
5–6 cardamom pods
300 ml(10 fl oz/1¼ cups) double or whipping cream

SAUCE
125 g (4 oz/½ cup) caster sugar
2 generous tablespoons (3 tbs) orange-flower water
1 tablespoon lemon juice
30 g (1 oz/¼ cup) walnuts

Remove the orange juice from the freezer well in advance to thaw: it need not be completely liquid. Whisk the eggs with the salt, using an electric beater. Dissolve the sugar with 6 tablespoons (½ cup) of water, and when clear bring to the boil and boil hard for three minutes without stirring. Pour the syrup on to the eggs, continuing to beat. Leave the beater switched on until the mixture has thickened and increased in volume.

Remove the seeds from the cardamom pods, and crush them. Add them to the egg mixture with the slushy orange juice. When perfectly amalgamated, whip the cream and fold it into the mixture. Freeze in the usual way (p. 473).

To make the sauce, dissolve the sugar in 4 tablespoons (⅓ cup) of water, stir until clear. Bring to the boil and boil hard for one minute. Take from the heat and add orange-flower water and lemon juice. Pour boiling water over the walnuts and remove their fine skin. Chop and add to the strained sauce. When you turn out the ice cream, pour the sauce over it just before serving.

TANGERINE ICE

MANDARINES GIVRÉES

After the French Cookery School series appeared in the *Observer* Magazine, there was a competition for the best dinner party menu chosen from Anne Willan's dishes. I think this was the favourite dessert. By the time we had looked through the hundreds of entries, we felt like screaming at the words *mandarines givrées*. But it really is a delightful way of presenting orange ice of any kind, and a most useful recipe to know so that you can adapt it to other citrus fruit.

The quantity of ice below is enough to fill 8 or 10 tangerine shells. The whole thing can be frozen for storage.

20 tangerines
juice of half a lemon
100 ml (3½ fl oz/good ½ cup) water
icing sugar (optional)

SYRUP
250 (8 oz/1 cup) sugar
150 ml (5 fl oz/⅔ cup) water
juice of ¼ lemon

First make the syrup. Heat the three ingredients over a low heat, until they are dissolved together and clear. Bring to the boil and boil for 2–3 minutes. Cool.

Grate the zest from half the tangerines, and squeeze the juice from them. Cut the remaining tangerines so that they each have a lid. If they are wobbly, shave a little peel from the base. Squeeze these tangerines too, being careful not to damage the skin: I find it easiest to scoop out the sections with a small spoon and then press them. This leaves the skins in good nick. You should end up with 750 ml (1¼ pt/generous 3 cups) juice. Add the grated zest, the lemon juice and the 100 ml (3½ oz/scant ½ cup) water to the juice, with the syrup. Taste and add icing sugar or extra lemon juice, if more sweetness or sharpness is required. Freeze until firm.

Attend to the shells next. If you squeezed them directly, scoop any pulp out and push them gently back into shape. If you scooped out the flesh, check to make sure no debris has been left behind. Chill the shells and fill them with the ice, which should have been well beaten. Replace the lids and store in the freezer, wrapped in film.

Filled oranges, tangerines and lemons look pretty served on a bed of bay leaves, on a white or ivory dish.

ORANGE HALVA CAKE

I cut this recipe out of a newspaper years ago, and find that it gives a better result than versions found in current books of Greek cookery. Make the cake two days in advance, so that the syrup can soak right in.

175 g (6 oz/¾ cup) caster sugar
175 g (6 oz/¾ cup) butter
2 teaspoons grated orange rind
5 tablespoons(6 tbs) orange juice
3 eggs, beaten
275 g (9 oz/2¼ cups) fine semolina
3 level teaspoons baking powder
125 g (4 oz/1 cup) ground almonds

SYRUP
275 g (9 oz/good cup) caster sugar
8 tablespoons (⅔ cup) water
5 cm (2 in) piece cinnamon stick
3 tablespoons (¼ cup) lemon juice
6 tablespoons (½ cup) orange juice
2 heaped tablespoons (3 tbs) chopped candied orange peel

Cream sugar and butter together, add rind, juice and eggs. Then mix in the semolina with the baking powder, and the ground almonds. Put into a large buttered ring mould – there should be plenty of room for the mixture to rise. If you do not possess a ring mould, use a large cake tin with a small jam jar stuck in the middle (butter the jar as well as the mould).

Bake for 10 minutes at gas 7, 220°C (425°F), then lower the heat to gas 4, 180°C (350°F) for a further 30 minutes. Test the cake in the usual way.

When the cake is nearly cooked, make the syrup. Simmer the first three ingredients together for 5 minutes, add the juices and peel.

Turn the cake out carefully – ease it first – on to a warm plate. Reheat the syrup to boiling point, and after a couple of bubblings, pour it slowly over the cake. It will slowly be absorbed. Cover the whole thing with foil and leave in a cool place for 2 days. If the cake seems on the dry side, make up some more syrup in half quantities. Serve with clotted cream and a scatter of toasted almonds.

ORANGE SYRUP CAKE

One of the best variations on the pound cake theme. The crunchy top and clear orange sharpness take away all need to dress up the cake with butter cream or glacé icing. For busy people, it is the perfect solution – speed with honour.

rind of an orange (see *recipe)*
200 g (6 oz/¾ cup) sugar (see *recipe)*
250 g (8 oz/1 cup) soft butter, cut in pieces
250 g (8 oz/2 cups) self-raising flour
100 g (3½ oz/scant ½ cup) candied orange peel, chopped
1 level teaspoon baking powder
good pinch salt
4 large eggs
2 tablespoons (3 tbs) orange liqueur or orange juice

SYRUP
juice of an orange
150 g (5 oz/⅔ cup) caster sugar
slice of candied orange or peel for the centre

If you have a processor, put the curls of peel into the bowl with *granulated* sugar. Whizz at top speed until the peel turns to bright orange specks.

If you use an electric beater, grate the rind from the orange with a zester or grater, or cut it off thinly and chop it fine. Put into the bowl, with *caster* sugar.

Put the butter in with the sugar and rind. Using 1 tablespoon of flour, toss the candied peel in it. Set the peel aside, returning any surplus flour. Sift the flour, baking powder and salt into the bowl. Break in the eggs and pour in the liqueur. Switch on or beat until the mixture is smooth. Fold in the peel.

Turn into an 18-cm (7–8-in) cake tin with Bakewell paper. Smooth the top down in the centre. Bake at gas 3, 160°C (325°F) for about 1½ hours. Test after 1¼ hours. The cake should be a light orangey brown on top, and contracting a little from the sides of the tin.

Have ready the juice from the syrup ingredients mixed with the sugar. Remove the cake wrappings and put it into a shallow dish that will hold the syrup at first (it soon disappears into the cake). Spoon the syrup over the hot cake evenly and leave it to cool. Put orange slice or peel in the centre.

ORANGE AND GINGER CAKE

A cake to have ready well before Christmas, then you have time to feed it with alcohol. An important item of the ingredients is candied orange peel: the only way to be sure of getting it, if you live outside London, is to buy those boxes containing large pieces of mixed peel – two should be enough. Pick out the orange sections, and cut them up, keeping one piece aside to decorate the cake when it has been iced. If you candy your own peel, recipe page 198, you will have no problem.

6 knobs of ginger in syrup, drained and chopped
125–175 g (4–6 oz/⅔–1 cup) candied orange peel, chopped
125 g (4 oz/1 cup) blanched, slivered almonds
60 g (2 oz/½ cup) shelled walnuts, coarsely chopped
about 625 g (1¼ lb/3⅓ cups) mixed seedless raisins, sultanas and currants
300 g (10 oz/2½ cups) plain flour
1 rounded teaspoon ground ginger
1 level teaspoon grated nutmeg
grated rind of an orange and a lemon
250 g (8 oz/1 cup) lightly salted butter
250 g (8 oz/1 cup) soft light brown sugar
4 tablespoons (⅓ cup) ginger syrup
5 size-3 eggs

½ level teaspoon bicarbonate of soda
1 tablespoon of milk
Cointreau or other orange liqueur

Put the ginger on the scales with peel and nuts. Add mixed fruit to produce a total weight of just over 1 kg (2¼ lb). Put them into a large bowl and add half the flour, the spices and rinds. Turn everything about with your hands, breaking up lumps of fruit, and making sure that the flour is thoroughly distributed.

Using an electric beater if possible, cream butter, sugar and syrup. Add the eggs one by one, then the rest of the flour and the mixed fruit. Dissolve the soda in the milk and stir it in, adding orange liqueur to produce a dropping consistency.

Have ready a 20-cm (8-in) cake tin, lined with greased Bakewell or other vegetable parchment. Round the tin tie a treble thickness of brown paper, coming about a centimetre above the top of the tin, or slightly more. Fold a wodge of brown paper, or newspaper, and place it on a baking sheet. Stand the tin on top.

Pour in the cake mixture, making a central depression. Bake at gas 1, 140°C (275°F) for 3 hours. Test with a skewer, and be prepared to give the cake another 15 or 30 minutes. A warm skewer pushed in diagonally, to the bottom, should come out clean.

Leave the cake to cool in its tin. Turn it out and peel away all paper. Put the cake upside down on a sheet of greaseproof placed on a sheet of aluminium foil. Wrap the greaseproof round the cake, then enclose in foil. Every week or so, open the parcel to expose the under side of the cake. Make holes with a skewer and pour in 2 tablespoons (3 tbs) of orange liqueur.

Four or five days before Christmas, cover it with almond paste and then ice it.

ALMOND PASTE: sift 250 g (8 oz/1 cup) icing sugar and mix with double its weight in ground almonds (4 cups). Beat a large egg, add 3–4 teaspoons of lemon juice, and use to mix the paste. Turn it out on to a smooth surface sprinkled with icing sugar and knead it to coherence. Roll out half of the almond paste and cut a circle rather wider than the top of the cake. Roll out the rest and cut wide bands to cover the lower part.

ORANGE GLAZE: bring 2 tablespoons marmalade (use medium rather than coarse cut) to the boil with a tablespoon of water. Sieve, and brush over the cake top and sides – cut off the top of the cake first if it needs to be evened. Alternatively, keep the cake upside down, to benefit from the flat surface.

Turn the top of the cake down on to the circle of almond paste, and press up the extra rim. Turn the cake over, and press again to fix it. Roll the sides of the cake on the almond paste strips, cutting away any surplus. Push all edges together to make as smooth a join as possible. Leave two days.

TO ICE: turn to p. 468 and make the glacé icing, with orange juice to moisten the sugar. Stand the cake on a rack and spread the warm icing over it smoothly. Allow to set or almost set, before decorating with pieces of ginger and orange peel.

WHOLE ORANGE MARMALADE

The simplest, easiest and best-flavoured marmalade. You can adapt the recipe to a mixture of three fruits, lemon, orange and grapefruit for instance. And if you have a pressure cooker, you can cook the fruit in two separate batches and still save time.

Scrub $1\frac{1}{2}$ kgs (3 lb) Seville oranges. Put into a large pan with $3\frac{1}{3}$ litres (6 pt/15 cups) water. Simmer until skin is tender and easily pierced – this takes about $1\frac{1}{2}$ hours, and the water reduces in quantity. Take out the oranges, cool and halve. Leave water in the pan. Remove pips from the oranges and tie them into a piece of muslin or stockinette, leaving a string so that the bag can be hung from the pan handle and bob about in the cooking marmalade.

Either cut the halves into strips of the size you like, or put the halves batch by batch into the liquidizer or electric shredder. Be careful not to overdo the liquidizer or you will end up with pulp. Put back into the pan of water, and add 3 kg (6 lb/12 cups) warmed sugar. Stir until the sugar dissolves.

Bring to the boil, and boil vigorously for about 10 minutes until setting point is reached (drop a little on to an ice cube – a quicker and more efficient test than the usual cold saucer).

Leave for 15 minutes for the peel to settle. Stir the marmalade, discard the bag of pips, then fill hot sterilized pots and seal in the usual way.

Sometimes I pour in Cointreau, when the marmalade comes off the heat. It does give an extra zest, but I am not in favour of adding whisky. It's too strong.

Note If you use a pressure cooker, put half the whole oranges into the pan with 900 ml (1½ pt/3¾ cups) water. Bring to steaming, put on 15 lb weight and pressure cook for 15 minutes. Cool rapidly under tap, remove weight and lid. Tip out and repeat with remaining oranges and same quantity of water.

ST BENOÎT THREE-DAY MARMALADE

As we drove into St Benoît-sur-Loire one evening on a long journey home, we realized in panic that not only was it Saturday but that next day was May Day, 1 May, France's most total holiday. Luckily there was one last room at the Hôtel Madeleine. We woke to a brilliant May Morning, thought of our daughter listening to carols under the tower of Magdalen College at Oxford, and went down to breakfast. It was served in a little courtyard room full of air and sun, very ugly, but with a view of oleanders in tubs and lilies-of-the-valley on the table. With the croissants, they served marmalade, a most delicious marmalade. The peel was chewier than usual, not tough, but more resolutely there. The flavour was light, yet fully and sweetly orange. A cheerful marmalade. People passed by delivering trays of sandy-stalked asparagus. The proprietor hurried by and we praised the marmalade. She stopped and smiled. A minute later she came back with the recipe.

Sustained – as it happened, fortunately – by that marmalade, we sauntered round the sculptures in the church. Crowds flowed in for mass. We stayed and endured the longest, most inaudible sermon we have ever tried to listen to. People round us at the back rustled, then they waved their service papers. They booed genteelly. On went the distant voice. They melted away. So did we, in the direction of home. Perhaps the sermon continues, crackling through a dozen microphones, battering at the resigned staring creatures on the stone capitals.

6 large oranges
½ large lemon
sugar

Cut off and discard the thick ends of the oranges, then slice them. The same with the lemon. Soak for 24 hours in ½ litre (18 fl oz/2¼ cups) water. Bring to the boil and simmer gently for 25 minutes (in fact, I cook it for an hour, to soften the peel a little more). Take off the stove and leave for another 24 hours.

Weigh the fruit and juice and put back into the pan with 650 g (1 lb 5 oz/2⅔ cups) sugar to each 500 g (1 lb). Boil until setting point is reached, about an hour. If the oranges were full of pips, they will rise to the surface during the cooking time and can be skimmed away. If the oranges were seedless, you have only to look out for the lemon pips.

Pot in the usual way.

CLEMENTINES IN ARMAGNAC

The best of Christmas preserves, to be eaten reflectively at the end of a special meal as dessert – with cream if you like. Tiny clementines come in at the beginning of the winter season – later they are nearer a satsuma in size – and you need to keep an eye open for them. They are called after Père Clément, a French priest in North Africa, who developed this fine-flavoured cross between the mandarin and bitter orange.

1 kg (2 lb) tiny clementines
600 g (1¼ lb/2½ cups) sugar
1 vanilla pod
Armagnac

Prick fruit four or five times each with a darning needle. Put 1 litre plus 100 ml (2 pt/5 cups) water into a pan with the sugar and vanilla pod. Stir until sugar is dissolved over a low heat, then bring to the boil and boil steadily for 4 minutes. Put in the clementines and bring back to simmering point. Cover and leave to simmer for an hour. Remove any clementines that appear to be cooked and soft – try to haul them out before the skins crack (this doesn't matter from the taste point of view, but it slightly spoils the appearance). After another 30 minutes, all the clementines should be ready.

Put the hot drained clementines into hot dry jars of a reasonably attractive appearance that can be put on the table. Pour in enough Armagnac to come three-quarters of the way up the fruit. Boil the syrupy orange liquor hard to concentrate the flavour and sweetness, but stop before it becomes too bitter. Use to top up the jars. Divide the vanilla pod between the jars and push down among the clementines, which must be covered by the liquid. Keep any remaining orange liquor.

Close the jars and leave in a cool dark place for at least a fortnight. As you eat the clementines, you can replace them with unsoaked prunes and top up with the reserved orange liquor and more Armagnac.

ORANGE APERITIF

Here is the most delicious aperitif for winter, refreshing, and an equivalent to summer's Kir of white wine and cassis. Remember that with this kind of recipe, quantities are a guide, not the Tables of the Law. Adjust them to suit your taste.

1 kg (2 lb) oranges
1 litre ($1\frac{3}{4}$ pt/scant $4\frac{1}{2}$ cups) dry white wine
250 g (8 oz/1 cup) sugar
150 ml (5 fl oz/$\frac{2}{3}$ cup) brandy

Cut oranges down into six pieces each, without peeling them. Put into a large jar and pour over the wine. Cover closely and leave for 7–10 days. Strain into a saucepan, pressing the oranges gently before discarding them. Add the sugar and heat gently stirring all the time to dissolve the sugar. Do not allow the wine to come anywhere near boiling point. It should not become more than warm.

Cool, add the brandy and strain through a double muslin into clean bottles. Cork tightly and leave for a further week to mature. Serve chilled with a little soda water.

Note A stick of cinnamon and 4 cloves can be put in with the oranges and wine to add a spiced flavour.

ORANGEADE

See Lemonade p. 220.

KUMQUAT SWEETMEATS

The tart crispness of these tiny oranges, their oval completeness, makes them ideal for *petits fours*. The colour glows through the fondant coating. The firm skin means that the sweetmeats will last until the next day without their juice ruining the sugar.

Break up and melt a 250-g (8-oz) packet of vanilla fondant with 2 tablespoons (3 tbs) rum or orange liqueur or bitter orange juice, over a pan of barely simmering water. Stir until perfectly smooth and liquid.

Wash and dry 24 kumquats carefully, and discard any that are soft. Spear each one through the stalk end with a cocktail stick, without piercing the skin at the far end. Cut a slice from the base of a couple of large sweet oranges so that they stand firmly – alternatively, use a flower vase with many holes in the top to dry off the kumquats once they are dipped.

Dip and spoon sugar over the kumquats, so that each one is evenly coated. Push the opposite end of the cocktail stick into the large oranges, or into the holes of the flower vase. Work quickly. If the fondant gets thick and muddy, add a little extra liquid, alcohol, juice or water according to taste. Once you become used to coating the kumquats and develop the right *tour de main*, you may well be able to coat 30 kumquats with one packet of fondant.

PICKLED KUMQUATS WITH ORANGE SLICES

These two make an attractive pickle together, but there is no reason why you should not use all kumquats or all orange slices. I make this recipe because the only kumquats I ever see are inordinately expensive.

For each 250 g (8 oz) kumquats, provide one large orange which has been well scrubbed. The kumquats only need rinsing.

Slice away and discard the peel ends of the orange, then cut the rest into slices and put them in a wide pan with the kumquats and enough water to cover generously. Bring to simmering point, and leave until the orange slices are tender. If the kumquats show signs of over-cooking and collapse, remove them.

Meanwhile dissolve 300 g (10 oz/1¼ cups) sugar in 250 ml (8 fl oz/1 cup) wine vinegar. Add a 5-cm (2-in) cinnamon stick, 8 whole cloves and 2 blades of mace. Once the liquid is clear and reaches boiling point, stop stirring.

Drain the cooking liquor from the oranges and kumquats into a bowl. Pour the syrup on to them, adding enough cooking liquor to cover the fruit. Simmer until the orange slices look transparent and slightly candied, adding extra cooking liquor as required.

Arrange the fruit in a wide glass jar, rinsed and dried upside down in a low oven. Cut the slices in two, three or four if you like. Pour on the boiling vinegar syrup, making sure that the fruit is covered. Fasten the lid tightly and leave in a cool dark place for at least a month to mature.

Serve with cold ham, pork, duck, and with duck pâtés. Or use the kumquats and some of their juice to flavour a *sauce bigarade*: with small bunches of watercress, they look delightful as a garnish.

AN ORANGE PORT IN RUSSIA

> At that time – in the early twenties – Batum was visited by a great number of feluccas with oranges and tangerines from near-by Turkey – from Rizeh and Trapezund (in Batum they said Trebizond). These

aromatic fruit were stacked in pyramids on the decks of feluccas which were as multi-coloured as Easter eggs.

I often saw one and the same scene: old Turks reclining on mat-covered oranges, drinking thick, aromatic coffee and smacking their lips.

The smell of coffee spread not only from feluccas, but also from the shingle on the beach. It was edged with coffee-grounds. Torn yellow shreds of tangerine peel were strewn conspicuously among them.

In the evening the sailors on the feluccas prayed sitting on their heaps of oranges, turned to the south-east, in the direction of Mecca, occasionally lifting up their hands or pressing their foreheads against the cold oranges. . . .

On the morning of my arrival . . . Lucienna fried a plaice. We ate the plaice served up with tangerines, washed it down with Rukhadze's vodka and felt happy.

Konstantin Paustovsky: *Southern Adventure*

Un orange sur la table,
Ta robe sur le tapis
et toi dans mon lit.
Doux présent du présent,
Fraîcheur de la nuit,
Chaleur de la vie.

Jacques Prévert

PAPAYA, PAPA, PAPAW or PAWPAW

The poet Edmund Waller writes of the Summer Islands – the Bermudas – in the mid-seventeenth century:

> There a small grain in some few months will be
> A firm, a lofty, and a spacious tree.
> The *palma-christi*, and the fair papa,
> Now but a seed, (preventing Nature's law)
> In half the circle of the hasty year
> Project a shade and lovely fruit do wear.

Indeed, the papaya is one of the hastiest trees to grow, fruiting within the first year. It is more a giant plant than a tree, seeding itself, a rampagious and vast Brussels sprout, producing a cluster of papayas round its stalk under a plume of leaves.

The names we use are from the Spanish *papayas*, which is a corruption of the Carib word *ababai*. The Caribs are the people who domesticated it, the original inhabitants of the islands when Columbus arrived. 'Perhaps,' said one plant writer, 'if we could retrace exactly the migrations of that remarkable people, we should find them marked by surviving populations of pawpaw trees.' Nowadays, they grow right round the world, among the most important of tropical fruits.

The problem with the papaya was the unpredictable size of its fruit. Some weigh 5 kg (about 10 lb). As the world fruit trade developed at the end of the nineteenth century and the beginning of this century, it was obviously a good idea to try and develop a reliable papaya of convenient size. In 1919 in Hawaii – a great place for papaya – the agricultural station came up with the handy sort of fruit that we now see in the greengrocery, and called it the Solo (because it was just right for one person). In South Africa about thirty years ago, at Letaba in the Letsitele valley, a grower called Len Hobson developed the papino varieties, golden-skinned fruit with golden to melon-pink flesh. The booklet I have, produced by his company, came out in 1972, a year which I suppose marked the beginning of large arrivals of papaya in this country. I would not say it has caught on as well as the mango – it is not such a stunning fruit – but when it is on sale in small towns like Marlborough or Cirencester, in branches of Waitrose, one might reasonably conclude that it is a success.

HOW TO CHOOSE AND PREPARE PAPAYA

From Bangladesh, a friend wrote to me of her pleasure in having papaya for breakfast, sprinkled with lime juice. That seems to be the favourite time and way of eating them right round the world: in his *Cook's Encyclopaedia*, Tom Stobart remarks that papaya and lime make the best breakfast at the start of a hot day. And the most beautiful, I would say. As you cut open a papaya, the particular intensity of the yellowish-pink flesh parts easily, and the glistening grey-black seeds spill out like a heap of caviar, large caviar. The seeds do you no harm, but they are usually removed, with perhaps a few saved for garnish if the fruit is to be served in some special way. As you dig into the fruit and taste it, you can see why sometimes it is called the Tree Melon. Most melon recipes can be adapted to papaya with little trouble, though you might think of substituting lime for lemon juice, and coconut cream for double or whipping cream.

Green papaya, which you might see in West Indian and Indian shops, is treated like a squash and made into soups and curries, or into a gratin with béchamel sauce – not especially exciting. The ripe fruit is best. Try a papaya fool, made with coconut cream, and sharpened with lime juice – sugar to taste – a few of those beautiful seeds on top; a speciality of Hawaii.

We make use of papayas in another way, without perhaps realizing it. The plant contains an enzyme which has a dramatic ability to tenderize meat (when an animal is killed in the tropics and there is nowhere – no cold room or butcher's shop – to hang it, the pieces are wrapped in green papaya skins and become remarkably tender). This enzyme, papain, comes from the latex which is drawn from the trunk and processed into a white powder. It forms part of the branded meat tenderizers you buy at the grocery.

The thing to remember when you buy papaya is that it will ripen in a warm kitchen, like an avocado, though it will feel a little softer. Sometimes the fruit develop a softish patch: use them immediately without waiting any longer. Another thing to remember is that they keep their flavour well when they are cooked. If you have to cut up a fruit that is still on the firm side, pour boiling syrup over it, or bring it up to boiling point in cold syrup and then leave it to cool (flavour the syrup with lemon or lime zest, thinly peeled). With mango and pineapple, with peaches or strawberries, it makes a good fruit salad – although in such mixtures, I would describe it as a background fruit.

HAWAIIAN PAPAYA CHOWDER

Green unripe papayas are cooked and treated like squashes, e.g. like courgettes, chayote, little gem squash and so on. Recipes belong more to a book on vegetables, but here is a simple Hawaiian soup recipe in case green papayas suddenly come your way.

Peel and seed the fruit, before dicing and measuring it.

3 thin rashers streaky bacon
lard
1 medium onion, chopped
½ litre (generous ¾ pt/2 cups) diced papaya
1 medium carrot, peeled, diced
375 ml (12 fl oz/1½ cups) milk
salt, pepper

Cut the bacon into strips and fry in a saucepan in its own fat. Add a little lard if necessary, some bacon these days does not have much fat to it. Push the bacon to one side and put in the onion. When it is lightly browned, add the papaya, carrot and ½ litre (18 fl oz/⅔ cups) boiling water. Add salt and simmer until papaya and carrot are tender. Pour in the milk, bring back to the boil, check the seasoning and serve.

PARMA HAM WITH PAPAYA

For six people, halve two ripe papayas. Scoop out the centre seeds and cut each half into three wedges. Peel off the skins and dice neatly. Sprinkle with a little sugar and lemon to bring out the flavour.

Arrange 375 g (12 oz) of thinly-sliced Parma ham – or more if you can afford it – on a serving dish. Leave a gap in the middle to take the papaya. Serve with brown or granary bread and butter.

VARIATION Substitute Westphalia or Bayonne ham for Parma ham (*prosciutto crudo*), or any of the numerous French country hams that are eaten raw in the same way.

STUFFED PAPAYAS WITH LASSI

If papayas are plentiful where you live, or if ever they become cheap and abundant, this is a good way of presenting them for a change. It is not the pinnacle of high eating, but for summer lunch when hot food needs to be slightly piquant and dry yet juicy, it is a good dish, especially if you also put a jug of *lassi* on the table as well.

Although raw meat is given in the ingredients, you could use the pink part of left-over veal or beef. It may seem a nuisance to chop it, but with a heavy sharp knife, it is quickly done and it tastes much better. Getting this kind of mixture right, dry but moist and meltingly spicy, is not too easy: you need to give yourself every help you can, and chopping rather mincing the meat makes the right start.

6 ripe papayas
2 limes, or 1 lime and 1 lemon
1 large onion, finely chopped
oil
2 large cloves garlic
1 small seeded red chilli
2.5 cm (1 inch) piece ginger root, peeled
½ kg (1 lb) veal or beef, chopped
2 pinches turmeric
60 g (2 oz/½ cup) blanched, slivered almonds
150 ml (¼ pt/⅔ cup) veal or beef or chicken stock
salt
6 rounded tablespoons (1 cup) cooked rice
2 level teaspoons garam masala
1 tablespoon each chopped green coriander and parsley
3 heaped tablespoons (1½ cups) fresh breadcrumbs
2 spring onions, sliced
melted butter

Halve and seed the papayas. Remove most but not quite all of their flesh, so that the skins have a nice pink lining. Chop the flesh into rough dice and put into a bowl with the finely-grated zest of the limes or lime and lemon. Squeeze in the juice. Range the papaya shells on a buttered baking dish.

Cook the onion slowly in 3 tablespoons (scant ¼ cup) of oil until it begins to soften. Raise heat to brown it nicely, stirring to get as even a colour as possible.

Meanwhile chop together garlic, chilli and ginger. Add to onion, and fry for 2 more minutes. Then put in the meat, turmeric and almonds. Raise the heat to brown the meat rapidly. Pour in the stock, lower the heat and cover. Simmer for about 15 minutes until the meat is tender and the liquid more or less vanished, with just a little juice remaining. Season. Stir in the rice and *garam* and add to the papaya dice. Taste and correct the seasoning. Add the herbs and taste again.

Stuff the papayas. Sprinkle them with breadcrumbs and pour on some melted butter. Bake in a fast oven, gas 6–7, 200–220°C (400–425°F) for about 20 minutes until the top has a golden crusty look. Sprinkle with the spring onion just before serving.

To make the *lassi*, put equal quantities of yoghurt and iced water – 600 ml (1 pint/2½ cups) of each is about right for six people – into a liquidizer or processor. Add some cream, if you like, and also 6 ice cubes. Whizz together until the mixture has a head of foam. Taste and add salt, then whizz again. Pour into a jug and stand in the refrigerator until required: the foam lasts for several hours. Put a pinch of cumin in the centre just before you serve, if you like, by way of decoration.

With a salad and some cheese to follow, this makes an agreeable meal. The *bonne bouche* is coming to the end of the spiced meat and rice, and finding the tender pink layer of papaya underneath.

BAKED PAPAYA WITH GINGER

A good dish to know – with slight adjustments, it can be produced in a savoury version for a first course, or to go with grilled chicken, barbecued spare ribs, duck cooked and cut up in the Peking style (p. 265), served with rice.

Ginger goes well with papaya, better than with melon. I would say.

3 papayas, halved and seeded
6 tablespoons (¾ cup) butter
5 lumps preserved ginger, chopped
zest and juice of a lime
6 tablespoons (½ cup) preserved ginger syrup

Place the papayas in a buttered baking or gratin dish of presentable appearance. Mash together the butter, chopped ginger, zest and half the juice of the lime. Divide between the six cavities. Sprinkle with the rest of the lime juice, then pour a tablespoon of syrup over each papaya half. Bake in a moderate oven, gas 4, 180°C (350°F) until tender, basting occasionally with the juices.

SAVOURY VERSION: replace the sweet ginger with a 2–3-cm (1-inch) piece of green root ginger, sliced, peeled and chopped. Use a very little sugar to emphasize the fruit's natural sweetness, not to make it pudding-sweet. Omit the ginger syrup.

PILAFF VERSION: Cook two sliced papayas in a little butter. Mix carefully with a rice pilaff, adding lime and preserved ginger to taste.

PAPAYA UPSIDE-DOWN PUDDING OR CAKE

A pretty and unusual version of upside-down pudding. The green sharp lime, coral papaya and golden cake mixture make a harmony of colour as well as of the taste.

2 limes
2 papayas, ripe but not too soft
60 g (2 oz/¼ cup) softened butter
30 g (1 oz/2 tbs) granulated sugar

SYRUP
lime juice (see *recipe)*
150 g (5 oz/⅔ cup) granulated sugar

SPONGE
175 g (6 oz/¾ cup) butter
150 g (5 oz/⅔ cup) caster sugar
175 g (6 oz/1½ cups) self-raising flour
½ level teaspoon baking powder
1 heaped tablespoon unblanched almonds, ground
90 g (3 oz/½ cup) candied citron peel, cut into thin shreds (optional)
3 large eggs, beaten

Using a zester or grater, remove the thin outer peel of the limes and set aside. Squeeze their juice on to a large plate. Peel, halve and slice the papaya thinly and put on to the lime juice-covered plate. Turn the pieces over and leave for several hours.

Spread a 25 cm (10 inch) shallow cake tin with the softened butter and sprinkle the sugar over the base. Arrange the drained papaya slices over it.

Put three tablespoons (scant ¼ cup) of the lime marinading juice into a bowl, and stir in the granulated sugar from the syrup ingredients.

Any lime juice remaining should go into a large mixing bowl. Add the butter and caster sugar from the sponge cake ingredients, and cream them together. Mix the flour, baking powder, ground almonds and candied peel. Add alternately to the butter mixture with the beaten egg. Mix in the grated lime peel. Do this in one of the following ways. If you use a processor, the almonds can be ground coarsely first, then the lime juice and other ingredients, apart from the candied peel and zest, tipped in. Peel and zest are mixed in last.

If you use an electric beater, you will have to grind the almonds separately, but everything can be mixed at one and the same time, peel and zest apart.

Spread the mixture carefully over the papaya slices, trying not to dislodge them. Bake at gas 4, 180°C (350°F) for 50–60 minutes until the top is nicely browned and the cake pulling away slightly from the tin.

Stir up the sugar and lime juice and pour over the cooked cake which will rapidly absorb it.

If you are eating the cake warm, ease round the edge with a knife and invert the whole thing on to a plate. If you are eating the cake cold, leave it to absorb the syrup for about 5 minutes, then ease and turn out on to a plate.

THE PAPAYA TREE

> In the great heat we often dined in the shade of a rock, from which we could see the waterfall, and feel its coolness and enjoy the murmur of its waters without being deafened by them . . . One day Virginia ate some papayas by this rock and then planted the seeds there for the future benefit of bird or traveller. Several papaya trees soon flourished by the rock, one of them a female, that is to say a fruit-bearing tree. When Virginia went away this female tree was scarcely knee-high. But so quickly do papaya trees grow that it had soon reached 20 feet, and was encircled aloft by row on row of ripe papayas. Coming on it one time by accident, Paul was delighted to observe that the papaya tree his beloved had planted had grown so tall, but then his delight gave way to deep melancholy at this evidence of all the time she had been away. . . . Sometimes this made him think of cutting the tree down, but sometimes he would put his arms round the trunk, and kiss the tree and talk to it, recalling to himself Virginia's kindness of heart. . . .
>
> Bernardin de Saint-Pierre: *Paul et Virginie* (1787)

PASSION FRUIT or PURPLE GRANADILLA

Are passion fruit – or purple granadillas – the most fragrant of all the eatable fruit of this world? I think they must be. The scent and sharp-sweet taste survive picking, packing, transport from their tropical vines (they are native to Brazil, but are now grown in many other hot regions, Central America for instance, Hawaii and parts of Australia). Visit Fauchon's marvellous food shop in the Place Madeleine in Paris when the passion fruit are in: an open box or two scents the whole shop – with a little help from the guavas – and the scent hits you as you push open the door. Suddenly an air of paradise.

They are still there, scent and savour, in a passion fruit ice cream.

Passion flowers delighted Spanish Jesuit missionaries in South America because in each of the complex flowers they could see the Signs of the Passion, the Three Nails, the Five Wounds, the Crown of Thorns, the apostles too (minus Judas and Peter). They believed the Creator had thoughtfully arranged the Passion Flower, the *Flos Passionis*, in this way, and had planted it in the New World ready to help in the conversion of the Indians.

The Signs of the Passion are all there in the blue flowers of the species we in the north can grow outdoors – if our luck holds. In a good summer the flowers of this *Passiflora coerulea* may ripen with fruit, but don't be too excited because they won't be the thrilling dimpled or wrinkled dark purple scented edible indispensable passion fruit, which are berries of the not so beautiful *Passiflora edulis*.

Seeing how very definitely these are worth the money you have to pay for them, I feel myself that the Spaniards of the sixteenth century might have found a better name for them than 'granadill' which means 'little pomegranate'.

Having said that passion fruit are worth the money you pay, I still feel that the price we have to give for the fruit in this country is little short of scandalous. I take tnat back, it is scandalous. In France, the trade knows that the way to sell fruit is to encourage a fruit-eating habit, the more you sell the more people buy. And the way to sell is to keep the price down. With something new, a little energy and intelligence is required as well to show customers that it is worth buying. The English habit is to do nothing whatsoever; if the Israelis or Dutch mount a campaign to sell avocados or peppers, they will go along with it, but I have never yet come across a greengrocer who has said to me, 'Here's something new, I'm selling it cheaply to get it going. Why don't you try it for a change?' In France, it happens all the time; the greengrocer knows his customers and, what is more important, knows his fruit, what to do with it, what it tastes like and so on.

HOW TO CHOOSE AND PREPARE PASSION FRUIT

You will have no choice most of the time, but go for the dark dimpled fruit, and choose the largest you can, as in this country they will be priced individually. Cut the top off to get at the seedy pulp. They can be served like this, with a spoon, and a jug of cream to pour into the cavity and mix with the pulp. The seeds are pleasantly crunchy.

To use passion fruit in cooking, scoop out the pulp into a small pan and warm it very slightly, with some of the sugar from the recipe to make a little extra bulk. Then sieve it vigorously, pushing through everything but the seeds. You can sieve the pulp without heating it, but it comes away from the seeds more reluctantly which makes for harder work. Keep some of the seeds, if you like, for decoration. Prepared in this way, passion fruit can be used to make a curd, as in lemon curd. Delightful as a tart filling or on bread, or as a sauce.

Chefs in France use prepared passion fruit juice and syrup, but I have not seen it on sale here. It is of high quality, and can be used to add extra flavour, but the fresh fruit is best.

PASSION FRUIT SOUFFLÉ
SOUFFLÉ AUX FRUITS DE LA PASSION

To launch the *Observer* magazine's French Cookery School series in 1980, we had a party in the kitchens of the Dorchester. Anne Willan, head of La Varenne, who wrote the cookery series, and Anton Mosimann, *maître-chef des cuisines* on his home ground, demonstrated their ways of making a soufflé. Afterwards there were twenty-six different soufflés, an A to Z, each with its sauce, made in the great revolving oven. A magnificent *tour de force* – and none of them overcooked.

Amande – almond
Banane
Chocolat
Dorchester – passion fruit
Epinards – spinach
Fromage – cheese
Grand Marnier – orange liqueur
Homard – lobster
Indienne – fish with curry
Joinville – crayfish
Kummel – aniseed liqueur
Lemon
Marrons – chestnut
Noisette – hazelnut
Orange
Poisson au curry – fish with curry
Quatre fruits – four red fruits
Rothschild – gold leaf dissolved in Kirsch
Strawberry
Taillevent – pear
Uval – grape
La Varenne – mushroom
Walnut
Xeres – sherry
Yorkaise – ham
Zingara – tongue

This was the first time I heard Anton expound his system of standing soufflés in a bain-marie, for anything up to half an hour, then baking them off in the oven. In his book *Cuisine à la carte*, I notice he only commits himself to 7 or 8 minutes with this particular elasticity, which is important if you plan a soufflé at the end of a meal. Anne Willan taught me that the completed unbaked mixture, in its dish, can wait for up to four hours in the refrigerator without coming to harm.

With these two suggestions in mind, you will be able to serve this passion fruit soufflé in perfection:

3 egg yolks
200 g (7 oz/good ¾ cup) caster sugar
150 ml (5 fl oz/⅔ cup) passion fruit juice
6 egg whites
5 g (¼ oz/½ tbs) butter and 2 teaspoons sugar for the dish
icing sugar
whipped cream

SAUCE
20 g (¾ oz/1½ tbs) sugar
juice of half a lemon
teaspoon cornflour

Beat together the yolks, half the sugar and two-thirds of the passion fruit juice, sieved free of seeds before measuring.

Whip the egg whites, add the rest of the sugar and beat until creamy. Fold into the yolks.

Butter the soufflé dish, sprinkle with sugar and tip out any surplus. Pour in the mixture. Poach in a bain-marie with barely simmering water for 7–8 minutes, then take the dish from the water and put into the pre-heated oven, at gas 3–4, 170–180°C (325–350°F) for 25–30 minutes. Sieve icing sugar over the top, and serve.

The sauce can be made in advance, or while the soufflé cooks. It should be served warm, rather than hot. Heat the last of the passion fruit juice with the sugar, lemon juice and about 4 tablespoons (⅓ cup) of water. Mix the cornflour with another 2 tablespoons (3 tbs) of water and stir into the passion fruit liquid. Stir over the heat until it clears and thickens very slightly. Add extra water to taste if you like, but the sauce should be strong in flavour and short in quantity, but in no way gluey from the cornflour.

Serve as well with whipped cream.

Note This soufflé is enough for 4 people. Should you wish to make it for a party of 6 to 8 people, bake two soufflés rather than one, putting the second one into the oven as you take the first one out – an extra five minutes in the bain-marie will not hurt it, and you can turn the oven up to gas 5, 190°C (375°F) if you want to serve the second one fairly rapidly.

To serve the soufflé, put two spoons at an angle into the dish, so that each person gets a fair share of the crisp outside and the creamy centre.

PASSION FRUIT SOUFFLÉ GLACÉ
SOUFFLÉ GLACÉ AUX FRUITS DE LA PASSION

I think that the standard *soufflé glacé* is much improved by serving with it a clear sauce of the same fruit, lightly thickened with arrowroot or cornflour but very lightly so that it pours easily.

8 passion fruit
caster sugar
4 egg yolks
packet (½ oz/2 envelopes) gelatine
300 ml (10 fl oz/1¼ cups) double cream, whipped
3 egg whites, whipped

SAUCE
2–3 passion fruit
caster sugar

juice of a large orange
1 tablespoon lemon juice
1 heaped teaspoon arrowroot or cornflour

Halve the fruit and scoop out the seedy pulp into a small saucepan. Add 125 g (4 oz/½ cup) sugar and a tablespoon of water. Stir over a moderate heat, until very hot but not boiling. Sieve into a large pudding basin, pushing through as much pulp as possible. Save some of the black seeds for decoration and throw away the rest.

To the basin, add the yolks, beating them in vigorously. Stand over a pan of barely simmering water and whisk with a rotary or electric beater until the mixture is bulky and creamy looking. Remove basin from the pan. Dissolve gelatine in 5 tablespoons (6 tbs) of very hot water, and mix into the passion fruit custard. Whisk until barely tepid. Add extra sugar to taste, then chill.

When the passion fruit mixture begins to set, fold in the whipped cream, then the stiffly beaten egg whites. Pour into a soufflé dish – collar it if you like – and return to the refrigerator to set.

Meanwhile, make the sauce. Put the pulp of 2–3 passion fruit into a pan with 60 g (2 oz/¼ cup) sugar, and the orange and lemon juice. Bring almost to boiling point, then stir in the arrowroot or cornflour dissolved in 100 g (3½ fl oz/½ cup) water. Cook until the sauce clears and thickens, then push through a sieve and taste. Add extra sugar, or more lemon juice to bring out the flavour.

Brush some of the sauce over the top of the souffle, and scatter a few seeds as decoration. Serve the rest of the sauce in a jug.

Note The soufflé can be set in individual glasses, or small white custard cups, with the same bright glaze patterned with black seeds.

PASSION FRUIT ICE CREAM
GLÂCE AUX FRUITS DE LA PASSION

The first time I made this ice, I used the yellow passion fruit of a different variety from the black ones, and sold in France under the name of *grenadilles*. They were the size of a medium apple, and not quite so intense in flavour.

Scoop the pulp of three to five passion fruit into a basin. Simmer together 125 g (4 oz/½ cup) sugar and 125 ml (4 fl oz/⅔ cup) water for 3 minutes, then pour on to the passion fruit. Leave to cool slightly, then sieve. Freeze until the edges are firm, the middle slightly mushy, then beat and fold in 150 ml (5 fl oz/$\frac{2}{3}$ cup) whipped cream.

If the passion fruit were not too squashy, keep the shells and serve the ice cream in them, using egg cups as the support or small custard glasses. This quantity makes plenty for four or five shells.

PASSION FRUIT APERITIF

MARACUDJA

At the Tours food fair in the spring of 1980, we came to the Guadaloupe stall and saw a large glass jar full of a pale yellow liquid with grey seeds in golden flesh floating around in it. It was labelled *maracudja, 10 francs a glass*. The man in charge told us that maracudja was the Guadaloupe name for passion fruit, and as he ladled out the liquid, he also told us how to make it. Very simple. Fill a glass jar one-third to a half full of passion fruit seeds and flesh. Mix together light rum and light cane sugar syrup (use heavy sugar syrup instead, made with soft pale brown sugar) in the proportions of 4 to 1. Fill up the jar. Close and leave for at least 6 weeks. Longer is better. When you open and taste it, you can add more syrup if you want it sweeter, but it should not be too sweet.

PEACH and NECTARINE

'Renoir used to say to young artists longing to paint – like him – the pinkish-brown tones of an opulent breast: "First paint apples and peaches in a fruit bowl." My greengrocer sold me worthy models for painting. Their curves were perfect and graceful. Their grooves were in exactly the right place. Their gradations of colour were an art school exercise, and as for the velvety down of their skin . . . "It's the left cheek of my girl", as a Japanese poet has written.'

So began an article by James de Coquet at the height of the peach season some years ago. Peaches for painting, peaches for writing poems about – but peaches for eating? His beauties were hard. When cooked, they were tasteless and still not tender. Horticulture has come into perilous times. Everything is selected, trained, peaches are now the children of arranged marriages. Perhaps they would have more taste and succulence if they were love-children? People understand the problems in France, where in every heart there sleeps a peasant. They are on the grower's side. They know he has to decide which currency he is growing for, the mark, the pound or the Swiss franc. They know he has worked hard to produce strains that will be roadworthy and can earn a living in the markets of Europe. They appreciate that he has produced the peaches of Renoir, with exultant curves, pretty enough to eat. But should they not also be eatable?

We can appreciate James de Coquet's sentiments, and feel an added crossness that our beautiful, unsatisfactory peaches cost nearly twice as much as his, although coming from the same growers.

Somewhere, surely, there must be the perfect peach? I thought I would try and find it, and started off with the famous Montreuil peach, because on every classic French menu Montreuil means peaches, just as Montmorency means cherries and Argenteuil means asparagus.

According to legend – perhaps it is true? – the peach began its career at Montreuil to the east of Paris in the seventeenth century. A retired soldier living there grew his peaches on the espalier system, which was a new idea, and took off most of the tiny fruit from the trees, leaving only a few to ripen on each one. This man, Monsieur Girardot, was a friend of de la Quintinie, Louis XIV's gardener, and one day confided in him that he would like to ask the king a favour. Knowing the king's fondness for fruit, de la Quintinie suggested that he presented a basket of his wonderful peaches at Versailles, anonymously. This he did. The king was entranced by the quality of the fruit and by their mysterious arrival. Next time he was hunting over in the direction of Montreuil, de la Quintinie told him that the peaches had been grown by a friend of his not far away. Of course the king had to go and see for himself, the man asked his favour and it was granted. After that, year by year until the Revolution, the family and other growers of Montreuil who were helping to supply the increasing demand for the fruit, sent a basket of them to the royal family.

Louis was not the only king who was captured by these peaches. Edward VII was dining incognito with a young woman in Paris. When the bill came, he was astounded at the price of the Montreuil peaches, and called the waiter over.

'Peaches must be very rare this year,' he said.

'Ah yes, sir, but not as rare as kings,' came the reply.

I rang up the Hôtel de Ville at Montreuil. No one knew, but passed me on to a local college of agriculture. No, we do not grow Montreuil peaches here, I was told, Never heard of them. When I went to see the district, I found that the old back gardens – once divided by espalier lines of peach trees – have quite disappeared into tower blocks and concrete. The big road to Charles de Gaulle airport goes through it.

Stumped. I went and told a friend about my search. 'Why didn't you ask me before?' he said. 'I remember eating Montreuil peaches after the war at the Escargot restaurant in Paris, in the rue Montorgeuil. They were sensational. And each one was presented in a special box as if it had been an expensive brooch, a black or navy box, I think, with the name of the restaurant printed on it in gold. You paid, but they were worth it.'

Off again, to the rue Montorgeuil with a friend, and the Escargot. There under the sign of the golden snail, in all that cherished decor of the past, surely I would find the answer? But no. Even the oldest cleaner was 'new', from the last eighteen years and less. I remembered that it was over thirty years since the war ended. We had an excellent meal – ending with strawberry fritters, *see* p. 422, which reminded me of my failure since strawberries were another speciality of Montreuil in the past. Does anyone anywhere tend a last surviving Montreuil espalier? It seems not. The French are more careless of their past than we are. Those magnificent peaches have gone for ever.

My journey was nothing, though, to the undertaking of Professor

Edward Schafer, of Berkeley University. His quest for the finest possible peaches has produced one of the best and most thorough books about food that I know, *The Golden Peaches of Samarkand*, a study of T'ang exotics. Now peaches are native to China. They have the kind of place in Chinese myth and poetry that apples have in ours.

When golden peaches, the size of goose eggs, were sent from Samarkand, T'ang courtiers must have thought they were seeing the mythical peaches of immortality that fruit every six thousand years, so that the gods can eat them and continue to live for ever.

The peaches were sent twice in the seventh century from Samarkand. Trees came with them. They must have been cherished, as T'ang emperors had the same kind of passion for fruit and vegetables as Louis XIV, ten centuries later. No one recorded what they tasted like.

For me, the exotic peach is the *pêche de vigne*, covered with a fine grey down, and deep-coloured through to the stone. In the past, every vineyard had its peaches, and the small ones still do. The fruit is ripe at vintage time, and the trees renew themselves, growing true from the fallen stones. A friend gave us a small peach tree from his vineyard and we planted it in Wiltshire. For years and years it did nothing, then suddenly it had thirty-nine peaches. The year was not warm enough to ripen them properly, but cooked they had a superb flavour. Even better were the purple vineyard peaches we found by accident, in Vendôme two years ago. They were purple right through, even and perfect in size. Someone was growing them commercially, for the following week I saw some in Paris and bought them. After the first lot, they were disappointing – the hand of commerce, presumably. We went back to Vendôme, but the peaches were finished, and last autumn there were none at all.

If anybody cares to grow fruiting trees and writes to the Ministry of Agriculture for a permit to bring in plants, they might on a journey through the wine districts of France be able to pick up a young *pêchier de vigne*. For the non-gardeners, look out in the markets for these peaches at the end of the summer, late August and September. They may also be marked 'sanguines', blood peaches, like blood oranges, on account of their rich coloured flesh.

Nectarines are smooth plum-skinned varieties of peach. They come white and orangey-yellow, just as peaches do and, as with peaches, the white ones taste better. The difference is quite marked, nectarines being both sharp and richer at the same time. In Italy, the peach is *pesca* – the nectarine *pesca noce* – because the Romans thought it came from Persia (hence, too, our peach and French *pêche*). In France, the *nectarine* may also be labelled *brugnon*, because the fruit was a great speciality around

Brignole in the Var. And should you ever come across the word *pavie*, it means a clingstone peach, grown originally at Pavie near Auch, a part of France where vast orchards now stretch kilometre after kilometre.

HOW TO CHOOSE AND PREPARE PEACHES AND NECTARINES

If you are housekeeping in Italy or France with a family to feed, buy peaches and nectarines in trays and boxes, rather than by the kilo. They will be a little cheaper and a continuing source of contentment in hot weather. In August, peaches, melons and late strawberries dominate the markets of both countries with their colour and smell.

An infinitesimal number of readers may live near an English peach orchard or big house where wall peaches and nectarines are still grown. All I can say is that if they can buy them, they are to be envied. My parents once said that the finest they had ever eaten anywhere had been grown in Mr Perrin's – or was it Mr Lea's? – garden in Worcestershire. But that was a long time ago, longer ago than the last Montreuil peaches.

Being on the whole so remote from the orchards, and being so used to the Golden Delicious apple which never changes, we can be taken by surprise that peaches from the same suppliers vary from year to year. Even from tree to tree it would seem, judging by the way a kilo of peaches will vary from superb to tasteless. This means that you are gambling rather than choosing. All you can do is make sure your horse is not lame or blind, by refusing peaches that are bruised or soft. Softness indicates woolliness and cannot be redeemed, as hardness sometimes can, by cooking.

To skin peaches and nectarines, pour boiling water over them. If they are very ripe, they will be peelable after about 15 seconds. Extra time will be needed for the rest. If you are poaching the fruit, leave the skins on – they improve the colour of the syrup and can easily be strained out or peeled off at the end of the cooking time.

The easiest way to cut up a peach, is to run a small sharp knife down from top to bottom, going through to the stone. Turn the peach slightly in your hand and make another cut, so that the wedge falls out, on to a plate, or better still into a glass of wine. Continue like this all round the peach. If you are doing a bowl of peaches in wine, do not prepare them more than an hour before the meal. The grandest form of what is basically a rustic affair is peaches in champagne, with a few wild or wood strawberries added. Myself, I prefer the simple version; at least you can add some sugar to the wine if the peach is not as sweet as it should be, without feeling that you are desecrating a noble example of the wine-maker's art. We open a bottle of a reasonable local white or red wine, whatever we have, just as our neighbours do.

If the peaches you buy are not quite as ripe as they should be, pour some boiling light syrup over the slices. It cooks them as it cools down. If they are slightly harder than that, slip them into some boiling syrup and bring the liquid back to a thorough boil before taking the pan from the stove. This kind of technique is ideal for cooking many foods that are tasteless and which spoil if they are overcooked.

PEACH OR NECTARINE SAUCE (for poultry)

Soften a large chopped onion in a generous tablespoon of butter (2 tbs) in a wide shallow pan. Deglaze with 2 tablespoons (3 tbs) wine vinegar. When it has almost bubbled away, pour in 150 ml ($\frac{1}{4}$ pt/$\frac{2}{3}$ cup) giblet or poultry stock, 125 ml (4 fl oz/$\frac{1}{2}$ cup) dry white wine and 2 tablespoons (3 tbs) sweet fortified wine.

When the mixture is simmering, put in 3 peaches or 4 nectarines, peeled and cut in slices. When they are tender, remove a few of the best pieces for a final garnish. Taste the sauce and judge whether it needs further concentration by boiling it down: it should be on the strong side. When it is right, liquidize the whole thing.

Return to the rinsed out pan, add salt, pepper, cayenne, a pinch of sugar as required. Pour in 4 or 5 tablespoons (5–6 tbs) of whipping cream, or beat in a large knob of unsalted butter.

Carve the bird and put it on to a serving dish. Pour over the sauce and arrange the pieces of peach or nectarine on top.

BIRCHER MUESLI

Peaches and strawberries are my idea of luxury for breakfast. Especially when they are used in this raw fruit porridge, more appetizingly known as Bircher muesli. The dish was invented by Dr Bircher-Benner, the pioneer of food reform, for patients at his Zurich clinic.

He explained that he felt muesli 'corresponds with the laws of the human organism', its composition of protein, fat and carbohydrate calories being identical with mother's milk. This may be so, and I do not see why mother's milk should be an ideal food for adults in any case, but the fact is that Bircher muesli tastes delicious. It's the great contribution of the vegetarian movement to the pleasures of European eating.

Unfortunately, we have become over-exposed to muesli and its commercial imitations. Generally they are far too sweet, and contain too many ingredients. The doctor's recipe was simple. Here are the quantities for one person:

1 level tablespoon rolled oats
3 tablespoons (scant $\frac{1}{4}$ cup) water
1 tablespoon Nestlé's sweetened condensed milk or single cream and honey
juice of half a lemon
1 peach or a few strawberries
1 small apple
1 tablespoon chopped nuts, almonds, walnuts, hazelnuts, cashews

In the original recipe, the oats and water were soaked together overnight. I prefer to mix them not long before the meal, so that the oats are still crunchy.

Immediately before the meal, stir in the milk, or cream and honey, and the lemon juice. Chop the peach or halve the strawberries, grate the apple and quickly mix into the oats before the apple has time to discolour: grate in peel and core. Add nuts and stir.

Raisins make a good addition. Other fruit, including soaked dried fruit, can be substituted for peach or strawberries, but I find that grated apple is essential to the texture.

PICNIC SALAD

The day before a picnic, swish out a wide-mouthed thermos with crushed ice, or ice cubes and water. Drain quickly and put in skinned, quartered peaches and a few raspberries. Sprinkle on lemon juice to prevent the peaches discolouring (happens all too easily if they are white ones). Tip in 125 g (4 oz/$\frac{2}{3}$ cup) caster sugar for every $\frac{1}{2}$ kg (1 lb) prepared fruit. Add 7 tablespoons gin ($\frac{2}{3}$ cup) and the juice of 2 oranges. Close and leave until required.

If you think of it, turn the thermos upside down when you go to bed. Then right way up in the morning.

PÊCHES MELBA

The most famous and travestied of all fruit puddings, I would say. Escoffier invented it in honour of the great singer, and his original idea was to celebrate her performance in *Lohengrin*. To this end he had a swan carved in ice and covered with sugar: between its wings was a dish containing vanilla ice cream and peaches. Later on, he improved the dish by adding a sauce of sieved sweetened raspberries, which is now known as sauce Melba.

Allow a peach per person, a white peach for preference. Halve, stone and poach them in a vanilla-flavoured syrup, p. 437. When cool, arrange them on a bed of vanilla ice cream, p. 473, and coat them with a purée of raspberries sweetened with icing sugar. Place the bowl in another bowl of crushed ice.

PÊCHES CARDINAL

For this variation of *pêches Melba*, Escoffier added a flavouring of kirsch to the raspberry sauce, and scattered the whole thing with shredded blanched almonds.

PÊCHES AURORE

Put a layer of strawberry ice cream into a bowl, and on top place peach halves, poached in a kirsch-flavoured syrup. Coat the peaches with a sabayon sauce, p. 469, flavoured with an orange liqueur.

This and the two peach recipes above can be served in individual bowls, of course, though they look more attractive in a large glass bowl, set in another bowl of cracked ice when the weather is hot.

PÊCHES CHINOISE

Cook the peaches in syrup from the preserved ginger jar in a covered pan. Cool and serve on vanilla ice cream. Reduce the syrup to a coating consistency and, when cold, pour over the peaches. Scatter with chopped preserved ginger and either chopped walnuts or chopped toasted almonds.

GRILLED PEACHES

Cut peaches in half and remove stones. Brush the cut sides with butter and sprinkle with sugar generously. Grill cut side up, gently at first to heat the peaches through, then more fiercely to brown the sugar. Serve hot with vanilla ice cream, or with whipped cream.

NECTARINE BAVAROISE

Not a correct classic bavaroise, but a lighter mixture that allows the flavour of nectarine to come through particularly well. My original version, and the most successful, contained *orgeat*, an almond and orange-flower syrup tasting very much of the past that you can buy in high-class groceries and foodshops in France. Worth looking out for when you're on holiday. Kirsch can be used instead, or an orange-based liqueur.

6 nectarines, approximately ¾ kg (1½ lb)
125 g (4 oz/½ cup) sugar
1 packet (½ oz/2 envelopes) gelatine
juice of an orange
2 tablespoons (3 tbs) orgeat *or appropriate alcohol*
100 g (3½ oz/½ cup) fromage frais *or yoghurt, drained weight*
100 ml (3½ fl oz/scant ½ cup) double cream, whipped
lemon juice, see *recipe*

Into a wide shallow pan, section nectarines into eight wedges each. Sprinkle with the sugar and 2 tablespoons (3 tbs) water. Cook on a low heat until the juices run, then raise the heat and finish the cooking. The red skins of the fruit will dye the syrup a glowing pink. Remove, drain and skin 12 sections. Put them in a whirligig star shape in the base of an oiled and sugared charlotte mould or other plain dish (use almond or a tasteless oil).

Put the rest of the fruit, skins and all, with the syrup into liquidizer or blender. Sprinkle on the gelatine, orange juice, and *orgeat* or alcohol. Whizz or process until very smooth. Strain into a basin, mix in the *fromage frais* or yoghurt, then fold in the cream. Taste and see if any lemon juice is needed to bring out the flavour. Add extra sugar if you like – icing sugar will mix in best at this stage, caster next best.

Carefully spoon some of the bavaroise over the nectarine sections in the dish, so as not to disturb them. Put in the coldest part of the refrigerator, or into the freezer, until almost set. Then spoon in the remaining mixture, starting round the edge so that it runs gently into the middle. Put back to set and chill.

To turn out, run a knife blade round the edge. Put a dish on top, then turn the whole thing over, and give a shake or two. If it is reluctant to turn out, put a cloth wrung out in very hot water over the metal mould, and shake again.

PEACH OR NECTARINE RIBBONS

BAVAROISE AUX PÊCHES/NECTARINES RUBANÉE

Turn to the recipe for Queen Claude's ribbons on p. 360, and substitute peaches or nectarines for greengages. A firm *coeur à la crème* mixture can be used instead of the yoghurt cream, *see* p. 471.

Rather than pistachio nuts, use almonds very lightly coloured in a low oven, gas 2, 150°C (300°F).

PEACH, OR PEACH, APRICOT AND GREENGAGE TART

TARTE AUX PÊCHES, OU AUX FRUITS

Every summer, the pastrycooks in France come out in a rash of fruit tarts. Very seductive, the circles of nicely browned pastry with circles of fruit inside; the peaches, apricots, greengages, alone or mixed, are often arranged in quarters or eighths so that the points catch the heat and the whole thing bristles and shines in the most appetizing way. I asked our pastrycook in Montoire how he managed to make such unsoggy fruit tarts. Here is his recipe.

Use a plain shortcrust, or a very slightly sweetened pastry, so that you can cook the tart at a high heat without the pastry burning (as a normally sweetened *pâte sucrée* would do). Over the bottom, scatter the pastry trimmings chopped into coarse even crumbs.

Quarter or cut the fruit into eighths, retaining the skins. Set the pieces, skin side down, tightly together in concentric circles. Peaches outside, then apricots, then greengages, with a peach or apricot half in the centre. Add no sugar. Bake at gas 7, 220°C (425°F) until the tips of the fruit are lightly caught with brown. Check from time to time. If the pastry looks like burning, protect it with a strip of foil.

While the tart is still warm, brush it with thick sugar syrup or apricot glaze.

OTHER WAYS OF REDUCING SOGGINESS:

1) Brush the pastry with lightly beaten white of egg.

2) Scatter the base with macaroon or sponge finger biscuit crumbs or with cake crumbs.

3) When making the pastry, set aside about a quarter of the rubbed-in mixture and use that instead of crumbs or cake crumbs.

PEACH DUMPLINGS
PFIRSICHKNÖDEL

An Austrian friend told me that in Austria, 'in the Wachau – the vineyards along a certain part of the Danube – they grow peaches. Small, very tasty, much better than the imported Italian ones: I'm sure there must be the same thing in France. And so you find in the marvellous restaurants there *Pfirsichknödel*.'

For anyone who lives in France, or who stays there at the end of the summer into autumn, it would be worth trying these Austrian dumplings, made to the recipe for fruit dumplings, p. 458. Replace the stones with sugar lumps.

PEACH AND ALMOND CRUMBLE

Allow a large peach per person, plus an extra one or two if you can afford it. Peel and slice and put into a pan. Pour on a very little water and sprinkle with sugar. Cook gently until almost tender and the juices have run out. Drain and cool.

Turn to the recipe for apricot and almond crumble, and follow that: p. 40.

The reason for cooking the peaches first is that if you use them raw, they become so juicy that the crumble is drowned.

The peach juice can be added to cream and heated through to make a sauce. Or you can use it instead of milk to make a sauce with eggs, on the custard sauce principle, page 470. Strengthen the taste, if need be, with orange juice and grated orange rind, or an appropriate fruit brandy or liqueur.

STUFFED PEACHES
PESCHE ALLA PIEMONTESE

A well-known Piedmont recipe, which is most successful if you use good quality macaroons or tiny *ratafias*. Sabayon sauce is usually served with it, but Melba sauce does well, too, especially if you need to make the dish in advance.

6 large peaches
150 g (5 oz/1¼ cups) macaroon crumbs

1 large egg yolk
2 tablespoons (¼ cup) caster sugar

Halve and stone the peaches. Enlarge the cavities and mash the peach you have removed. Mix it with macaroon crumbs, egg and caster sugar. Pile this mixture back into the cavities, mounding it up and smoothing the top.

Set the peach halves in a lightly-buttered dish and bake at gas 3, 160°C (327°F) for 45–50 minutes. The filling should crackle apart on top.

Either serve warm with sabayon sauce (p. 469); or serve cold in a dish into which you have sieved a bed of Melba sauce (p. 302).

PEACHES ON BUTTERED TOAST
PÊCHES EN CROÛTE

The first thing to think about is the bread. If you cannot buy a brioche loaf or good white milk bread, you must make your own. I use the brioche recipe on p. 463, and bake it in a tall round coffee tin. Sometimes I make it plain, sometimes I add hazelnuts.

Cut one or two slices per person, according to the diameter of the loaf, and cut them a centimetre (good ¼ in) thick. Trim away the crusts to make neat slices. Butter them on one side, and put them butter side down on a metal baking tray. Then butter the top sides, right to the edge.

Skin, halve and stone a ripe peach per person, and cut it into 8 wedges. If the peaches seem very juicy, with the risk of making the bread too soft, poach them lightly in some syrup or bake them briefly on their own. This will have the effect of drawing out the juice. They can then be drained and the juice used to flavour cream.

Arrange the fruit wedges on the bread, sprinkle them with sugar and pour over a little melted butter. Place in a hot oven, gas 6–7, 200–220°C (400–425°F) and leave until the bread is brown and crisp underneath. Baste the fruit with the juices that flow out (butter principally). The fruit may or may not caramelize. If it doesn't, don't feel anxious. The point is the contrast of melting fruit and crisp bread; serve ice-cold cream, mixed with any juices from pre-cooking the fruit if you like, but it's not essential. I prefer the buttery flavour and crunch unmasked by the softness of cream.

Apricots, apples, pears, Chinese gooseberries, can all be cooked in the same way. Spice may be mixed with the sugar, which is a good idea if you want to use fruit that has been canned without added sugar. Altogether one of the nicest puddings there is.

SPICED PEACHES AND NECTARINES

Choose 1½ kg (3 lb) small fruit and rinse them. Turn to the recipe on p. 449, and substitute 3 blades of mace for the ginger and chilli. Cinnamon and cloves as specified.

As with apricots, you should test them the moment that the skins begin to show signs of wrinkling, and the fruit takes on a lop-sided look. It is a wise precaution to put in an extra peach, to allow scope for testing. You can always put the fruit back to cook a little more, if necessary.

PEAR

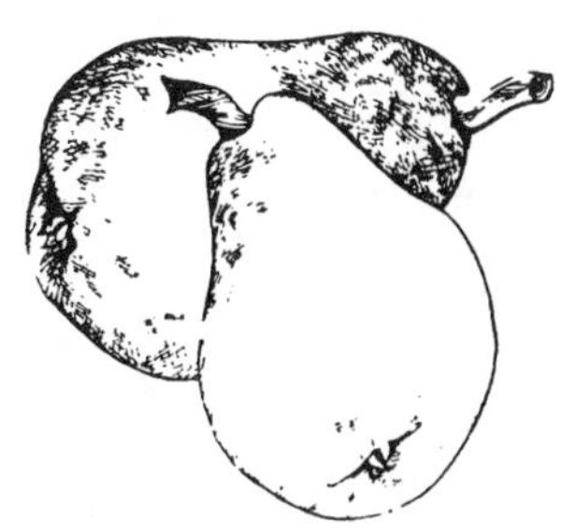

A favourite book of mine is Edward Bunyard's *Anatomy of Dessert* (now out of print, but worth pursuing in second-hand bookshops and public libraries). He doesn't bore you with gracious cadences and profound emptiness. He knew that wine and food do not spring fully bottled and wrapped in plastic from the head of some divine supermarket manager. He was a great horticulturist and historian of plants, a fine enjoyer of skill and knowledge.

Take the chapter on pears. Did you know that the Williams' Bon Chrétien, Williams' for short, or the Bartlett pear, was raised in 1770 in Aldermaston in Berkshire? That we have to thank a schoolmaster, John Stair, for it, and not Mr Williams or Mr Bartlett who were the distributors in this country and the United States.

The original Bon Chrétien was brought to France to the château of Plessis-les-Tours, now on the outskirts of Tours, by Saint François de Paul from his home in Calabria. He had been pressed into leaving Italy by Louis XI who had a superstitious hope that the saint's prayers would improve his health. St François told him, with robust and uncourtly good sense, that the life of kings has its limit like the life of ordinary men. In the end, unlovable Louis died in the saint's arms in 1483. At court, the virtue of Saint François' life was such that he was called the good Christian, *le bon chrétien*, and the pear was known, too, as Bon Chrétien, since he had loved it.

An even earlier pear continues famous, although we no longer see it in the shops. That is the *poire d'angoisse*, which was originally an instrument of torture (a pear-shaped metal contraption was pushed into people's mouths and then expanded). Poires d'angoisse were called after this abomination, as they were sharp in the mouth too (hay was put into the cooking water in an attempt to soften the flavour). In the thirteenth-century streets of Paris, sellers went round shouting

poires d'angoisse crier haut

which was I suppose a grim reminder of the connection, 'Cry loud the pears of anguish'. The phrase 'to swallow the pears of anguish' means to suffer humiliations and distress.

By the seventeenth century, France had become the great country for pears. Pears were the supreme fruit of the age. Three hundred varieties were known of which about twenty-five were worthy of the attention of the discerning man. The discerning man was de la Quintinie, Louis XIV's gardener, though perhaps gardener is not quite the right word. De la Quintinie started off in the law, but had a passion for gardening. He was a man of the highest skill and intelligence, and his exact title was *Intendant général des jardins fruitiers et potagers de toutes les maisons royales.* No sinecure. Not only did he have to create the gardens at Versailles – from a pile of rubble, and soil which had all the qualities that one would least have chosen – he had to keep Louis XIV happy. The king's passion for fruit and vegetables was so great that he demanded asparagus in December, strawberries in April, *petits pois* in May and melons in June. And he loved pears. He had a habit of peeling fruit in such a way that the skin could be put back into place: the work of art was then handed to whoever might be in his good graces.

His ways may sound appalling in these times of constitutional monarchy (though a little keen greed on the part of one of our princes might do wonders for the fruit and vegetable trade in this country even now). They gave an immense push to the general interest in fruit growing. It became the smart thing to do. Introducing a fine new pear at court, as the Marquis de Chambert did with the *virgouleuse* from his estate at Virgoulé in the Limousin, was one of the ways to get on. A favourite pear of Louis' was called La Robine: de la Quintinie rechristened it La Royale, declaring that the pear which has least faults should bear the title of the man who has most merits. Other favourites on the list were bergamot varieties, the Crassane, Louise Bonne, Cuisse-madame (my lady's thigh – nice name for a pear, just as Cuisse de Nymphe is exactly right for a delicate pink and white rose), Bon Chrétien d'hiver and Beurré grise.

What had transformed pear growing around Paris – and peach growing, too – was the new cultivation of espaliered trees. In open orchards, the fruit was blown about, and ripened unevenly. With espalier trees, the whole business could be effectively controlled, with shelter, warmth and attention. 'Now we are no longer obliged to go to Touraine to eat a Bon Chrétien,' said one writer. Everything grows on the edge of Paris, and abundantly. Other towns may have a lot of one thing, Paris has a lot of everything. In those pre-railway days, cities were dependent for most of their supplies on land within a radius of about thirty kilometres. Zola gives the last picture of the system at the beginning of *Le Ventre de Paris*: the market carts, with their sleepy drivers propped against the cauliflowers or apricots, leaving their horses to take the well-known road to Les Halles.

By Zola's day, pear-raising had come firmly down to gentleman-gardener level. Which means that the gentleman did the thinking, supplied the money and enthusiasm, and his gardener did the actual work assisted by a team of underlings:

In this garden
was raised in 1849–50
the celebrated pear Doyenné du Comice
by the gardener
Dhommé and by Millet de la
Turtaudière
President of the Comice Horticole

This inscription, in gold letters on a nice piece of local slate, is set discreetly on an old stone wall in Angers, the capital of Anjou. You may go and pay your respects – but take care, for you will find it between the crashing traffic of the boulevard du Roi René and the bicycles of students. They come to eat at the canteen built on the site of the very old garden which belonged to the Comice Horticole in the nineteenth century (it goes back to the fourth century, this garden which is a garden no more: Saint Hilaire, a lover of good things, built a chapel on the spot in 356 and called it Notre Dame du Verger because it was surrounded by fruit trees).

To be fair, Monsieur Millet de la Turtaudière was far more than a gentleman with an agreeable hobby. He deserved that pear. He got catalogues and plants and advice from all over Europe, from the great de Candolle and from our Thomas Andrew Knight, founder member of what is now the Royal Horticultural Society, and its president from 1811 until he died in 1838. How the town of Angers did not have the piety to keep that garden going, I do not know. It is a shocking thing. Other distinguished pears were raised there as well, or around Angers, and approved by the Société d'Horticulture under its lively and exacting President.

Edward Bunyard reflected that two thousand years had been necessary to arrive at the supreme pear, to bring man to this state of skill and appreciation. He laments that he was not there to taste the first fruit of that tree, and that the world's great gourmets – by which he meant, I suppose, Brillat-Savarin, Talleyrand and so on – had died 'still unacquainted with the perfect Pear'. I agree with Bunyard, and would like at least to have seen Millet de la Turtaudière handed a first Doyenné du Comice by his gardener Hilaire Dhommé, to have been there when he took the first bite, oblivious, in his pleasure, of the juice running stickily down his chin.

Angers is circled these days by orchards in which the espaliered trees run, in parallel rows, to the horizon. In spring, they are covered with white blossom so regular that the orchards seem covered with yard after yard of lace. In the slightly less amiable climate of Normandy, but from another Angers pear, Monsieur Boisbunel raised the large, almost round Passe Crassane at Rouen a year or two after the Comice pear. We now import it, labelled Passacrassana, from Italy, where the warmth cherishes the delicate tree – Bunyard describes it as a pear for the gardener who likes to live dangerously – making it less of a hazard to commercial growers.

Monsieur Boisbunel raised an even rounder pear, but a tougher one this time, a small russet-green fruit. Look out for it if you take autumn holidays in the west of France. We pass a tree on a favourite walk and have nearly cracked our teeth on its fruit. I have tried simmering the pears without success. Now I know why. Bunyard identified it as the Olivier de Serres, named 'in honour of the Father of French Agriculture'. It should be stored until February or March. By then it has developed a 'flavour which makes it a worthy epilogue to the pear season'.

Towards the end of the summer holidays, which is almost the beginning of the pear season, look out for local varieties in the markets of France if you are shopping for a picnic. Earlier, the pears are mainly the rather dull Guyot, which improves with cooking. Then come the Williams' pears. Look out, for among them you may see a red pear of a deep yet vivid coral-toned red. Moreover, it is entirely red, not just gold with large red splashes. We made its acquaintance last year, in September; the grower, from Ronsard's village of Couture, was proud of his fine pear and made us taste it. He gave the impression that it was a happy mistake, something to do with his Williams' pears, but what? He could not account for it, but weren't we lucky, Madame, to have it at all? What do names matter?

HOW TO CHOOSE PEARS

Extremely difficult. Pears should be picked firm and then ripened in the house. A ripe pear gives very slightly round the stem, but should be in no way squashy. All this provides problems for the shopkeeper and supplier. The result is that most people have never eaten a decent pear in their lives. And anyone who read the pear introduction may well be wondering what on earth the fuss was about when the Doyenné du Comice was first produced.

When commerce was less efficient, I suspect that many more people grew pears. My husband remembers being taken by his mother to the Midlands before the First World War, and visiting house after house where pears were produced from back garden trees, as a special treat. They were sampled and compared as knowledgeably and thoughtfully as wine at a tasting.

Perhaps we shall return to this pleasant occupation of pear growing, now that family trees are on sale. In one bush, three varieties are grafted and the size makes them suitable for the smallest garden. Then at least you may be able to experience perfection. Doyenné du Comice is bound to be one of the three varieties (Williams' Bon Chrétien and Conference the other two). 'The tradition is to pick them on or soon after 1 November and then watch them as they come to perfection (the old legend that towards the end it may be necessary to get up at 3 a.m. to find absolute perfection is not a great exaggeration)' according to R. H. L. Gunyon, who wrote and sent me much information on pears.

Meanwhile, as your family tree grows, you will have to continue with shop supplies. Buy pears firm, and keep them in the kichen until they are ripe. Eat them as soon as possible, storing them if you must in the bottom of the refrigerator wrapped in newspaper as protection.

One thing I learnt in Italy, in Florence, was that fine pears should be set off by fine cheese. That was thirty years ago. Now the idea has spread. French pear salads are dressed with Roquefort vinaigrette, and at Locket's in Westminster, they serve the pear and Stilton savoury described later in the section.

I read the other day that the pear in cooking always needs help from another fruit, even if it is only lemon juice to prevent its flesh discolouring as it is peeled and quartered, or lemon peel in the poaching syrup. On reflection, I suppose it is true – think of pears with sauce Melba or simmered in a white wine syrup, finished with orange juice, or mixed with black currants or the red juice of quinces. A pear sorbet requires lemon sharpness, even if the finishing touch is a dosage of Williams' pear *eau de vie*.

A LESSON IN EATING PEARS

At Trôo, we had a good friend who helped with the hard labour of the house. One autumn day at lunch we were eating some beautiful pears, when suddenly he laughed, and went on laughing until his eyes were rimmed with tears. '*Une de ses poires*,' he would say, then begin laughing again. At last he quietened down and told us the tale. His father, Lucien, had been coachman to a wealthy *curé* in the small market town where we had seen the bright red pears. The *curé* was very fond of Lucien, and they often had their meals together in the presbytery.

> One such time the lunch included two fine pears which Lucien had brought in, by way of the stables, from the *curé*'s garden. The *curé* took one of the two pears from the fruit stand and bit into it with great enjoyment, without peeling it. When he had finished, Lucien took the other pear and started to peel it with some attention. The *curé* suddenly realized what he was doing.
>
> *'O, Lucien, qu'est-ce que tu fais là? Pour déguster un bon fruit, il faut manger avec le peau.'*
>
> *'Mais oui, M. le Curé, je le sais bien, mais il y a une de ces poires qui a tombée dans la merde, et je ne sais plus laquelle.'**

Of course, the *curé* was right – you should eat a pear, skin, core and all. But wash it first, thoroughly.

HAMBURG EEL SOUP
HAMBURGER AALSUPPE

I gave a simple recipe for Hamburg eel soup, made with fresh pears, in *Fish Cookery*. Since that book was published, I have taken to using dried pears which give a better flavour, and using them, too, for this version of Alan Davidson's from *North Atlantic Seafood*. He points out that North Germans have long made a Saure Suppe in spring and early summer when the first vegetables arrive. They put in a ham bone, and to add a balancing sweetness, the last of the winter's dried fruit is also added, the new apples, plums, apricots and pears not yet being available.

To this recipe, cooks in Hamburg began to add eel in the nineteenth century. To me, this is what makes the soup.

1 smoked ham bone
1½ kg (3 lb) prepared vegetables, peas, carrots, leek, celery, cauliflower, string beans, kohlrabi etc
200–300 g (about ½ lb) dried fruit, pre-soaked
½ kg (1 lb) eel, cleaned and cut into sections
2 bay leaves
bouquet garni
summer savory
dash of vinegar
sugar, salt

**Notes From An Odd Country*, Geoffrey Grigson.

DUMPLINGS
250 g (½ lb/2 cups) flour
½ teaspoon salt
1½ tablespoons (2 tbs) marrowfat, suet or butter
1 egg
150 ml (5 fl oz/⅔ cup) water

Simmer the bone in plenty of water for 20 minutes, removing scum as it rises. Add the vegetables in batches, according to the time they are likely to take to cook. Then put in the drained fruit. Simmer until everything is tender.

Poach the eel until cooked in water to cover, with the herbs and vinegar.

Extract ham bone and remove any bits of meat. Chop them and put back into the soup, adding sugar and salt to taste. Mix the dumpling ingredients, rolling them into little balls. Poach them in water, with a lid on the pan, for 5 minutes, then put into the soup with the eel. Simmer briefly, then serve.

LOCKET'S SAVOURY

Here is a reminder of the splendid Locket's savoury (I usually serve it as a first course), which I included in *English Food.*

Toast two small slices of white bread per person. Cut away crusts. Top them with watercress, then slices of Doyenné du Comice, and shavings of Stilton cheese. Put into an ovenproof serving dish and bake in a moderate oven for about 5 minutes, until the cheese starts to melt. Grind over plenty of pepper.

If I can buy good pears, we always have this dish for supper over the Christmas holiday. Refreshing and sharp and crisp.

PEAR SALADS

Pears make a good salad, to be served on its own as a first course, or with any of the cold meats that combine well with fruit.

The basic simplicity is peeled, sliced pears dressed with vinaigrette, and a good sanding of pepper.

1) You can build on this by using walnut or hazelnut oil and cider vinegar, instead of olive oil and wine vinegar or lemon juice. Scatter the top with chopped walnuts, or chopped toasted hazelnuts, and surround the pears with a ring of watercress sprigs.

2) Pears and cheese go well together. Start the dressing by creaming some Roquefort, Stilton or Gorgonzola. Add sunflower oil, a little wine vinegar and seasoning to taste.
3) Peeled, halved and cored pears can be filled with curd cheese or drained *fromage frais*, beaten up with a little cream. Add salt, pepper, parsley and tarragon, or sugar and some chopped nuts, according to the stage of the meal they are intended for. Serve with walnut granary bread, p. 462.

Note Don't forget to rub over the peeled and sliced or cut surfaces with lemon or orange juice, to prevent them turning brown. It is a good idea to have a deep plate ready and waiting, with the juice squeezed into it, then the pieces can be dropped in as you prepare them.

AUSTRIAN PEAR SALAD
KAISERBIRN-SALAT

We do not grow or import Kaiser pears, but Comice or Williams' pears can be used instead for this Austrian dish. The essential extra ingredient is Maraschino – in Austrian homes, 'it is kept almost only for this reason. Every keen housewife and cook comes back from Italy with a bottle of Maraschino for fruit salads.'

Choose perfectly ripe pears (this is a dish for showing off fine pears at their finest point).

Peel, quarter and core the fruit. Roll each piece in caster sugar and put it into a dish. Squeeze over the juice of lemons in appropriate quantity, and add Maraschino. Leave for an hour until sugar, lemon and liqueur turn to syrup. Taste and add more lemon or more Maraschino if you think fit. Serve chilled.

If you have a zester, grate some very fine shreds of lemon peel over the whole thing before taking it to table. Serve chilled as a dessert.

BUTTERED PEARS
POIRES AU BEURRE

According to the size of the pears, allow one or slightly less for each person. Choose them ripe but still firm. If you have some Williams' pear brandy, choose Williams' pears. Otherwise Doyenné du Comice, Passa-crassana or any other fragrant eating pear will do just as well. Even the first rather tasteless Guyot pears of summer are interesting when treated this way. This simple dish is so good that I am amazed that only French children seem to know it. Until we went to Trôo with a small daughter of our own, we had never tasted it.

Peel, core and cut the pears into even wedge-shaped slices. Sprinkle them with lemon juice as you work, to stop them turning brown.

Fry them gently in clarified butter until they are an attractive golden brown underneath. Turn them and sprinkle them with 2–3 tablespoons caster sugar. Gradually the juices will caramelize.

Put the tender pieces on to a dish, or plates, and deglaze the pan with water mixed with a little pear brandy if possible. Boil down to a good flavour. Mix in some extra butter, unclarified this time, still stirring, and pour over the pears.

Serve with custard sauce, or just on their own.

For a more detailed description of the method, turn to Buttered Apples on p. 13.

COMPOTE OF PEARS WITH QUINCE AND VANILLA

A beautiful red dish for autumn dinners, the colour set off by the bloomy blackness of a vanilla pod.

4 quinces
1 whole vanilla pod
250 g (8 oz/1 cup) sugar
8 large pears

Rub any grey fluff from the quinces. Wash them and remove peel, then quarter and core them. Put the vanilla pod and quince debris into a pan and cover generously with boiling water. Simmer steadily for half an hour, covered. Strain off the liquid into another pan, extricate the vanilla pod from the quince debris, and put it into the pan with the liquid, plus the sliced quince and sugar. Simmer with a lid on the pan, until the fruit begins to become tender and the liquid looks deep pink. Peel and core the pears, cutting them into wedges. Put them carefully into the pan, cover it and leave to simmer until the pear slices are tender.

Remove the fruit to a bowl with a slotted spoon. Taste and consider the liquid: if it is copious and watery, boil it down hard. It should have a syrupy consistency. Pour over the fruit, putting the pod on top, and leave to chill.

Serve with cream and little biscuits: shortbreads or almond biscuits are the best kind.

CHINOISERIE
POIRES À LA CHINOISE

A delightful and unusual recipe which, I am sure, is thoroughly French. That is why I've translated the title as Chinoiserie.

100 g (3½ oz/good ½ cup) sultanas
100 g (3½ oz/1 cup) pine nuts
3 level dessertspoons honey
50 g (1½ oz/3 tbs) butter
8 large ripe pears
175 ml (6 fl oz/¾ cup) dry white wine
3 dessertspoons preserved ginger syrup
150 g (5 oz/½ cup) red currant jelly

Chop the sultanas and nuts, then mix them with the honey (if the honey is very stiff, melt it to measure out the quantity needed). Use the butter to grease a deep oven pan which will take the pears comfortably.

Peel the pears. Cut a thin slice from the base so that they stand upright. Cut a cap from the top. Scoop out the core and pips, trying to leave the cavities closed at the bottom: this is not always possible.

Stuff the cavities with the honey mixture, replace the caps and stand the pears in the pan. Pour in the wine. Cover with foil and bake at gas 6, 200°C (400°F) until they are tender.

Remove the foil. Put the pears in bowls, and return to the switched-off oven to keep warm. Add the ginger syrup and jelly to the pan juices, and boil them vigorously, stirring, until smooth and syrupy. Pour over the pears and serve at once.

POIRES BELLE-HÉLÈNE

Escoffier became the great master of delicious puddings, and for a particular reason. At the time when he first met César Ritz in 1883 at Monte Carlo, women were just beginning to dine out in restaurants – something they had not done before, unless they had no reputation to lose. To encourage – and profit from – this civilizing trend, Ritz designed luxurious hotels with spacious, tactfully-lit restaurants, serving Escoffier's simplified menus of exquisite food.

These menus often ended with some special dish to compliment the women – beautiful, rich, famous, royal, notorious – who had brought him success. They had to leave the restaurant with a last impression of delicious flattery. So he invented a repertoire, based largely on fruit, including *pêches Melba* and *poires Mary Garden* for the two singers, *mousseline d'oeufs Réjane* and *poires Belle-Hélène* for Gabrielle Réjane, the star of Offenbach's *Belle-Hélène* and other light operas produced at the Paris Vaudeville.

Poires Belle-Hélène consists of pears poached in vanilla syrup, arranged on slices of vanilla ice cream and sprinkled with crystallized violets: a chocolate sauce, hot, is served separately. It's a dish that has had much criticism from some food-writers, for the combination of delicate pears with chocolate sauce, but when properly made it can be good. The secret lies in taking very little sauce.

In advance, make the vanilla ice cream on p. 473.

PEARS
6–8 fine dessert pears, just ripe, still firm
juice of a lemon
6 heaped tablespoons ($\frac{3}{4}$ cup) vanilla sugar
1 vanilla pod
300 ml ($\frac{1}{2}$ pt/$1\frac{1}{4}$ cups) water
crystallized violets

SAUCE
200 g (7 oz/good cup) bitter or dark dessert chocolate
4 tablespoons ($\frac{1}{3}$ cup) strong black coffee
3 tablespoons (scant $\frac{1}{2}$ cup) sugar
175–200 ml (6–7 fl oz/generous $\frac{3}{4}$ cup) whipping cream
2 tablespoons (3 tbs) kirsch (optional)

Peel, halve and core the pears, sprinkling them with lemon juice. Bring the next three ingredients slowly to the boil in a shallow pan, stirring to dissolve the sugar. Simmer for 5 minutes. Slip in the pears, cut side down. Immediately raise the heat, so that the syrup boils up and covers the pears. Maintain the heat for 30 seconds, then lower it, cover the pan and leave the pears to simmer until barely tender and slightly transparent. Turn them over after 4 or 5 minutes. Transfer the pan, covered, to a cool place. The pears will be perfectly cooked by the time they are quite cold.

To make the sauce, use a double boiler or a basin over a pan of simmering water. Put in the chocolate, broken up, with the coffee. As the chocolate begins to melt, stir it together, then gradually add sugar and cream; when smooth, take off the heat and store until required. Reheat over simmering water, adding kirsch at the last moment.

To assemble, put slices of vanilla ice cream into six or eight dishes. Arrange two well-drained pear halves on top of each. Sprinkle with a few crystallized violets. If the assembly has to be done before the meal, keep cold in the refrigerator. Reheat the sauce while you serve the cheese course, pour into a sauce boat – preferably a metal one that can be heated before you pour in the sauce.

ICED PEAR SOUFFLÉ
SOUFFLÉ GLACÉ AUX POIRES

The lightest, whitest and most poetic pudding in this book, the best way of ending a special meal. It can be made with other fruit purées and an appropriate alcohol, or even gin if all else fails. Tiny mirabelle or cherry plums do very well: they should be lightly cooked like the pears, then sieved to get rid of stones and tough skin. Raspberries, strawberries, kiwi fruit, Cape gooseberries – in fact most soft fruit – need only be processed and then sieved if they have pips or stones.

If this soufflé stands around, it can separate very slightly at the base. This does not matter, but it spoils the cloud-like appearance. Once you have made it, put it into the freezer, and aim to serve it in a frozen but soft condition, so that it is halfway between a fairly firm cold soufflé and an ice cream. If you have no freezer, and want to get ahead, make the Italian meringue up to 6 days in advance, and store it in a covered bowl in the refrigerator. Fruit and cream can be added a couple of hours beforehand, or less, and the whole thing returned to the cold.

250 g (8 oz/1 cup) sugar
4 large egg whites
500 g (1 lb) Williams' or Doyenné du Comice pears
1 lemon
300 ml (10 fl oz/1¼ cups) double cream
4 tablespoons (⅓ cup) pear eau de vie

First make the Italian meringue. Dissolve the sugar in 125 ml (4 fl oz/ ½ cup) water over a low heat. When clear, bring to the boil and boil hard until you reach the hard ball stage, 120°C (248°F). If the sugar crystallizes above syrup level, wash down the sides of the pan with a brush dipped in water.

Meanwhile whisk egg whites electrically. Pour on syrup straight from the stove, and continue whisking until the meringue swells to a cloudy mass. Or use a rotary beater.

Peel, core and chop pears coarsely. Cook with a tablespoon lemon juice until just tender. Process or crush to a smooth purée. Taste and add more lemon, but no sugar.

Fold whipped cream and *eau de vie* into the Italian meringue, then the fruit purée. Turn into a soufflé dish of a generous litre (2-pt/5-cup) capacity, or a slightly smaller one collared with oiled paper. Serve chilled with sugar thins (p. 461) or almond biscuits.

ITALIAN PEAR PIE

Make shortcrust pastry with 375 g (12 oz/3 cups) flour and use half to line a 23-cm (9-in) deep pie plate. Put in peeled and cored pear slices, uncooked if properly ripe, lightly cooked if on the hard side. Scatter over them quarter their weight of sugar, and some chopped candied peel (p. 198 for making your own, which is nicer). Add a little cinnamon, too. Cover with the rest of the pastry and bake first in a hot oven, gas 6, 200°C (400°F) for 15 minutes, then at a moderate temperature until the pears are cooked.

Put a level tablespoon (2 tbs) of sugar, 50 g (2 oz/¼ cup) butter and 3 tablespoons (scant ¼ cup) of rose-water into a pan and bring to a boiling point. Pour syrup into 125 g (4 oz/1 cup) icing sugar, sifted, beating thoroughly until smooth. Spread over the warm tart, leave for a moment or two, and serve tepid or cold.

PEAR TART
TARTE AUX POIRES

A beautiful recipe. Rich enough for an elegant dinner, simple enough to please the youngest child at Sunday lunch. To me, this is perfect food, with each good ingredient making its point in a pleasant harmony.

creamed pâte sucrée, *p. 456*

FILLING
4–5 ripe pears
1 tablespoon lemon juice
300 ml (10 fl oz/1¼ cups) double or whipping cream
2 large egg yolks
vanilla sugar

Roll out the chilled pastry and line a large tart tin, about 25 cm (10 in) in diameter, or slightly larger. Bake blind – *see* p. 453 – at gas 7, 220°C (425°F) until very lightly coloured, about 15 minutes.

Meanwhile peel, core and slice the pears. This can be done longways or across, keeping the pear halves in shape. Sprinkle with lemon juice to prevent discolouring. Beat together the cream and egg yolks, sweetening to taste.

When the pastry comes out of the oven, arrange the pear slices on it. Long slices go in concentric circles, the sliced halves should be put on the pastry and spread out slightly so that they are flattened but still in the shape of the halves, with the pointed ends towards the centre. Pour over the cream mixture, and return to the oven. Close the door and turn the heat down to gas 5, 190°C (375°F) until the cream is nearly firm. It tastes best when not quite set in the centre. Serve warm.

Note Apples can be used for this tart, eating apples. It is a good idea to cook the slices in a little butter first, so that they begin to soften. Should your pears be on the hard side, cook them gently in a little syrup with lemon juice and lemon zest.

SPRINGFIELD PEAR CAKE

Every autumn, our fruit-farming neighbour at Springfield gives his friends the fine large almost-ripe pears that he cannot sell to the shops because they have slight blemishes. This is a favourite way of using them, to make an upside-down cake flavoured with ginger.

TOP

90 g (3 oz/⅓ cup) lightly-salted Danish butter, cut up
90 g (3 oz/⅓ cup) granulated sugar
2 tablespoons (3 tbs) syrup from preserved ginger
3 or 4 large firm pears
juice of a lemon

CAKE

125 g (4 oz/½ cup) softened butter
125 g (4 oz/½ cup) caster sugar
100 g (3½ oz/scant cup) self-raising flour
1 level teaspoon baking powder
30 g (1 oz/¼ cup) ground almonds
2 large eggs

3–4 tablespoons ($\frac{1}{4}$–$\frac{1}{3}$ cup) syrup from preserved ginger
4 knobs preserved ginger, coarsely chopped

Take a shallow cake tin or *moule à manquer* that measures 23–25 cm (9$\frac{1}{2}$–10 inch) across and at least 3$\frac{1}{2}$ cm (1$\frac{1}{4}$ inches) deep. Set it over a low heat and put in the Danish butter. Stir about with a wooden spoon, pushing the butter up the sides of the tin to grease them. Add sugar and syrup and stir about until you have a rich creamy-fawn bubbling mixture, a pale toffee mixture. Remove from the heat.

Peel, core and thinly slice the pears, turning them in lemon juice on a plate so that they do not discolour. Arrange them in a sunflower effect on the toffee base. Put the larger pieces round the outside, curved side down and overlapping, so that the base is evenly covered with two rings of pear slices, and some central pieces.

Tip all the cake ingredients, except the chopped ginger, into an electric mixer bowl or processor, and whizz to smoothness. Or beat everything vigorously together with a wooden spoon. Add the ginger and spread over the top of the pears. Bake at gas 5, 190°C (375°F) for 45 minutes. If the top is richly brown and well risen, turn the heat down to gas 4, 180°C (350°F). Leave another 15 minutes, or until the edges of the cake have slightly contracted from the tin, and a skewer pushed in almost horizontally comes out clean. Leave to cool for a few minutes on a wire rack, then run a broad knife blade between the tin and the edge of the cake to make sure there are no sticking patches.

Put a serving plate with a rim on top, upside-down, then turn the whole thing as rapidly as possible – use a cloth to protect your hands. A certain amount of juice will flow from the cake, but this only adds to its deliciousness.

Serve hot, warm or cold, on its own, or with cream as a pudding. It's a spoon-and-fork cake, being far too messy to eat with your fingers at afternoon tea.

ANJOU PEAR CAKE
FONDANT AUX POIRES

When the Doyenné du Comice was presented to the Comice at a meeting on Thursday 10 October 1850, in Angers, two more pears were mentioned. One from the year before, buttery in texture and called after the President, the Beurré Millet; and one which was obtained at the same time as the Doyenné and worthy to be ranked with it, the Fondante du Comice, the melting Comice pear. These two were prized very highly at the time, but they have been overshadowed in spite of their 'exquisite qualities'. At least the two words, buttery and melting, remind us of the ideal qualities of the pear, which in its best manifestations is quite without grittiness on the tongue.

$\frac{1}{2}$ kg (1 lb) pears
juice of a lemon
6 rounded tablespoons (1$\frac{1}{2}$ cups) flour
1$\frac{1}{2}$ level teaspoons baking powder
pinch salt
75 g (2$\frac{1}{2}$ oz/$\frac{1}{3}$ cup) caster sugar
2 large eggs
generous 60 g (2 oz/good $\frac{1}{4}$ cup) butter, melted
1 tablespoon water

GLAZE
2 tablespoons (3 tbs) apricot jam
2 tablespoons (3 tbs) water

Peel, core and cut up the pears into a basin containing the lemon juice. Turn them in the juice so that they do not darken. Mix remaining cake ingredients in the order given, then fold in the pears with any juice. Turn into a lined 20-cm (8-in) cake tin, which has been brushed with extra melted butter. Bake for an hour at gas 4, 180°C (350°F), or until cooked.

Make a glaze by boiling together the apricot jam and water. Sieve out the lumps.

Remove the cake from its tin and brush it over with apricot glaze. It can be eaten hot or warm as a pudding, with cream if you like, as well as cold.

TARTE DE CAMBRAI

On arriving in France, we usually go straight to the *droguerie* to buy paraffin for the stoves. We are always greeted enthusiastically – being the first of the returning cuckoos of summer – but one year was exceptional. No preliminaries of health, weather and absent friends, straight to business. 'I've been waiting for you,' said Madame Guilbaud. 'I've a recipe for your readers. Easy. Quick. Delicious. Take it down.' I obeyed, and have been grateful ever since.

4–5 large ripe pears
1 tablespoon lemon juice

BATTER
10 level tablespoons self-raising flour
6 level tablespoons vanilla sugar
4 tablespoons oil
8 tablespoons milk
2 whole eggs
pinch salt

PLUS
60 g (2 oz/¼ cup) butter
extra sugar

Peel, core and slice the pears. Sprinkle with the lemon juice, turning the slices over.

Mix the batter ingredients in the order given. Grease a 25-cm (10-in) flan or shallow cake tin with a butter paper and pour in the batter. Arrange the pear slices on top. Dot with butter, and sprinkle evenly with extra sugar. Bake for about 50 minutes at gas 6–7, 200–220°C (400–425°F), or until golden brown and risen, with bits of pear showing through. Eat warm, and serve with cream, if it's being eaten as a pudding.

Note Tarte de Cambrai can also be made with apples, and with peaches (they should be lightly cooked first, to draw off most of the juice). Bananas work well, too. To make the top extra good, scatter 30 g (1 oz/ ¼ cup) blanched slivered almonds over it before putting in the oven.

PICKLED PEARS

Spiced pickled pears to serve with ham or salt pork, but also with cold duck. If you prefer a deep red appearance, use red wine vinegar instead of white.

6 large ripe but firm pears
500 g (1 lb/2 cups) sugar
250 ml (8 fl oz/1 cup) white wine vinegar
1 teaspoon whole cloves
1 teaspoon whole allspice
small piece nutmeg or 7 cm (3 inch) piece cinnamon

Peel, core and cut pears into 8 slices each. Cover with water – about 750 ml (1¼ pt/generous 3 cups). Boil sharply for 5 minutes. Strain off and measure liquid. To 600 ml (1 pt/2$\frac{1}{2}$ cups) add sugar, vinegar and spices. Pour over the pears and simmer until the pieces are cooked and translucent – 20–30 minutes. Pour into a bowl and leave overnight. Drain off liquid next day into a wide shallow pan and boil hard for 5 minutes to reduce it slightly. Pack pears into warm sterilized jars with the spices. Pour in boiling syrup to cover. Seal while warm. Store for at least a month before using.

Note Any leftover syrup can be used for cooking pears to serve hot, German style, with roast venison, or casseroled wood pigeon, or roast duck. Sometimes, in a good French restaurant, a neatly-cut fan of pickled pear slices will be laid on your plate with a piece of hot duck, usually the breast steak grilled and sliced, and a discreet vegetable or two on either side.

WILLIAMS' PEAR LIQUEUR

Two of life's mysteries – how does the ship get into the bottle, and how does the pear get into the bottle of pear *eau de vie*?

The first, I am told, is done by strings. The second is a gardener's trick. Once the pear blossom kerns into tiny fruit, select a likely one and push it carefully on its branch into a large bottle. Keep the bottle in place with forked sticks placed firmly in the ground. Stop up the end with tape or wax, to keep insects out. The bottle acts as greenhouse and the pear swells. When it is ripe, the bottle is cut from the tree.

To make the liqueur, you pour in a litre (1$\frac{3}{4}$ pt/scant 4$\frac{1}{2}$ cups) of white fruit spirit (use vodka in this country – it will give a better result anyway). Cork the bottle and leave it as much as possible in the sun for a month.

Open the bottle and add 300 g (10 oz/1$\frac{1}{4}$ cups) caster sugar. Close it again and leave for three months until the sugar has dissolved. Finally add a liqueur glass of *eau de vie de poire*, and it is ready for drinking.

If you do not grow pears, put a ripe Williams' pear into a wide-mouthed bottling jar and pour in the vodka to come about two-thirds of the way up the jar and to cover the pear. You will need less than a litre (1$\frac{3}{4}$ pt/scant 4$\frac{1}{2}$ cups), so scale down the sugar proportionately when it is time to add that. It will do no harm, however, to put in the same quantity of pear *eau de vie*.

Note I believe you can do the same trick with a bunch of grapes, though I have never seen it. Choose a rounded bottle of the old-fashioned Chianti type, to allow for the larger width of a fine bunch of grapes.

PERSIMMON and SHARON PERSIMMON

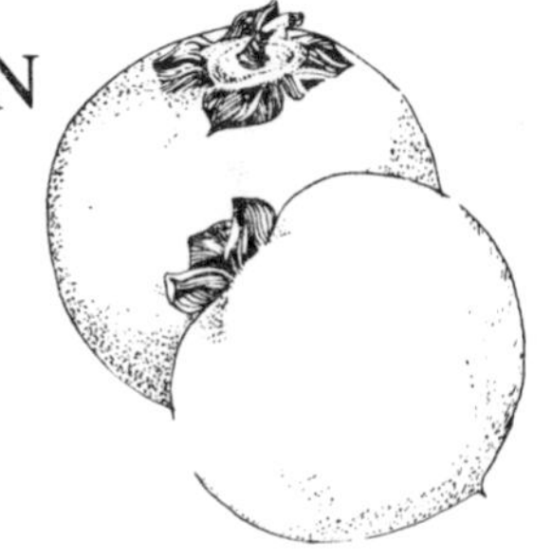

The first persimmon we ever heard of in this country grew in Virginia, *Diospyros virginiana*, the wild persimmon that was soon known to the early settlers, and described by travellers to the colony. William Strachey wrote, in his *Historie of Travell into Virginia Britania* (1612), that 'they have a plum which they call Persimmons like to a medlar in England, but of a deep tawny colour they grow on a most high tree, when they are not fully ripe, they are harsh and choky, and fur a man's mouth like alum, howbeit being taken fully ripe it is a reasonable pleasant fruit, somewhat luscious, I have seen our people put them into their baked and sodden puddings, there be whose taste allows them to be as precious as the English apricot I confess it is a good kind of horse plum.'

It is the 'simmon of various American folk songs and rhymes:

> Possum in a ''simmon tree,
> Raccoun on de groun',
> Raccoun ask de possum
> To shake dem 'simmons down.

and Strachey describes it well. When the fruit is full and nicely firm, apparently ripe, it is far too astringent to eat. Like a medlar or a sorb, it must be left until it is really soft, pulpy, you might think on the verge of rotting. Then its translucent glowing skin does indeed hold a luscious fruit. I have tried the 'baked and sodden', i.e. stewed puddings given in American cookery books, and cannot say I liked their heavy sadness. The only reason for making them would be an excess of persimmons, which is something we are never likely to be able to afford.

The persimmon we buy here, the glowing tomato-like fruit with its dry, neat brownish fawn calyx, is originally from Japan, *Diospyros kaki* (the French have taken up the Japanese name of *kaki*). You might see it labelled Sharon or Sharon fruit, meaning the persimmon developed by the Israelis in the Sharon valley: this can be eaten with pleasure at the firm stage, skin and all, seeds and all. No need to wait for the pulpy, plummy softness to develop before you eat it. With Sharon persimmons, greengrocers do not run much of a risk of losing stock. They do not have to contend with complaining customers who ate their persimmons too soon. It is good that persimmons can be popularized and made available, that they have a chance of becoming as familiar a part of our diet as avocados are now. But choose them when they are brown under the skin, or they taste too bland, by comparison with the soft ripe persimmon and the rich flavour that succeeds the bitterness. I hope that we shall be able to have both kinds in increasing quantities, the Sharon persimmon which slices so neatly and prettily for salads – try it alternating with avocado, and French dressing, served with smoked chicken – and the deeper-tasting soft persimmon for dessert, served with cream and nothing else.

Naturally the Japanese understand persimmons well. They have long been a major fruit in that country. The poet Issa used them to symbolize the maternal self-sacrifice:

> Wild persimmons
> The mother eating
> The bitter part.

The German surgeon, Kampfer, who was attached to the Dutch East India Company in Nagasaki from 1690–1692, was dazzled by the strange autumn beauty of the tree in full fruiting. The leaves fall off before the persimmons are ripe. Seen from a distance, they stand out on the tree like tiny tomatoes sewn all over a crinoline. Bashó, the poet, who died two years after Kampfer left Japan, described an old village in a *haiku*, each house with its twisted persimmon tree. A sight for Samuel Palmer who might have thought he really was looking at magic apple trees. The great poet for persimmons, Shiki, lived nearer our time and suffered from eating too many of his favourite fruit. He was sent thousands of *haiku* by hopeful readers and conscientiously read through them all to choose the best. As a reward, he would eat two persimmons every three thousand *haiku*. He wrote himself a poetic epitaph:

> Write me down
> As one who loved poetry,
> And persimmons.

Every child in Japan knew the origins of the persimmon tree from one of the miraculous tales about the childhood of the great Samurai leader of the twelfth century, Yoshitsune. It tells how the great giant Benkei was knocked over by a blow from Yoshitsune who was no bigger than a boot, but braver than a tiger. The noise of his fall was so tremendous that the earth shook and split open. And out of the crack came a tree covered with these beautiful squashy orange-red fruit that were full of juice for the thirsty fighters. From then on they were inseparable, and Benkei became Yoshitsune's champion fighter – he has been described as the Little John of Japanese legend.

HOW TO CHOOSE AND PREPARE PERSIMMONS

Always enquire about the persimmons you are buying or check with the box labels. You will then know whether they are the Sharon kind that can be eaten straightaway, sliced through to show the beautiful star shape of the seed pattern, or whether you should keep them in the kitchen for a few days until they have that translucent glow that in many fruit would indicate over-ripeness.

With Sharon fruit, lemon juice helps to bring out the flavour. It can be served as a first course salad on its own with a light vinaigrette, or with other fruit (avocado and watercress, as I have mentioned, or grapefruit) or cream cheese and toasted nuts. It can be arranged with cold poultry and smoked meat, as you might serve melon. Or it finishes the meal, with sugar and soured cream mixed with double cream. The skin is edible.

With persimmons of the normal kind, slice off the top and eat the pulp with a spoon, adding cream if you like. If you cannot afford one persimmon each, the pulp can be scooped out and turned into a fool or water ice. You could put it into little tartlet cases, using the mango pie recipe on p. 233 as a guide. A mousse or a bavaroise or a fruit curd or custard can be flavoured with good persimmons.

PERSIMMONS IN CHINA

According to one thirteenth-century play, they sold candied persimmons in the street, crying 'supple-supple-soft, quite-quite-white, crystal-sweet, crushed-flat candied persimmons from Sung-yang'.

PHYSALIS or CAPE GOOSEBERRY or GOLDEN BERRY

The prettiest of fruits, a prettiness we are all familiar with – in the garden, at least, where orange Chinese Lanterns, *Physalis alkekengi*, are left to dry on their stalks, to be cut eventually and brought in for winter decoration. Another variety with similarly shaped fruits, *Physalis peruviana*, is grown especially for eating: the orange berry inside the greenish calyx swells to a plump cherry size.

This physalis, from South America, was already a major fruit crop in South Africa at the back of the Cape of Good Hope at the start of the nineteenth century. So to us, they became famous as Cape gooseberries. Nowadays, the cans are labelled Golden Berries. They taste well enough from tins, but this way you lose the beauty which is so important a part of the fruit, the orange-red berry glimmering through its dried out, gauzy calyx.

For this reason, and for the fresh piquancy of the flavour, physalis was bound to attract the attention of growers in Europe. The fruit had been known for sometime, but it was not until the penitentiary of Dumbea in New Caledonia sent some seeds to the Societé d'Acclimitation in Paris, in 1878, that things got going. Two important members of the Société, Monsieur Bois and Monsieur Pailleux (the one who first grew Chinese artichokes in his garden at Crosnes) did their best to push its cultivation. They saw its usefulness as a winter soft fruit. From their book, *Le Potager d'un Curieux*, we can see how they set to work, and how successful they were.

They described the potential of the crop for a Mediterranean grower. He could plant out of doors. If the fruit could be protected from frosts, he would be able to pick them all winter, leaving them on the plant – 'berries picked in Antibes on March 7th were perfect when received by a member of the Societé d'Acclimitation. Their ripeness and their freshness were irreproachable.' If the fruit could not be protected, it could be picked once the calyx was perfectly dry, and then stored in a cool dry place for four months. This meant, either way, that growers could supply Paris with spaced and regular deliveries. Not only is this nice for the Parisians, but it means that a stable price can be maintained for the growers. The gentlemen do not spell this out, but it is implied. The plant is also extremely productive, plants having been seen by the authors with over 100 physalis.

This storability is an enormous advantage. That delicate-looking calyx is, in fact, quite a tough protection, from insects as from the bumps and drops of transportation. I have bought physalis in London at the end of January (for a disgusting price, admittedly) and they were perfect.

As the second edition of the *Potager d'un Curieux* was being prepared, physalis were much in demand. Paris confectioners and pastrycooks could never get enough of them for their tarts and *petits fours*. In one season, a popular *pâtissier* had sold 30,000 physalis berries, wrapped in fondant or caramel. The two gentlemen were delighted; and we should be grateful to them.

Nowadays, the Curious Gardener can buy seeds for the edible as well as the garden variety from Thompson and Morgan of Ipswich, and grow them in his vegetable garden – his *potager* – with the kind of effort and care demanded by outdoor tomatoes. The name on the packet is Golden Berry, which is a good if unromantic description (physalis goes better with the delicate appearance). And as my berries were 15 pence each, I would think that any small gardening effort they demand was worthwhile.

Other physalis varieties are edible – the berry of *Physalis alkekengi* in our gardens can in fact be eaten though it is very sharp and not particularly succulent on account of its small size. There is the dwarf Cape gooseberry, *P. pruinosa*, or ground cherry, which is also good to eat, either raw or cooked, with its slightly acid sweetness. The large tomatillo or jamberry (*P. ixocarpa*) is important in Mexican cooking, and can be bought here in cans: a point to watch when you are using Mexican recipes, as you can get the impression that the required item is just another form of tomato (sometimes it is called the Mexican husk tomato). The berry is sticky and the calyx sits closely round it, instead of acting like a cage. I have never eaten it raw, but it is said to be dull that way: it needs a little cooking to bring out the flavour (when canned, it has of course already been cooked). As Elisabeth Lambert Ortiz points out in *Latin American Cooking*, when recipes tell you to skin a 'green tomato', it means that you should cut away the calyx that surrounds it, not that you should remove the skin of the fruit itself.

GOLDEN BERRIES IN DISGUISE
PHYSALIS DEGUISÉS

I first saw these enchanting sweetmeats at the Café Royal in 1972, when

Guy Mouilleron was chef there at the Relais and making his name in London. He gave me one to try, picking it up by its light wings from the tray of *petits fours*.

It never occurred to me to make them myself – fondant being a matter for confectioners – until Whitworths sent me sample packets of prepared fondant that are now on sale at every supermarket. Do not be put off by the pixillated wrapping, pink elephants and green sugar mice, or by the play-dough nastiness of the brown and pink kinds. Go for the vanilla fondant, which is quite good and can be improved by reheating with flavourings of your own. A plate of *petits fours* with coffee now becomes a possibility. A few golden berries for beauty, kumquats or large firm strawberries for elegance, and some tiny Danish shortbread knobs, p. 460, or discs of date nougat make a good ending. *Petits fours* always seem to give great pleasure.

The catch is that fruit fondants, fruits in disguise, must be made the same day you intend to serve them. If they hang around, juices escape and ruin the sugar coating. For this reason, avoid rinsing soft fruit and make sure you dry kumquats well. Some books suggest using canned fruit or cherries in brandy, but I find them too sweet. Pleasure comes in biting through sugar sweetness to a tart crisp fruit.

With the fondant all ready, the process is simple. All you need for success is organization.

First prepare the golden berries. Snip down the Chinese lanterns in three or four places, not right to the stalk but to where they curve in sharply towards it. Very carefully bend the sections back, so that they curve away from the berry. Trim them with scissors if they look ragged. These wings are tougher than you might think, and make efficient handles for dipping the berry.

250 g (8 oz) of fondant is enough for 24 berries, perhaps a few more once you get the knack and work more quickly. Break it up into a small basin, and add 2 tablespoons (3 tbs) of rum and 2 of water. Set the basin over a pan of barley simmering water and stir until the fondant turns to a smooth liquid. Crush any tiny blobs so that you end up with an evenly-coloured mixture.

Quickly dip in the berries, spooning the icing over them. If you have someone to help, it makes things easier. You can hold your dipped berry until it dries, while the helper dips hers. In fact, fondant sets remarkably quickly so long as it is not too liquid in the first place. You soon manage to get an even covering with a tiny point at the end.

Put the dried berries into little paper cases, with their wings nicely disposed.

GOLDEN BERRY CAKE
GÂTEAU AILÉ

If I had a mid-winter birthday, this is the cake I would choose for its charming delicacy and prettiness. The rum colours the fondant a pale fawn, the berries are bright orange, the wings greenish fawn. The nuts on the apricot glaze are tawny brown. The alternation of wings and brightness, the contrast of sweet and sharp with rich genoese and toasted nuts, make it a special delight.

As things are at the market, I make the quantity below for eight people. Gardeners with a supply of golden berries can raise the quantities according to their good fortune. Only the cake can be made in advance and frozen. Assembly and decoration should be done the day the cake will be served.

Provide yourself with a genoese baked in two sandwich tins of 16–18-cm (6½–7-ins) diameter, and at least 30 golden berries – 40 to 45 will be better, plus:

4–5 tablespoons (5–6 tbs) rum
sugar
apricot glaze, p. 477
30 g (1 oz/¼ cup) chopped toasted hazelnuts
250 g (8 oz) vanilla fondant
icing sugar

Remove all the berries from their Chinese lanterns except five. Cut and bend back the lanterns of these five to make wings. With them, put four of the finest berries and set aside for the time being.

Halve the rest into a heavy pan with enough water to prevent them sticking but no more. Cover and cook, not too fast, until they dissolve into a pulpy mass. Remove from the heat, stir in 2 tablespoons (3 tbs) of rum and a very little sugar. The purée should be tart, rather than sweet like jam, as there is plenty of sugar in the rest of the cake. Taste and add another spoonful of rum, if you like. When cool, spread on the base of one cake and put the second cake on top, base side down.

Heat the apricot glaze and brush it over the top of the cake and round the sides. Roll the sides in the nuts.

Over simmering water, melt the fondant in a small bowl with the last of the rum and a tablespoon of water. When it is evenly liquid, dip into it four of the five winged golden berries. They dry rapidly and should be left to finish hardening on a plate sprinkled with icing sugar. Spread enough of the remaining fondant over the cake to cover the top.

Finally, alternate the four remaining plain berries with the four fondant berries round the top of the cake. Put the last bright winged berry in the centre. Leave in the refrigerator until required.

CAPE GOOSEBERRY JAM

Hilda's Diary of a Cape Housekeeper, by Hildagonda Duckitt, was published in 1902, and reprinted in facsimile by Macmillan in 1978. The pleasant life of Dutch and English settlers is enchantingly described. It sounds like paradise before the Fall, when God was white and saw it was good. The section on servants is enlightening, but hardly different in tone from books for English housewives of the same sort of date. It is difficult to accept a world where everyone knew their place and was happy – unless of course it was a bad place. What then?

> Those who have Cape gooseberry bushes and have trailed them over a fence (where they like best to grow), can now by picking the ripe ones every day and collecting them for a week make a nice jar of jam . . . and for afternoon tea I don't know anything that is more appreciated by visitors from England, or nicer to send home.
>
> Cape Gooseberry Jam: – If you have, as suggested, collected the ripe gooseberries (the pods turn yellow when ripe), they may even be kept for 10 days till you have 3 or 4 lb [1½ or 2 kg] for a cooking. Shell them, and give one prick with a steel pin. Wash if dusty, then put the fruit in an enamelled or copper saucepan rubbed with olive oil, adding just enough water to moisten the gooseberries, and set it to boil pretty briskly for 7 or 10 minutes, then add the sugar (equal to weight of fruit); let it boil for another 10 or 15 minutes. Test if it is good by dishing a little in a saucer and letting it cool; if the syrup has a crinkly or creamy surface it is right and the syrup must be oily and thick. This jam will keep very well, and is one of our best Cape jams.

See also Old-fashioned Apricot Tart (p. 38) and use physalis.

PINEAPPLE

At Ham House near Richmond, there is a painting of John Rose, the royal gardener, handing Charles II the first pineapple grown in England. 'Pines' are mentioned in John Evelyn's *Gardener's Calendar* of 1664. Such exotic fruit were among the first products of the 'stove houses' buildings heated by Dutch stoves; with the development of these buildings – the stoves replaced by steam and piped hot water, the brick walls and roofs gradually replaced by glass – exotic fruits were more and more grown for the delectation, and at the expense of the landed gentry of this country.

If you stop and imagine yourself limited to the fruit you can grow and store in one way or another, you can understand what the first months of the year were like without fresh fruit. For the poor and middling, the lack was made up by dried fruit of which we once consumed quantities that were the amazement of the rest of Europe including the people who produced it. The rich, however, imitated tropical Edens in which pineapples, bananas, mangosteens, guavas, passion fruit might be attempted, or – more successfully and more cheaply – glassed in their south-facing brick walls to force 'wall fruit', apricots, peaches and so on.

The great stimulus to the production of fruit was Louis XIV's passion for it, and the demands he made on his long-suffering but successful gardener, de la Quintinie (whose book on gardening Evelyn translated). With all the excitement that exploration of America had provided, it is not surprising that people should have longed to have for themselves the pineapple, the King of Fruit.

The pineapple may have acquired its English name – the French stuck to *ananas* from the Tupi-Guarani language of Brazilian Indians – from its resemblance to pine cones, but it was a great deal more difficult to produce. Mr Rose may have been able to present his royal boss with a pineapple but it was another fifty years before Henry Tellende, gardener to Sir Matthew Decker at Richmond, was acclaimed as the first person to grow a decent pineapple. In 1714, he managed to bring the pineapple to perfection, 'to rejoice in our climate'.

This success pushed other landowners into a pineapple craze. They began to crown their gates and the corners of their roofs with stone pineapples, a symbol of hospitality, while teams of gardeners were busy on the real thing at ground level. William Speechley raised ten thousand in pots at Welbeck for the Duke of Portland, and published a book on pineapple growing in 1779. English gardeners led Europe. Jean-Sébastien Mercier, the French playwright, visited the grand establishment of the Duc de Bouillon near Evreux in Normandy, and gasped at the extent of pineapples there – four thousand pots, soon to be six, of this excellent fruit now 'naturalized in England'. If only the French would learn the simple procedures carried out by the English gardener who directs the operation, they would soon overtake the English. On his table every day, the Duc has eight or ten pineapples. Think of it! Fewer, presumably than the Duke of Portland.

The pineapple, and the other hot-house fruit and early vegetables, demanded effort and space. Do you remember Catherine Morland being shown the kitchen-gardens at Northanger Abbey? 'The walls seemed countless in number, endless in length; a village of houses seemed to arise among them, and a whole parish to be at work within the enclosure.' General Tilney explained that the gardens were his hobby-horse. Though he did not much care about food, 'he loved good fruit; or if he did not, his friends and children did. There were great vexations, however, attending such a garden as his. The utmost care could not always secure the most valuable fruits. The pinery had yielded only one hundred in the last year . . . '

The force of this ironical passage did not strike me until I read J. C. Loudun's account, in his *Encyclopaedia of Gardening* (1827), of the management of the pinery. His twenty-six large pages of small and smaller type, with diagrams, described just such a place as General Tilney had designed, no doubt in accordance with the instructions of William Speechley.

When you learn that each of the thousands of pots produced only one pineapple at a time over a period generally of two to three years, when you read of the nursery, the succession houses – General Tilney was solicitous to know how Mr Morland's succession houses were worked – the tan bark pits, dung-heat and fire-heat and steam-heat and oak leaves, shifting and potting, the insects and the cleansing, you begin to see that pineapple production was an extraordinary battle between the English gentry and the English climate. And when you think that as well as the pineapple battle, there was the guava skirmish, the banana engagement, the mangosteen flurry, you can understand that country life for the very grand was exciting and full of incident. The rewards were displaying better fruit than your neighbours, or graciously lending them an early pineapple or two as decoration when they were having a party.

Steam ships must have taken the heart out of so much industry. By the 1870s, they were bringing pineapples from the West Indies in 'pretty good condition'. This at least extended their enjoyment from the very rich to the well-off. And canned pineapples began to make an occasional treat for the rest of us. It is only in the last few years that pineapples have come down to a price and availability that puts them within the reach of almost everyone. A fine pineapple, enough for four to six people, even more if it is dressed up or used for ices, costs less than a packet of twenty cigarettes. What would Mr Speechley, or Mr Loudun, have thought of that?

HOW TO CHOOSE AND PREPARE A PINEAPPLE

Go by the plume of leaves. If it is fresh and lively, the pineapple is in good condition. Once you can pull out one of the leaves, with a neat tug, it is ripe enough to eat. By this stage, it will be smelling of pineapple, too. Occasionally, as you wait for a pineapple to ripen, a soft patch may develop. It can be cut away: the rest of the pineapple will be fine. With an over-ripe pineapple, make sauces, either sweet or savoury. Simmer the best parts, throwing away any that are obviously bruised and turning brown. Pineapples are around for most of the year. I have the feeling that they are best in February, March and April, but this may be an entirely subjective judgment reflecting the smaller range of fruit after Christmas which makes me extra thankful to have pineapples.

The easiest way of preparing a pineapple is to twist off the plume and then slice it across. Each slice can then be peeled, and the centre removed, if it is very hard, with an apple corer or tiny round cutter. Do this on a plate so that none of the juice is wasted.

If you wish to keep the pineapple whole, so that you can later cut it into long wedges like a melon, take hold of the plume (cut off the sharp ends first) and cut away the peel in downward strips. Remove the base as well. Stand the pineapple up again and cut diagonal channels with a small very sharp knife so that the 'eyes' come away with the tiny triangular strips. Try using a canelle knife or a mushroom cutter, but usually the eyes are too deeply embedded for them to be very effective.

If you want to turn the pineapple skin into a container (for fruit salad, or a pineapple ice), slice off a lid carefully, preserving the plume intact. Leaving a firm wall, cut down and round inside the pineapple so that you can remove it in chunks. It is likely that whatever you do with these pieces will provide you with enough to replenish the skin for a second helping or another meal. Empty or full, the skin can be frozen for future occasions. When you fill the skin with pineapple ice – pineapple liquidized with light syrup or Italian meringue, plus alcohol such as kirsch or gin – mound it up above the rim and perch the lid on top.

Another way of preserving the skins is to slice the pineapple down through the centre, leaves and all. You can mould an ice cream inside, and serve the two halves rounded side up, then turn them over for serving. Or you can mound up the ice cream in each half, and display it as if you had two pineapples. An American fancy, which children may find fun, is to use the flesh from inside the two halves and spear it on cocktail sticks, alternating the pineapple with cheese, bacon bits and so on; the halves are then placed round side up on a dish and the cocktail sticks pushed into them.

A warning – fresh pineapple cannot be set with gelatine, and it is none to easy to freeze either. Simmer it first, and all will be well.

A disappointment – I had imagined in my youth that pineapples grew magnificently on enormous cactus-like trees, in heraldic dignity, untroubled by wind. In reality, they grow directly from the ground and are as dull to look at *en masse* as cabbages in a field.

PINEAPPLE AND CUCUMBER SALAD

Diced pineapple and cucumber – equal quantities of each. Just before serving, drain well and mix with whipping cream and soured cream sharpened with a little lemon juice, salt and pepper. Or use a light mayonnaise. Chopped fresh mint can be added.

Serve with lunchtime drinks in the garden, and provide small salty biscuits to scoop it up with. The friend who gave me the recipe decorates it with blue borage flowers which look right on the cool, pale salad.

HAM OR GAMMON WITH PINEAPPLE

There are many versions of this dish, most of them horrible. Cross-Channel ferries, cheap catering have all had their way with the idea until their victims are sickened by merely the words on a menu sheet – sweet sweet pineapple, obligatory cherry, over-salted tough gammon. The point of the dish was to soften the fibres of salted meat with the juices of fresh pineapple, which makes sense of the whole thing. And of course fruit goes well with cured pork as everyone knows – figs or melon with Parma ham, or the sweet mustard fruit pickles of Cremona (*mostarda di Cremona*, an ancient product but difficult to find now in good quality), the range of plum and raisin sauces that are made from China to Great Britain. The point is the slight sweetness of the fruit and its acidity, not too much added sugar.

In this simple version – which I like because it can be adapted to thick slices of cooked ham or gammon, or to uncooked gammon steaks – Madeira provides the main sweetness apart from the natural sugar of the fruit.

Provide a thick slice of ham or gammon for each person, soak it if it is uncooked and salty.

Also provide a slice of peeled fresh pineapple for each person, plus one or two extra according to whether you are feeding six or a dozen people.

Melt a generous knob of butter in a shallow dish. Turn over the ham or gammon slices in the butter, then put the pineapple on top. Chop the extra slices roughly, and cook them in about 150 ml ($\frac{1}{4}$ pt/$\frac{2}{3}$ cup) water. When tender, remove from heat and add 3 tablespoons (scant $\frac{1}{4}$ cup) Madeira. Pour with any pineapple juice collected when cutting the slices into the ham dish.

Mix a tablespoon (2 tbs) of sugar with a teaspoon of cinnamon and sprinkle a little over the pineapple slices. You may not need it all. Bake in the oven at gas 4, 180°C (350°F) until bubbling and tender, basting occasionally for 10–20 minutes according to whether the gammon or ham was cooked or uncooked. Spoon off the juice and boil it down. Pour back, and serve with watercress and fried potatoes.

CARIBBEAN CHICKEN WITH PINEAPPLE

If you cannot buy fresh pineapple, use a can of pineapple, sliced, in unsweetened syrup.

1 chicken, cut into pieces
1 lime
salt, pepper
oil or clarified butter
175 g (6 oz/1$\frac{1}{2}$ cups) chopped onion
175 g (6 oz/$\frac{3}{4}$ cup) chopped tomato
2 heaped tablespoons (3 tbs) raisins
1 teaspoon of seeded, chopped red chilli
1 large clove garlic
1 pineapple or can of pineapple rings
chicken stock

Marinate the chicken in the grated peel and juice of the lime, with salt and pepper, for at least half an hour, turning the pieces every ten minutes. Drain and brown to light gold in oil or butter, then put in onion, tomato, raisins and chilli (dried chilli can be used, but include the seeds). Crush the clove of garlic with the blade of a knife, remove the skin and put into the pan.

Peel, slice and cut up the pineapple, if using a fresh one. Pour pineapple juice, whether from fresh fruit or the tin, into a measuring jug and bring up to ¼ litre (8–9 fl oz/1 cup) with chicken stock. Tip into the pan, cover it and leave the chicken to cook, turning the pieces occasionally and removing the breast as it becomes tender.

Arrange the chicken on a hot serving dish and keep it warm. Pour off surplus fat from the pan juices, and tip in the pineapple pieces. Boil hard to reduce and correct the seasoning. Pour some over the chicken but don't swamp it. Serve the rest in a bowl or jug. Provide boiled rice to eat with the chicken.

DUCK WITH PINEAPPLE
VIT NÂÙ THOM

In France, Chinese cookery has become popular, as you would expect, via Vietnam and returned settlers as well as Vietnamiens who wisely left before the Americans took over. This duck recipe has the South-East Asian accent of pineapple that we also know from the Cantonese cookery of many of our Chinese restaurants. I have adjusted it slightly from a little book I bought at the splendid oriental shop near the station at Tours, *Petit Livre de Recettes de cuisine Sino-Vietnamiennes et exotiques*. Use a can of unsweetened pineapple rings, better still a fresh pineapple peeled and cut into six rings and simmered in a little water until tender.

It is worth making the effort to find cloud ears (or wood ears as they are sometimes called). Their crisp jellied texture adds a lot to this dish. Most oriental stores have them, although they are not so much in demand as the *shiitake*.

1 large duck with giblets
1 large onion, sliced
2 large cloves garlic, halved
bouquet garni
*10 g (⅓ oz) dried Chinese mushrooms (*shiitake*)*
10 g (⅓ oz) dried cloud ear mushrooms
60 g (2 oz/¼ cup) butter

1 large onion, chopped
1 rounded tablespoon (¼ cup) flour
1 large clove garlic, chopped
flour
1 can pineapple rings
1 teaspoon tomato concentrate
salt, pepper, cayenne, soy sauce
¼–½ kg (½–1 lb) string beans, half-cooked

First bone the duck and cut it into pieces about 3 cm (good inch) square. Break up carcase and put with giblets, sliced onion, halved garlic cloves and *bouquet garni* into a large pan. Cover generously with water and simmer for 2–3 hours. This can be done in advance.

Reduce stock to ½ litre (¾ pt/2 cups). Put rinsed mushrooms into separate bowls and pour very warm water over them, to cover. Leave to soak and swell, about 20 minutes. Cut them into 3 or 4 pieces each. Discard the stalks of the Chinese mushrooms. Keep and strain the liquor.

The duck can take 1½ hours to cook, so give yourself plenty of time (the dish can always be reheated, but keep fruit and beans for adding at this stage to preserve their freshness). Brown it in half the butter, putting it skin side down first: you will need to do it in two pans, or two batches. As you turn the pieces over, add the chopped onion and garlic. When the duck is nicely coloured, pour off the surplus oil. Amalgamate the duck in one pan. Sprinkle with flour and turn the pieces again for 3 minutes.

Pour in the pineapple juice, tomato concentrate, mushroom liquor and enough of the stock to come most of the way up the duck. You will need more or less according to the quantity of pineapple juice. Add the mushrooms, stirring them in. Cover and simmer until the duck is almost cooked, then remove the lid and allow the sauce to reduce.

Taste for seasoning when the duck is tender, adding soy sauce if you like it, a tablespoon or two (1–3 tbs). Put in the pineapple cut into cubes, and the beans. Simmer a further 10 minutes. Serve with boiled rice.

'A TART OF THE ANNAS, OR PINE-APPLE'

Richard Bradley, first professor of Botany at Cambridge, inventor of the kaleidoscope, friend of Sir Hans Sloane, and author of *The Country Housewife and Lady's Director** among other books on larger themes,

*Part I was published in 1727, Part II with the pineapple recipes in 1732. The two came out in one volume in 1736 from which the facsimile edition, beautifully edited and introduced and explained by Caroline Davidson, was published in 1980 by Prospect Books.

was the first cookery writer to give recipes for pineapple in England (and for green turtle soup, too). This tart came from a correspondent of his who knew Barbados. Here is the way he describes it:

> Take a pineapple, and twist off its crown: then pare it free from the knots, and cut it in slices about half an inch thick; then stew it with a little Canary wine or Madeira wine, and some sugar, till it is thoroughly hot, and it will distribute its flavour to the wine much better than any thing we can add to it. When it is as one would have it, take it from the fire; and when it is cool, put it into a sweet paste, with its liquor, and bake it gently, a little while, and when it comes from the oven, pour cream over it, (if you have it) and serve it either hot or cold.

To get full benefit of crisp contrast, I bake blind a 20-cm (8-in) tart case made from sweet shortcrust, p. 456. The pineapple is cooked in about 6 tablespoons (½ cup) of Madeira, with water to bring up the level at first. As it begins to soften, I raise the heat and add a tablespoon (2 tbs) of sugar, and cook it to a lumpy golden-brown marmalade. When it is ready, extra sugar can be added, but remember the pastry sweetness. Pour it into the hot pastry, and then add a liberal quantity of the best double cream – 250 ml (8 fl oz/1 cup) or even the whole pot. Serve immediately. The juice marbles and colours the cream to a beautiful mellow shade of brown.

If you leave the tart to get cold, the cream and juice set together to a custard-like consistency.

The tart cannot be cut neatly, especially when hot, but looks, smells and tastes wonderful.

PERFECT CREAM CHEESE PIE

I dislike cheesecakes made with a depth of filling that sticks to the roof of your mouth and clogs the teeth. A Canadian friend's recipe that she had from her mother seems to me exactly right, with the filling not much thicker than the crust. In the original, Graham crackers are used, wholemeal biscuits with a slight similarity to our digestives but less sweet; they can be bought at Panzer's of 15 Circus Road, London N.W.8, who also supply them to delicatessens in the rest of the country through their wholesale company W. & P. Foods Ltd.

If you have to make do with digestives, which are perfectly successful, cut the sugar in the filling. The friend now lives in the Var, where neither Graham nor digestive biscuits can be found: she crushes wholemeal *biscottes* with sugar instead and finds it works well. The quantity of butter may seem low. It works if you crush the biscuits finely in a processor. Crumbs produced by the brown paper bag and rolling pin method may need a little extra.

CRUST
20–22 Graham crackers
60 g (2 oz/$\frac{1}{4}$ cup) melted warm butter

FILLING
375 g (12 oz/1$\frac{1}{2}$ cups) full fat soft cheese
2 large eggs
up to 125 g (4 oz/$\frac{1}{2}$ cup) sugar
1 medium pineapple, peeled, cut up, lightly cooked or 1 can unsweetened pineapple rings, cut up

TOPPING
300 ml (10 fl oz/1$\frac{1}{4}$ cups) soured cream
2 tablespoons ($\frac{1}{4}$ cup) caster sugar
up to 1 teaspoon vanilla essence

Crush biscuits and mix with butter. Spread in a shallow oblong tin, 18 × 28 cm (7 × 11 in) or into a 23–25-cm (9–10-in) tart tin with a removable base. Cook for 15 minutes at gas 4, 180°C (350°F).

Mix together the filling ingredients, adding sugar to taste, and folding in the pineapple. Spread on hot crust. Bake a further 10 minutes, or slightly longer until filling is barely firm. Cool 10 minutes.

Beat topping ingredients smoothly together, going discreetly with the vanilla essence (I take it you use sugar from a jar containing vanilla pods). Pour over the pie. Return to oven for 10 minutes. Cool. Cover with foil loosely, so that it does not stick. Chill 24 hours and eat with spoons and forks, as top remains slightly liquid.

PASKHA

Easter is the great festival of Russian Orthodoxy, which is easy to understand. We rejoice when it comes to our grey climate; how much more must it mean as the long bitter winters of Moscow and Leningrad lumber to an end? There the feast will include two cakes, surrounded by candles – the yeast-raised *kulich*, a form of brioche and *paskha*.

The mould used for *paskha* is made in wood, in the shape of a truncated metronome, pierced so that the liquid can drain from the soft cheese. It is lined with butter-muslin which gives an attractive surface, and the decoration in glacé fruits starts with the intitials XB, the Cyrillic letters for the words 'Christ is risen'.

I include the recipe in this book for the sake of the glacé fruit, but also because without the glacé fruit it makes a good cheese cream to serve with soft fruit, a more solid version of the French *coeur à la crème*, p. 470. If you use *fromage frais*, weigh it after draining overnight in a muslin or sieve. The same applies if you use moist curd cheese. Petits suisses and similar cheeses can also be used.

125 ml (4 fl oz/½ cup) double or whipping cream
10 cm (4 in) piece vanilla pod, split
2 large egg yolks
90 g (3 oz/⅓ cup) caster sugar
100 g (3½ oz/good ½ cup) chopped candied fruit, including glacé pineapple, and peel
125 g (4 oz/½ cup) creamed unsalted butter
¾ kg (1½ lb/3 cups) curd or other medium fat cheese
100 g (3½ oz/scant cup) blanched chopped almonds

DECORATION:
about 60 g (2 oz/½ cup) toasted almonds and (⅓ cup) candied fruit and peel

Bring cream to the boil with vanilla. Beat eggs with sugar then pour on the cream, still beating. Return to the heat and cook without boiling until thick. Remove from the heat, dip base of pan in cold water and stir in the candied fruit and peel. Cool and take out vanilla pod. Mix in remaining ingredients.

Line a washed new flowerpot with butter-muslin or stockinette. Put into the mixture. Leave from 10 hours to 3 days in the refrigerator to drain (stand pot on dish). Turn out and decorate.

POLISH GLACÉ PUDDING

The great Joseph Conrad was not above contributing a preface to one of his wife's cookery books, and thoroughly applauding the enterprise. He thought that the English would be saner and happier if we gave up our national addiction to patent medicines and took care with food instead. A conscientious cook makes 'for the honesty and favours the amenity of our existence'. So a cookery book is 'the only product of the human mind altogether above suspicion'.

Mrs Conrad described the kitchens they experienced on their travels: 'In Poland the kitchen was a kitchen but also served as the cook's room; and chancing once or twice to time my visits at inconvenient moments, I would miss the bed and see instead innumberable dirty dishes adorning a long board that stretched the whole length of the bed, resting each end on the wooden frame. A girl seated on the floor without shoes or stockings would be busy preparing mushrooms or other vegetables.'

This particular recipe is a variation on the tutti frutti theme.

125 g (4 oz/$\frac{2}{3}$ cup) raisins
125 g (4 oz/$\frac{1}{2}$ cup) sugar
600 ml (1 pt/2$\frac{1}{2}$ cups) milk
1 large egg
1 tablespoon ($\frac{1}{4}$ cup) flour
125 g (4 oz/1 cup) slivered toasted almonds
125 g (4 oz/$\frac{1}{2}$ cup) chopped glacé pineapple (see p. 349) or chopped canned pineapple
125 g (4 oz/$\frac{1}{2}$ cup) well-drained chopped ginger
3 tablespoons (scant $\frac{1}{4}$ cup) Sauternes or other sweet white wine
900 ml (1$\frac{1}{2}$ pt/3$\frac{3}{4}$ cups) whipping cream

Bring raisins, sugar and milk slowly to the boil, then keep just below simmering point for 15 minutes. Beat egg and flour, then pour on some of the milk to make a smooth mixture. Gradually add the rest of the milk through a strainer and press on the raisins to push through their juice but do not sieve the raisins.

Return mixture to the pan and stir over a moderate heat to thicken the custard. Cool, stir in almonds, pineapple, ginger and wine. Whip cream stiff and fold into the custard.

Taste and adjust the flavouring items if you like. Turn into a charlotte or decorative mould and freeze at lowest possible temperature. Stir occasionally.

This is enough for 12 people. You might find it more practical to freeze it in two moulds, unless you plan a party. If you want to reduce the quantities by a third, keep the whole egg but cut down the flour to 2 level teaspoons.

PINEAPPLE SURPRISE

Take a large and beautiful pineapple. Slice off the top, and set it aside carefully, so that the leaves of the topknot don't get damaged.

Scoop out the pulp carefully and as neatly as you can, leaving a good firm pineapple wall. Cut up the pieces, ruthlessly cutting away hard bits and hard core (simmer these bits in a light syrup, then strain and use for other fruit salads, and for making ices or jelly). Weigh the pineapple pieces, and add an equal weight – more or less – of skinned peach slices, or peeled and pitted grapes, or orange sections freed from skin and pith.

You can also add small but perfect strawberries – *fraises de bois* are the ideal. Do not worry if you cannot afford an equal weight to the pineapple.

Sprinkle the mixed fruit with icing sugar and a little Cointreau, Grand Marnier or kirsch. Leave 2 hours. Taste and adjust sugar or liqueur quantities if necessary. Pile the fruit into the pineapple shell, keeping back the liquid until the end and then adding enough to moisten the salad. Top with the pineapple leaves. Chill for several hours, wrapped completely in a plastic bag (otherwise the smell will dominate the refrigerator).

Note After the salad has been eaten, you can freeze and keep the pineapple shell for another time. Use it when you have to resort to canned unsweetened pineapple which is not at all to be despised.

PINEAPPLE BLITZ TORTE

A splendid pineapple cake which a Canadian friend acquired from her mother. Canned pineapple is used for the filling, but fresh pineapple makes a sharper keener contrast with the cream and cake.

CAKE
100 g (3½ oz/scant ½ cup) sugar
100 g (3½ oz/scant ½ cup) butter
4 egg yolks, beaten
70 g (2½ oz/good ½ cup) sifted flour
1 level teaspoon baking powder
¼ teaspoon salt
60 ml (2 fl oz/¼ cup) milk

MERINGUE
4 egg whites
125 g (4 oz/½ cup) sugar
1 teaspoon vanilla essence

FILLING
250 ml (8 fl oz/1 cup) double cream
1 rounded tablespoon (¼ cup) icing sugar
250 g (8 oz/1 cup) crushed, drained pineapple, or chopped fresh pineapple
¼ teaspoon vanilla essence

Cream sugar and butter, add the egg yolks and beat thoroughly. Sift the flour again with the baking powder and salt, and mix into the bowl alternately with the milk.

Line a couple of 20-cm (8-in) cake tins with Bakewell paper, and brush it over with melted butter, if you like. Pour in the cake mixture and bake at gas 3, 160°C (325°F) for 15 minutes.

During this time, whisk the egg whites stiff, then add the sugar a tablespoon at a time, whisking until the white forms lustrous peaks. Add the vanilla. Remove the half-baked cakes from the oven and pile them with the meringue. Return them to the oven and cook for a further 15 minutes. Cool, then remove from the pan.

Have the filling ready – whisk the cream and icing sugar until thick. Fold in the pineapple and vanilla.

Put one of the cooled cakes, meringue side down on a plate. Spread with the filling, then put the second cake on top, meringue side up.

PINEAPPLE UPSIDE-DOWN PUDDING (CAKE)

A splendid pudding when made with fresh pineapple, which lifts it right out of the second class. Resist the temptation to stick glacé cherries into the holes of the pineapple slices: lightly toasted hazelnuts rubbed free of skin are more harmonious. I agree that you need something to bring the final appearance up to the taste.

For 6–8 people, I use an oval gratin dish of 2-litre (4-pt/10-cups) capacity, measuring 35 × 25 cm (14 × 10 in). It takes a large pineapple, peeled and not too thickly cut, to cover it. Remove the central core from the slices and halve all except three slices to go in the centre. Toast 60 g (2 oz/½ cup) hazelnuts in the oven that you have switched on to heat up for the pudding. Rub off their skins.

Into the base of the gratin dish pour 60 g (2 oz/¼ cup) each of butter and sugar that you have melted together. Brush the mixture up the sides, then let it fall back. Put the pineapple slices on the base, with hazelnuts in the holes and gaps. Now mix with an electric beater or processor the following ingredients

250 g (8 oz/1 cup) soft butter
4 large eggs
250 g (8 oz/1 cup) sugar
200 g (6 oz/1½ cups) self-raising flour
2 level teaspoons baking powder
50 g (2 oz/½ cup) ground hazlenuts, or hazelnuts and almonds mixed
liquid left from cutting up pineapple, or a couple of tablespoons (3 tbs) alcohol from the pineapple in brandy jar (see *below)*

When smooth, spread over the pineapple evenly. Bake at gas 4, 180°C (350°F) for about 50 minutes, or until cooked – if you use a deeper dish than mine, it will take longer. The top should be a beautiful golden-brown. Run a knife blade round the edge of the pudding and turn it out on to a hot dish. If any nuts have moved out of place, put them back where you can and don't worry about the rest. Absolute precision is not always appetizing.

Serve hot, warm or cold. Soft chilled *coeur à la crème*, p. 470, goes well with the hot pudding. Cold, as cake, it needs nothing with it.

PINEAPPLE IN BRANDY

This and the recipes for three-fruit marmalade and glazed pineapple are based on instructions from the admirable *Complete Book of Fruit* by Leslie Johns and Violet Stevenson. For once the title approximates to the contents – origins, cultivation, varieties, distribution, uses in cooking with a recipe or two, sometimes more, for each fruit. For the enthusiast who has a chance of travelling, or who just wants to dream about fruit he may never have a chance to eat, it's the best book on the subject. Fruit is taken in its wide sense, to include for instance avocado and cashew nut, whereas in this book I keep to fruit as opposed to vegetable in the ordinary culinary sense.

This recipe is intended for use with several pineapples, but it works perfectly well for one or two. To each fruit, you need:

1 clove
1 teaspoon brown sugar
450 g (15 oz/scant 2 cups) granulated sugar
brandy

Scrub fruit well, drain and twist off the top leaves. Put into a large enamel or stainless steel pan with the clove and brown sugar and enough water to cover the fruit. Boil until tender, testing with a skewer or larding needle. Drain the pineapple and leave it to cool. Keep 4 tablespoons (⅓ cup) of the liquid.

Slice, peel and cut the pineapple into slices. Take out the hard central core with a small cutter or apple corer: the hole should measure approximately 2½ cm (1 inch) across.

Layer the rings with the granulated sugar, ending with sugar, in a bowl and leave overnight in a cool place. Next day, drain off the syrup and mix with the 4 tablespoons (⅓ cup) cooking liquid. Boil until clear and skim. Tip into a measuring jug.

Put the pineapple rings into a bottling jar. Bring the syrup back to boiling point and pour over the pineapple. Add an equal quantity of brandy – the fruit should be covered. Seal and leave for at least a month. Extra brandy can be added if you want to drink the juice as a liqueur, rather than as an adjunct to the fruit.

GLAZED PINEAPPLE

In this confection, which can be used for several recipes in this chapter and for ices, decorating cream cheese tarts, cakes, or to flavour creams and custards, a whole large pineapple is reduced to about 250 g (8 oz) of sweet, sticky rings of concentrated flavour.

Cut a large peeled pineapple into eight rings, taking out the middle with a corer in the usual way. Cover with water in a wide pan and simmer until tender. Remove rings and reduce the liquor to 250 ml (8–9 fl oz/1 cup). After measuring, return to the pan with 375 g (12 oz/1½ cups) granulated sugar and 3 tablespoons (¼ cup) honey. Stir about until everything dissolves to a clear liquid and bring to the boil. Slip in the pineapple and simmer until translucent. Lift rings and place on a sieve to cool. Then put on a sheet of kitchen parchment or greaseproof paper and dry off in the plate-warming oven or a not-too-hot airing cupboard.

This pineapple can be used for the perfect cream cheese pie (p. 342), but cut down the sugar.

PINEAPPLE, GRAPEFRUIT AND LEMON MARMALADE

A pleasant version of three-fruit marmalade, in which the pineapple blends into a delightful harmony. Worth making a few pots to vary the year's round of fine Seville marmalade, see p. 278.

You need not use a whole pineapple; cut off, say a half-kilo (1 lb) twisting away the leaves first. Weigh out an equal quantity of grapefruit, and then of lemons.

Peel, core and cut up the pineapple. Halve the other fruit, take out their pips and tie them into a piece of muslin with a string long enough to suspend the little bag from the handle of the preserving pan. Squeeze the citrus fruits and put the juice into the pan, along with the pineapple and 1¾ litres of water to every ½ kg of fruit (3 pt/7½ cups to the lb). Cut up the citrus shells and add them, too.

Boil in the usual way until the fruit is tender and the liquid reduced by a half. This takes an hour approximately.

Measure the pulp and return it to the pan. Weigh out 500 g (1 lb) sugar to every 600 ml (1 pt/2½ cups) pulp, warm the sugar and stir it into the panful of fruit over a low heat so that it dissolves before boiling point is reached. Boil steadily until setting point, stand 10 or 15 minutes to allow the peel to settle, then pot in the usual way.

PINEAPPLE–GUAVA, see FEIJOA

PLUM, GREENGAGE and DAMSON

We all probably have some kind of romantic view of plum orchards, an old one in Gloucestershire or Hereford, the trees crooked and slightly mossy, the branches laden and bending down with bright red and golden drops. The long grass dotted with red and gold, over all the slow murmur of wasps. The plums from those trees warm, meltingly delicious.

I describe the plum-grower's nightmare. His trees are nicely organized, all of a height and shape. I bet a wasp doesn't dare show the beginnings of a buzz. Inevitably those trees will be producing Victoria plums, and yet more Victorias, the apotheosis of a long reign is a flood of bland, boring plums. Poor Victoria. She began life in 1840, a stray seedling found in Sussex and introduced by a nurseryman in Brixton. A Cinderella story if ever there was one, except that the bridegroom was not a beautiful prince but a multi-million fruit business. Useful, comfortable, but looking for a good cropper. Victorias are for canning. Victorias are for plums and custard, that crowning moment of the school, hospital, prison and boarding house midday meal: I reflect that Mr Bird invented his powder round about the time that Victoria plums were beginning their career. Indeed, Mr Bird could be seen as the discreet lover, the husband's best friend, the front man, in a Pooteresque *menage à trois.*

Looking at it in another way, Victorias might be seen as the King Edwards or Whites of the plum business. This is a shame for the varieties of plum are legion, and have been at least since Roman times when Pliny spoke of the *ingens turba prunorum*, the enormous crowd of plums.

Edward Bunyard in his splendid *Anatomy of Dessert* remarked of plums that there is little 'encased in red, black or blue' that is worthy of the serious plum-eater's attention. He was speaking of dessert fruit, uncooked fruit. He also made the point that 'the distinction between a "Gage" and a plum is only British. . . . Generally a plum with a good flavour is called a gage in this country' – and he made his selection of the best accordingly. Who would disagree that the finest eating plums all have yellow or green skins? This even applies to the winter plums we get from South Africa, bred I believe from Japanese and Chinese varieties. The red ones are sweet and acid and little more, but in March and April come golden plums of a beauty that holds your eye and a subtle charming flavour that has managed to survive the exigencies of transportation.

I do not understand why more greengages are not planted. Friends, family, neighbours always have Victorias to spare. I am always being asked for ways of using up Victorias. Nobody ever needs a recipe to use up greengages.

In northern Europe, the greengage had a royal beginning. It came from Italy to the banks of the Loire at Blois and was taken up by Claude, wife of François I. She gave her name to the new plum which has been known ever since in France as the *reine-claude*. In the marvellous gardens that her parents had made running from the château to what is now the railway station, she had many trees grown and gave them to anyone who might be interested in planting them in their own gardens. Now when the season comes round, the markets for many miles around Queen Claude's castle at Blois are piled with greengages. And after the first ones are over, the Bavay comes along – the *reine-claude de Bavay* – bred by one of Napoleon's retired generals, Major Esperen at Malines (he raised other, yellow plums as well, and a new Bergamot pear). A little plummier than the greengage, not quite as fine, but good.

Why green*gage*? The answer is that Sir William Gage, of Hengrave Hall, Bury St Edmunds, brought back *reine-claude* trees from France among many other items, in 1724. The labels were lost. The splendid new fruit was known as the green Gage's plum, I suppose, which settled down to greengage. From it, a nurseryman nearby began to hybridize new plums, Jervaise Coe. In 1800, he produced his masterpiece, the Coe's Golden Drop, which many people regard as the supreme plum of all: 'the skin is rather tough, but between this and the stone floats an ineffable nectar.' This and its two bud sports, Coe's Violet and Coe's Crimson Drop, are for gardeners only. No commercial grower would contemplate such a poor cropper.

In France, the greengage was used to raise a transparent plum – transparent, because if you hold one up to the light you can just see the dark shadow of the stone. It was called *reine-claude diaphane*, diaphanous Queen Claude, an enchanting name that pictures her in muslin robes in some Botticelli painting. In reality, the poor woman was more likely to be swollen and haggard from constant childbearing; it weakened her so much that she died young, at the age of twenty-five, in 1624, having had seven children in eight years.

Her diaphanous plum has survived rather better, both in France and this country. One of our great nurserymen, Mr Thomas Rivers of Sawbridgeworth, was visiting a rose nursery near Paris – he was a great raiser of roses – and was offered some plums. The one he ate was so good that he imported trees. They did not work well, so he raised seedling trees from some of the stones of this *reine-claude diaphane*. The result of this was the Early Transparent of 1866. The nursery still continues in his family: 'From 1850 to 1875 there was scarcely a season without some new fruits from Sawbridgeworth, and many of them have been decided acquisitions . . . thirty-one peaches, sixteen nectarines, twelve plums, six pears and some apricots, cherries, raspberries and strawberries' have originated there. 'As a raiser and introducer of fruits . . . Thomas Rivers stands pre-eminent – even greater than . . . Thomas Andrew Knight',* who was our first systematic plant breeder. Late Transparent (1888) and Golden Transparent (1894) were other contributions to the group of transparent plums that came from Rivers' nursery.

Damsons are so easy and familiar that I imagine nobody reflects very much on their romantic name which is a shortening of Damascene, i.e. the plum from Damascus. It seems likely that other strong-flavoured plums came to us from those parts. Perhaps only the damson kept the name because they breed true – which means that the modern fruit tastes like the damsons of Damascus in, for instance, the time of St Paul.

Such power in a fruit I find difficult to enjoy. Many people like damson cheese but I find it needs plenty of something else, whether bread and butter, junket, ginger-bread or roast lamb, to make it tolerably pleasant. Damson gin is all right, but in that it is only used as a flavouring.

PRUNES

In a masochistic and patriotic egotism of suffering, I had always thought that prunes and rice pudding were unique to Great Britain (like strikes). Alone in the world we suffered, or made our children suffer – 'Never did me any harm! Sit there till you eat it up . . . Starving piccaninnies . . . Wicked waste . . . Have you Been today?' But no! In middle middle-age, I at last learned from one of the Tartarin novels by Alphonse Daudet that they once dominated the *tâble d'hôte* at all Swiss hotels causing mournfulness and faction.

If our regime has not been too overwhelming, we can return to prunes as adults now that French and Mediterranean dishes have taught us that

*Edward Bunyard, *Anatomy of Dessert*

prunes have their virtues. They may intensify the flavour of meat dishes, make a contrast to the delicacy of some fish without submerging them, balance the richness of puff pastry in tarts, and promote reckless extravagance in the matter of sweetmeats, when stuffed with apricots and kirsch in a pale orange paste.

My one check in the matter of prunes has been the Algerian *L'Ham Lahlou*, which demanded almost equal quantities of mutton, prunes and sugar, and a glass of orange-flower water. We cut the sugar to 60 g (2 oz/ ¼ cup) from 430 g (over 13 oz/1½ cups), but the dish was almost as sweet as a mincemeat from the ½ kg (1 lb) of prunes. The orange-flower water vanished. I might have got away with it as a sort of Westmorland sweet pie or mincemeat, but even so it would not have been a dish worth making again.

The best prunes are reckoned to come from Agen, or the district around known as the Agenais. As this is Armagnac country, too, they are often put together in a most felicitous way. We went to Agen out of piety one Sunday, and found there the most rumbustious of markets – by this I do not mean that the market was a drunken riot but that the fruit and vegetables had a healthful exuberance, a perfection of shine and curve and good humour. In consequence, the market people were happy and full of jokes, anxious to tell you all about their stock. We bought some plastic, sealed containers of Agen prunes that needed no soaking, and asked the man who sold them to us if the market was always like this. 'You should come in the summer,' he said. 'It's marvellous then! A bit quiet today. Not much choice.' That was the last Sunday in May. The day before we had driven through mile after mile of orchards at that quiet stage between blossom and fruit. We had visited Marmande before Agen, and there the tomatoes were not yet ripe. Perhaps it was a dull stage of the year. And we just did not understand what a full season market could be.

The prunes were fine, and so were all the vegetables we bought, and the strawberries, and the little prune tarts and stuffed prune sweets. But I cannot say that at a blind tasting I could unhesitatingly distinguish between an Agen prune and a Californian prune dried from fruit of the same tree, the *prune d'Agen* which was taken to America in the middle of the last century by a Frenchman, Louis Pellier. The output of California is such that in most groceries and supermarkets in France, the prunes you buy will have come from the States. Look in the markets for the *pruneaux d'Agen*, at the nuts, glacé fruit, pickles and spices stall. Dodging the wasps is hazardous, so you have to make a quick choice of the size you want. The biggest for eating, the middle ones for cooking whole, the small ones for purées. Take the chance of buying angelica, too. It comes in foot-long stalks, not those pathetic little 2 oz packages.

In long-cooking dishes, there is no special need to soak prunes. If you have to soak prunes, and forget to put them in water the night before, pour boiling water over them to cover them generously and they will soon soften. Use the soaking liquor always, whether it is wine, water or tea: keep it to add to another dish, if it is not required with the prunes.

HOW TO CHOOSE AND PREPARE PLUMS

Avoid bruised plums, or plums with damaged skins. Tough skins can be peeled away easily from ripe fruit; for firm fruit, pour on boiling water as if they were tomatoes. When eating squashier plums, you can often suck out the flesh without the need to peel them at all: some people consider that it is the skin that makes them indigestible to people with sensitive stomachs.

When preparing plums for cooking, halve them and crack the stones to get at the kernels (or some of them). They improve plum and greengage jam, for instance, and plum dishes of all kinds.

The acidity of some varieties seems to be increased by cooking them in syrup. Simmer them instead in plain water, then drain and sugar them afterwards: if you use the least water possible, so that they partly steam, you get a reasonable result. Moreover, the skins can be removed easily afterwards.

Before you cook plums, decide whether you wish to keep them intact or let them dissolve into a purée. In the first case, leave the skin: in the second place, remove it – even if the purée is to be strained, it is a good idea because the skin seems to be the element that adds an unpleasant acidity. Of course, there may be occasions when the extra strength that the skins give is desirable. In some creams and mousses, for instance.

With the best plums, i.e. the gages and yellow plums, the skin is not so strong or so damaging. The whole effect of such plums, when cooked, is milder, just as it is when you eat them. When it comes to the tiny delectable mirabelle plum, and you want to make jam, leave the stones to rise to the surface and remove them at that stage: there is never any point in skinning them that I have found.

CHINESE PLUM SAUCE

A thick, chutney-like sauce, that keeps well in covered jars. It can be served on its own with roast duck or pork, or with barbecued spare ribs. Or in Russian style with lamb kebabs and grilled chicken.

1 large cooking or sharp dessert apple
250 g (8 oz) dark red plums
250 g (8 oz) apricots
150 g (5 oz/⅔ cup) sugar
125 g (4 oz/⅓ cup) white wine vinegar
1 chopped chilli, minus seeds (if dried, retain seeds)
extra sugar

Peel, core and cut up the apple. Cook with 4 tablespoons water in a covered pan, until soft. Add plums and apricots, sugar, vinegar and chilli. Cover again and simmer for ¾–1 hour. Sieve out the stones.

Taste and consider your sauce. If it is very thick, add a little water. If unbearably tart, add a little sugar. Serve hot or cold.

DANISH PLUM SAUCE

To go with boiled ham, salt pork and gammon.

Simmer a ½ kg (1 lb) yellow skinned red plums in 300 ml (½ pt/1¼ cups) dry white wine and a dessertspoon of tarragon vinegar. Sieve, then boil down with 3 tablespoons (¼ cup) of honey, until you like the taste.

If the sauce is too thin, stir in a level tablespoon of arrowroot mixed to a cream with a tablespoon of water.

GEORGIAN BEEF AND PLUM SOUP
KHARCHO

Slice ½ kg (1 lb) brisket of beef into 18 pieces. Put them in a pot and cover with 2 litres of water (3½–4 pt/8¾–10 cups). Bring quickly to the boil, skim and cover. Turn down heat to keep the water simmering. Leave for 1½ hours or until beef is very tender. Now add salt, pepper, 3 crushed garlic cloves, 2 sliced onions, 125 g (4 oz/½ cup) rice and 125 g (4 oz) stoned quartered plums – use dark red ones that are on the sour side.

Simmer 30 minutes. Fry two large skinned, chopped tomatoes and a teaspoon of tomato purée in 2 tablespoons (¼ cup) butter or beef dripping. Add to soup and simmer for another 10 minutes. Check the seasoning and dilute the soup if you feel it's too strong. Add a little sugar, if it's too tart. Scatter with chopped green coriander or dill and parsley.

LAMB WITH PLUMS

Brown a joint of lamb in butter. Put into a deep heavy pot with ¼ litre

(8 fl oz/1 cup) red wine. Cover and cook in a moderate oven until tender, or on top of the stove (turn the meat occasionally).

After half an hour, add to the pot 10 dark red skinned plums, a clove of garlic chopped small and a medium onion, chopped.

When the meat is ready, remove it to a serving dish and keep it warm. Skim fat from the cooking juices, then push them through a sieve into a clean pan. Season with cinnamon and either allspice or nutmeg, plus a little sugar. Do not make the sauce too sweet.

Pour a little sauce over the lamb and serve the rest in a bowl.

PORK IN THE VOSGES STYLE

PORC À LA VOSGIENNE

Tiny yellow mirabelle plums are used for this dish which is found in varying styles in the Vosges. You can adapt it, for instance, to thick slices of tenderloin cut on the bias, cooking it in sauté style on top of the stove.

Skin, bone, season and roll a roasting joint of pork, preferably the loin. Brown it in butter. Put it into a deep pot with 2 glasses (1½ cups) of dry white wine. Cover and cook either in a moderate oven, or on top of the stove. Allow 1½–2 hours for a 1½ kg (3 lb) piece.

Meanwhile, stew 2–3 large sliced onions in butter without browning them, and cook ½ kg (1 lb) mirabelles with the minimum of water and a thin sprinkling of sugar. Drain, retaining the liquor.

Make a bed of the onions on a serving dish. Season them. Arrange the sliced pork on top down the middle, with the mirabelles round it. Keep warm.

Skim surplus fat from the pork cooking juices. Add the mirabelle liquor and 100 ml (3½ fl oz/scant ½ cup) beef or veal stock. Boil up hard and season, putting in plenty of pepper. Whisk in a good knob of butter and strain over the pork. There should not be much sauce, just enough to moisten the meat.

PLUM AND WALNUT PIE

A most delicious pie, with walnuts, cinnamon and butter to counteract the fruit's acidity. Choose a mild plum: the Quetsche or Zwetschken is ideal.

shortcrust pastry
½ kg (1 lb) plums, halved, stoned, chopped

125 g (4 oz/½ cup) soft light brown or demerara sugar
125 g (4 oz/1 cup) chopped walnuts
2 teaspoons ground cinnamon
grated rind of ½ lemon and ½ orange
60 g (2 oz/¼ cup) melted butter
beaten egg or top of the milk to glaze

Line a 20–23-cm (8–9-in) pie dish about 2.5-cm (inch) deep with pastry. Mix plums, sugar, walnuts, cinnamon and grated rinds and put into the pastry. Pour over the butter. Cover with pastry, pinching the edges and making a central hole. Brush over with beaten egg or top of the milk. Bake at gas 5, 190°C (375°F) for about an hour. Serve warm rather than hot with cream.

This pie can also be made with greengages, almonds and white sugar.

PLUM JALOUSIE AND PLUM SHUTTLE

Puff pastry goes particularly well with plums on account of its butteriness. If you use the frozen kind, roll it out and spread it with melted butter; fold, chill and roll it out again; repeat once more. Just the thin coating of butter improves the flavour.

To make a jalousie, roll out two oblongs of pastry. Put one on a damp baking sheet, brush round the rim with water, and cover the centre with halved, stoned plums (Victorias will do), cut side down. Sprinkle with sugar and cinnamon, and walnuts as in the recipe above if you like.

Fold the other oblong in half, lengthways, cut a series of parallel slits in the centre, leaving a rim. Open out the pastry and put it over the plums. The slits should have a look of a slatted Venetian blind, or *jalousie*. Press down the rim of pastry. Brush with beaten egg, and bake at gas 7–8, 220–230°C (425–450°F) until golden brown. Lower the heat after 20 minutes, and leave for another 10 or 15 minutes until the plums are tender.

Plum shuttle is a country version of plum jalousie, in other words, a plum turnover. Again, it should be eaten soon after baking, to get the freshness of taste, plum juice and butter.

Cut squares of puff pastry measuring about 15 cm (6 inches) each side, or slightly larger. Halve and stone plums, and put them on the pastry slightly to one side. Brush the rim of the pastry with water, and then fold one side over the other to make a narrow oblong. Press the rims together, make a slash or two in the top and glaze. Bake as above.

LORRAINE MIRABELLE TART

TARTE LORRAINE AUX MIRABELLES

Make shortcrust pastry with 375 g (12 oz/3 cups) flour, and other ingredients in proportion, *see* following recipe.

Line a 23–25-cm (9–10-in) tart tin with three-quarters of the pastry, cutting the circle large enough to flap right over the sides and down to the table.

Mix ¾ kg (1½ lb) mirabelles, stoned, with 2 large eggs beaten up with 100 g (3½ oz/scant ½ cup) sugar and 100 g (3½ oz/scant cup) ground almonds – no need to blanch the almonds before grinding them.

Put into the pastry. Roll out the remaining pastry and cut a circle the size of the tart. Put the circle on top of the plums, brush round the edge with water and press the underneath pastry lightly over it, as neatly as you can (this is a bit tricky – if you are hopeless at this kind of thing, flip the pastry edges over the filling and put the circle neatly on top instead). Make a hole in the middle, and pierce the circle part with a fork in a decorative way. Brush with egg glaze, and bake 30–40 minutes at gas 6–7, 200–220°C (400–425°F), until nicely coloured.

GREENGAGE AND ALMOND TART

Make a shortcrust pastry with 250 g (8 oz/2 cups) flour, 125 g (4 oz/½ cup) butter, ½ teaspoon salt and a tablespoon (2 tbs) of sugar, with water to mix. Line a tart tin of 30 cm (12 in) with the pastry, and bake it blind until firm but not coloured.

Spread it with an almond cream, made by mixing 100 g (3½ oz/scant cup) ground almonds, 100 g (3½ oz/scant ½ cup) each of sugar and melted butter with a large egg and a level tablespoon (2 tbs) thick cream, and a tablespoon of kirsch if you have it.

On top, put 1 kg (2 lb) of stoned greengages that have been lightly cooked with 100 g (3½ oz/scant ½ cup) sugar and 150 ml (¼ pt/½ cup) water. Drain them well, and place them close together.

Put into the oven until the pastry is coloured and the almond cream just cooked: gas 7, 220°C (425°F).

Remove and brush over with a glaze made by mixing apricot jam with a little of the reduced greengage cooking liquor (see p. 477).

This tart can be made with apricots or peaches or cherries, as well. For family size, use a 20–23 cm (8–9 in) tin and reduce the quantities by a third, roughly speaking, and use a small egg for the almond cream.

QUEEN CLAUDE'S RIBBONS
BAVAROISE AUX REINE-CLAUDES RUBANÉE

When we were young, I preferred books from our parents' childhood: *The Gorilla Hunters* and *Coral Island* were more exciting than Biggles. *Jessica's First Prayer* stirred the tearful beginnings of political conviction, and an improbable school story gave me an image of elegance – the rest was soon forgotten – that there has never seemed any reason to change. At a fête to raise funds, the sixth-form girls all dressed in white muslin with green sashes: there was a picture of them, all length and style in the manner of Sargent, cool, in a hot garden.

White and pale or apple green are even better in summer food than in summer clothes. When I tried this bavaroise with a central band of apricot, as a change, not only did it look less attractive, but the apricot strength submerged the clear greengage and yoghurt creams.

GREENGAGE BAVAROISE
½ kg (1 lb) greengages
125 g (4 oz/½ cup) vanilla sugar
1 packet (½ oz/2 envelopes) gelatine or 5 leaves
250 ml (8 fl oz/1 cup) whipping or double cream

YOGHURT BAVAROISE
250 ml (8 fl oz/a good cup) yoghurt or fromage frais
125 g (4 oz/½ cup) vanilla sugar
1 packet (½ oz/2 envelopes) gelatine or 5 leaves
250 ml (8 fl oz/1 cup) double cream
plus *2 tablespoons chopped pistachio nuts*

Choose a mould, not too elaborate, that will hold just over 1 litre (2 pt/5 cups) and brush it either with almond oil or a tasteless cooking oil. Turn it upside down to drain. Or else choose a deep glass serving dish, if you are not keen on turning jellies and so on out of moulds. It will need no oil.

Make the greengage bavaroise: halve and stone the greengages, putting them into a heavy pan with a couple of tablespoons of water and a tablespoon of the sugar. Cook gently at first, then faster until the greengages are soft. Sieve and add them to the gelatine dissolved in 6 tablespoons (½ cup) very hot water. Add rest of sugar. (If you have a liquidizer or processor, put the gelatine in it with the water, then the hot stewed greengages and remaining sugar.) Whizz until very smooth. This will leave faint dark speckles, which I like. If you do not agree, sieve the mixture – a matter of seconds.

When the mixture is cool, whisk the cream until light and bulky, not too stiff, and fold into the greengage carefully. Pour a layer into the mould or serving dish, a centimetre thick is about right, or half an inch. Put into the ice compartment of the refrigerator or into the freezer until set firm.

Meanwhile make the yoghurt bavaroise. Beat the yoghurt (or *fromage frais*) with the sugar, add gelatine dissolved in hot water and the whipped cream, as for the greengage bavaroise; hurry things up if you like with a liquidizer or processor.

When the first greengage layer is set, pour on an equal layer of yoghurt bavaroise. Return mould or dish to chill. Repeat with the remaining creams; aim at 8 layers. Do not make the layers too exact, as if the bavaroise had been turned out on an assembly line in a factory. When you pour on the last layer, put mould or dish into the main part of the refrigerator until serving time. Should there be much delay, cover it with plastic film.

If you are making this bavaroise in the summer, with no delays, the creams will stay just liquid enough to be poured. In colder weather, they may set. Stand the bowls in hot water, and stir gently until they become just pourable again.

To serve, either turn out the pudding and scatter it with pistachio nuts, or scatter the nuts over the surface of the pudding in its glass dish. If by some lucky chance, you come into an inheritance of pistachio nuts, scatter them on every green layer of pudding as well as on the top.

ICED PLUM/DAMSON/MIRABELLE/ CHERRY PLUM SOUFFLÉ

SOUFFLÉ GLACÉ AUX PRUNES/PRUNES DE DAMAS/ MIRABELLES/MYROBALANS

Follow the recipe for iced pear soufflé on p. 320. Stone the fruit, cook it in a shallow open pan without lemon juice, then sieve out the skins. Plum purées do not need processing as they are so easily sieved.

PLUM STONE NOYEAU

If you have ever made cherry brandy, you will have noticed the faint almond flavour that comes from the kernels. Plum stones, or apricot or peach stones, have the same effect and can be used on their own to make a simple noyeau for flavouring puddings creams and cakes, fruit salads and so on. True *noyeau* – or *noyau* – was long made at Poissy on the outskirts of Paris, for the especial benefit of confectioners and pastry-cooks, before the days of almond essence. *Crème de noyaux* is still made in Holland.

Bruise and crack slightly with a hammer enough stones to fill a small glass bottling jar. Cover with a tasteless *eau de vie* or vodka. Close tightly and leave for several months, at least three, in a cool dark place.

PLUM, DAMSON OR GREENGAGE CHUTNEY

A good spicy traditional recipe, which is much improved by using wine vinegar rather than the usual malt. The fruit should be prepared before weighing.

1 kg (2 lb) stoned plums or damsons or greengages
250 g (8 oz) sliced apple
250 g (8 oz/2 cups) sliced onion
125 g (4 oz/⅔ cup) raisins
125 g (4 oz) coarsely shredded carrot
250 g (8 oz/1 cup) demerara sugar
1 level tablespoon salt
1 level teaspoon ground cloves
1 level teaspoon ground ginger
1 level teaspoon ground allspice
1 small red dried chilli
600 ml (1 pt/2½ cups) white wine vinegar

Mix the fruit, onion, raisins, carrot and sugar together in a basin (use your hands). Put salt and spices into a pan, and pour in the vinegar. Bring slowly to the boil and add the basinful of fruit etc. Stir up well, and bring again to the boil. Leave to simmer steadily until thick and chutney-like. Remember it will become a little thicker as it cools down. Pot and cover with plastic film or the usual jam pot covers, rather than metal lids which will turn very nasty in contact with the vinegar. Store for at least a month before using, in a cool dark place.

DAMSON CHEESE

During Jubilee year damson cheese was in the news with Women's Institutes and other organizations making it night and day – or so it seemed – to go with roast saddle of lamb served at a City banquet for King Birendra of Nepal. He had acquired a taste for English food at Eton, as a schoolboy, which is more than most of the English do at their schools.

Rinse and remove the stalks from your damsons. Weigh them, and for every 1 kg (2 lb), measure out 150 ml ($\frac{1}{4}$ pt/$\frac{2}{3}$ cup) water. Cook them together slowly, on top of the stove or in the oven according to your convenience. Sieve the panful when the fruit is very soft.

Weigh the pulp, and put it into a heavy preserving pan. Weigh out three quarters of its quantity in sugar, or slightly more. Some recipes give an equal weight, but I think that a little less sugar is a good idea.

Heat the two together, stirring, and when the sugar has completely dissolved, bring the pan to the boil. Boil steadily until you can make a clear track through the purée with a wooden spoon, showing the base of the pan. It is a good idea to wrap your stirring hand in a cloth towards the end of cooking time: then, if the splutters become over-enthusiastic you will not be burned.

Avoid the usual shaped jam jar, because damson cheese needs to be turned out for slicing. Use straight-sided jars or bowls. A tip I had from a friend was to brush out the jars with glycerine for easier turning out.

COCKIE-LEEKIE

Put a 1-kg (2-lb) piece of stewing beef, e.g. shin cut across to include the bone, into a pot. Cover generously with cold water. Simmer an hour. Add a boiling fowl and a tied bundle of trimmed cleaned leeks. Simmer until tender. Put in 500 g (1 lb) prunes, soaked, when meat is nearly ready. Just before serving, put in the white part of four more leeks cut into rings. Take from the heat so that the leek rings do not become soft.

Discard the bundle of leeks. Give everyone a bit of everything in a deep soup plate. Or serve the broth first and the rest afterwards.

STUFFED DANISH TENDERLOIN

Pork tenderloin, split and stuffed with apples and prunes, is a favourite Danish main course. Here is the version I like best.

12 large prunes, soaked in plenty of water
2 pork tenderloins
1 teaspoon each sugar and cinnamon, mixed
2 teaspoons salt
pepper

250 g (8 oz) cooking or sharp apple, peeled
butter
150 ml (5 fl oz/⅔ cup) cream
1 level tablespoon (2 tbs) each butter and flour
red currant jelly

Simmer prunes in their soaking liquor until you can easily remove the stones and halve them. Keep the liquor. Trim and slit the tenderloins lengthways without cutting them right through, so that they open like a book. Lay them out flat, skin side down, and bash them until they are nearly as thin as a veal escalope.

Mix the seasonings and sprinkle over the pork, grinding on plenty of pepper. Slice the apple thinly, discarding the core, and lay on the pork with the prune pieces. Roll up each tenderloin from one of the narrow ends and tie with string.

Brown the rolls in butter, and put them side by side into a heavy pan. Deglaze the frying pan with the prune liquor and pour over the pork. Pour in the cream. There should not be a great depth of liquid, though obviously this will depend on the fit of the pan, about a centimetre (½ in). Bring to simmering point, cover tightly and leave to cook gently for 30 minutes, turning the rolls at least once.

Remove pork rolls and slice them, arranging the slices on a serving dish. Don't forget to take away the string. Taste the pan juices, reduce or add a little water – this will depend on how well the pan was covered. Mash butter and flour together. Add to juice in little bits to thicken them – keep just under the boil, and stop when the sauce is creamy rather than very thick. Whisk in red currant jelly to taste, starting with a level teaspoonful. Pour some over the meat, serve the rest in a sauceboat. Put straw potatoes round the pork, and some watercress.

TZIMMES

A dish of the autumn Rosh Hashanah, the Jewish New Year, made with the fruit and vegetables of the season, sweetness being appropriate as the celebrants look forward with hope. All sorts of fruit can be used, apples, pears, a little quince, as in Moroccan *tajine*, to combine with the sweetness of the carrots. Everyone has their own variations. Here is the outline:

250 g (½ lb) prunes
½ kg (1 lb) carrots

1¼–1½ kg (2½–3 lb) brisket of beef
375 g (12 oz/3 cups) sliced onions
salt, pepper
½–1 kilo (1–2 lb) potatoes
1 heaped tablespoon goose or poultry fat
1 heaped tablespoon (¼ cup) flour
60 g (2 oz/¼ cup) sugar
juice of a lemon

Soak prunes. Peel or scrape and cut the carrots into batons. Put the beef on top of the onions in a deep close-fitting casserole. Cover with boiling water, put on the lid and simmer until the meat is tender – about 2 hours. Add prunes and carrots with seasoning after 1 hour.

Peel and slice the potatoes and arrange them in an ovenproof dish, with salt. Put beef on top surrounded by onions, prunes and carrots. Make a roux with the fat and flour, then add a generous ½ litre (¾ pt/2 cups) of defatted liquor from the casserole to make a sauce. Season with sugar and lemon and pour over the beef in the dish. Bake in a moderate oven at gas 5, 190°C (375°F) until the potatoes are tender, the meat browned.

Note The closer the fit of the casserole, the less water you will need to cover the beef and the better tasting your sauce will finally be. If you have beef stock to hand already, use that instead of water to cook the meat in.

OXTAIL WITH PRUNES AND CORNMEAL FRITTERS

QUEUE DE BOEUF À L'AGENAISE ET AU MILHAS

The idea of using prunes with meat survives throughout France, from Agen and the surrounding Agenais where the best prunes are produced, right to the north. The combination is improved sometimes by adding cornmeal, as they do in south-western France, the first part of the country to take to maize, in the mid-sixteenth century. You can serve it, as the Italians do, in a soft, coherent, yellow mass straight from the pan. Or you can leave it to cool, then slice and fry it – I much prefer this second way as it adds crispness to rich and tender meat. Although cornmeal ousted millet as a peasant food centuries ago, it did not oust the old millet names of *milhas* and *milhassou*.

When cooking oxtail, which freezes well, it is worth making more than you need for one meal. Left-over pieces can be reheated with grapes (*see* p. 183). Quantities below give enough for eight at least. When counting out the pieces, do not include the thin ends of the tail: they are good only for stock.

3 oxtails cut into 5 cm (2 in) pieces, or 16–20 pieces if already cut
seasoned flour
3 medium to large onions, peeled, quartered
3 large carrots, peeled, quartered
2 large leeks
bouquet garni
strip of dried orange peel (optional)
4 cloves garlic, halved
150–200 ml (5–7 fl oz/⅔–¾ cup) red wine
beef, veal or chicken stock, or water
pepper, parsley

PRUNES
16–24 dried prunes
4 tablespoons (⅓ cup) Armagnac or brandy

MILHAS
breakfast cup of yellow cornmeal
corn oil
salt (see *recipe)*

The snag about oxtail is the fattiness it adds to the stewing liquor. For this reason, it is essential to complete most of the cooking the day before you intend to eat it. Or two days, or even three. This gives the fat plenty of chance to rise and settle in a cake, on top of the liquor, which can then easily be removed.

It is also a good idea to cut away as much as you can of the fatty part of the tail, leaving each piece with only a thin layer. Press the cut sides into seasoned flour, but add no extra salt.

The fatty lumps should next be heated in a heavy frying pan, until they have given up enough clear fat in which to brown first the onions and carrots, then the oxtail pieces. Transfer the vegetables, then the oxtail to a wide casserole: if you cannot arrange the pieces in a single layer on top of the vegetables, put the largest pieces in first, the smaller ones on top, making a close fit.

Trim the leeks; remove their two outer layers carefully, slice the rest and add to the casserole. Place the *bouquet*, and orange peel if used, inside the leek layers, to make a roll: tie with thin string, and add to the casserole with the garlic, the wine and enough stock or water to cover completely. No salt should be added. Bring to simmering point and put on the lid. Simmer either on top of the stove, or in the oven gas 1, 140°C (275°F), until the meat parts from the bones visibly. Time will vary according to the oxtail and your success in controlling the simmer – check after 2½ hours, be prepared to allow 4 or even 5 hours.

Put the oxtail into a heatproof serving dish, and sprinkle lightly with salt. Strain liquid into a bowl. Pick out the carrot from the sieve, and cut it into neat centimetre batons. Add them to the liquid. Allow everything to cool, then cover and store in the refrigerator.

Meanwhile soak the prunes with enough water to cover well, and the Armagnac or brandy.

Bring two breakfast cups of water to the boil in a pan, and add the cornmeal slowly. Stir until very thick. Add more water if the cornmeal rapidly becomes too thick to stir. It needs 10–15 minutes. Turn into a buttered dish. Leave overnight.

To finish: remove the cake of fat from the oxtail liquor, which will have set to a jelly. Put a little of the jelly with the oxtail pieces, enough to provide moisture and steam. Cover with foil and reheat in the oven, or over a pan of boiling water.

Meanwhile cook the prunes in their own liquor, then stone them. Put any remaining liquor into a large frying pan with the rest of the jelly and the carrots. Boil vigorously, skimming away greasy bubbles, until the liquor reduces to a syrupy concentration. This strong reduction would be unpleasant if you started with a salty jelly, which is why no salt should be added to the oxtail earlier on in an absent-minded moment. When you taste the sauce, remember too that there will be some dilution from the oxtail juices. Put the prunes into the sauce to reheat gently, then scoop them out with the carrot and place them both among the oxtail pieces in their dish. Pour over the sauce, well peppered, and scatter with a pinch or two of parsley.

While the sauce is cooking down, cut the cornmeal *milhas* into neat fingers, allowing three or four for each person. Fry them golden brown and crisp in the oil. Tuck them round the dish just before you serve it, so that the yellow contrasts with the sombre prunes and meat.

Note If you do not wish to make *milhas*, or cannot buy the cornmeal, serve triangles of home-made bread fried in butter, or a few fried potatoes. People who do not mind soft textures will be perfectly happy with egg noodles or boiled potatoes.

PORK WITH PRUNES AND CREAM
PORC AUX PRUNEAUX DE TOURS

With the new fatless breeds of pig, I find that the boned loin that is the traditional meat for this famous recipe can be tough. Thickly-cut slices of tenderloin seem to work much better. And they are quick to cook. Quantities below are for 6 to 8, depending on the size of the tenderloins and the rest of the meal.

½ kg (1 lb) large prunes
½ litre (¾ pt/scant 2 cups) Vouvray or other medium-dry white wine
2 pork tenderloins
seasoned flour
clarified butter
1 tablespoons (2 tbs) red currant jelly
300 ml (10 fl oz/1¼ cups) double or whipping cream
salt, pepper, lemon juice
parsley

Soak and cook prunes in the wine, adding a little water if they seem in danger of drying out. Strain and stone the prunes; keep the liquid.

Trim the tenderloins and cut them diagonally into thick slices (a centimetre or just under half an inch). Turn them in flour and fry them quickly in a little butter until they are nicely browned on both sides. Pour in the prune juice and simmer a few moments until they are cooked through. Transfer to a serving dish, arrange the prunes round them and keep warm. Add jelly to pan juices, whisk and boil down if necessary to concentrate the flavour (this will depend on the quantity of prune juice): it should be a delicious syrupy essence. Pour in the cream and stir and boil vigorously to make a sauce. Season to taste, adding a little lemon juice (this dish is properly made with *crème fraîche* – *see* note below). Pour over the pork, scatter with a small amount of chopped parsley. Serve very hot, with a few homemade straw potatoes.

Note If you can buy soured cream, use 100 ml (3½ fl oz/scant ½ cup) with 200 ml (7 fl oz/scant cup) double cream. And if you are one of the lucky few who can buy *crème fraîche*, use that on its own.

CHICKEN PIE WITH PRUNES

PIE DE POULET À L'AGENAISE

Armagnac is not the only alcoholic speciality of the Agenais. Pineau des Charentes is made in the neighbouring region and much sold there: it consists of brandy and grape juice, blended with various refinements. Nowadays, it is becoming a popular drink in much of the west of France, and I advise bringing back a bottle or two as you will not see it here. Normally it is drunk as an aperitif, well chilled and with ice, but it is splendid for cooking with, being a little sweet.

Cut up a 1½-kg (3-lb) chicken – one of the yellow maize-fed chickens from France that some shops now sell, if possible. Make the pieces fairly small and flour them. Soak 350 g (11–12 oz) Agen prunes with 150 ml (5 fl oz/½ cup) of Pineau des Charentes.

Brown the chicken pieces in butter, and flame them with 4 tablespoons (⅓ cup) of brandy. Put in the prunes and their liquor, plus 150 ml (5 fl oz/½ cup) chicken stock, cover and simmer 15 minutes. Check the prunes occasionally, and the moment they are easy to stone, remove them. Replace the stones with cubes of cooked ham – you need about 100 g (3½ oz).

Put the chicken pieces into a pie dish, with the prunes. Season. Reduce the cooking liquor by half, pour off any fat or blot it away, and add to the dish. Cover with puff pastry or shortcrust if you prefer, and glaze with beaten egg, then cook in the oven preheated to gas 4, 180°C (350°F). Allow 40 minutes, but test to see if the chicken is tender after 30 minutes.

STUFFED PRUNES FOR ROAST GOOSE

This is one of my favourite prune recipes, which I came across years ago in Louis P. de Gouy's *The Gold Cook Book*. It makes me think of Christmas Eve when we all sit round in the kitchen talking and making the last preparations. People help and the fiddly part doesn't take long. If I were better organized, I should do most of the work in advance and store the prunes in the freezer.

¾ kg (1½ lb) giant prunes, soaked
300 ml (10 fl oz/1¼ cups) red wine
1 small lemon, thinly sliced
375 g (12 oz) pork, ⅓ fat and ⅔ lean, chopped
60 g (2 oz/½ cup) onion, finely chopped

12 large green olives, chopped
60 g (2 oz/¼ cup) butter
pinch each of nutmeg, mace, marjoram and thyme
salt, pepper
2 small eggs

Put the prunes into a large pan with any remaining water. Add the wine and lemon, plus more water until the prunes are covered. Bring to the boil, cover and simmer until tender but not too soft. Cool, drain and remove the stones. Any liquor left over can be used up in making giblet or carcase stock; don't waste it.

Cook the pork, onion and olives slowly in the butter in a large frying pan. Season with spices and herbs as the meat is cooking. When the pork is tender, taste and add extra herbs if you like. Remove to a basin and season well, then bind with the eggs. Leave to cool down. Stuff the prunes and put them in the refrigerator until you are ready to put them in the cavity of the bird. If any are left over, reheat them round the bird during the last half hour of cooking time.

ELIZABETH DAVID'S BLACK FRUIT FOOL

A recipe for devotees of prunes and dried fruit. To me, very much a winter dish of the cold northern counties of Britain, though I daresay that it would not be out of place around Agen and Bordeaux. There prune tarts have a filling which tastes much the same.

250 g (8 oz) prunes, unsoaked
150 ml (5 fl oz/⅔ cup) red wine or 75 ml (2½ fl oz/⅓ cup) port
125 g (4 oz/⅔ cup) raisins
60 g (2 oz/⅓ cup) currants
whipped cream

Put prunes into a pot with the wine and plenty of water to cover. If the pot has no lid, put foil closely over the top. Bake for 2–3 hours at gas 3, 160°C (325°F). In another pot, put the raisins and currants with water to cover, and cook for 1–1½ hours in the same oven. Sieve, liquidize or process the prunes with any juice left. Drain the raisins and currants, and sieve them, too, or add them to the prunes in the liquidizer or processor. See that both purées are well mixed together.

Serve well chilled in glasses, with sweetened whipped cream on top. Sponge or shortbread fingers go well with the fool, which can also be used as a filling for tarts and pies.

PRUNES STUFFED WITH WALNUTS

I gave this Middle-Eastern recipe ten years ago in *Good Things*; here is a brief summary for those who don't know it.

Soak a ½ kg (1 lb) large prunes in boiling tea, then simmer them in the tea remaining until the stones can be removed without too much trouble. Replace them with shelled walnut halves, and range the stuffed prunes in a single layer in a large pan. Bring 2 heaped tablespoons (3 tbs) of sugar and a tablespoon of lemon juice to the boil with 300 ml (½ pt/1¼ cups) water, simmer and then pour on to the prunes. Cover and cook gently for half an hour. Transfer carefully to a dish, pour on the remaining syrup and chill. Cover with sweetened whipped cream, and chill for a further hour or so. Serve with walnut biscuits, p. 459.

PRUNE AND WALNUT TART

Make 125 g (4 oz/1 cup) shortcrust pastry (*see* p. 452) and partly bake a 20-cm (8-in) pastry case blind (p. 453), then spread it with apricot jam. Fill it with a coherent not too stodgy purée of prunes – use about 375 g (12 oz/1½ cups) sweetened. Make a lattice of strips over the top with pastry, and put a perfect walnut half in each gap. Brush the strips and edge with beaten egg glaze. Complete the baking.

PRUNE TARTS AT MOISSAC

Moissac is worth visiting for its sculptured church porch, also for its pastry-cooks, who take full advantage of being near Agen with its prunes and Armagnac.

I saw three tarts in one shop near the church:

1) A very light, puffy double crust tart, with a big hole in the centre to show the prune purée filling, which was flavoured with Armagnac.
2) An open flan in *pâte sucrée*, with prunes set in a cream and egg custard.
3) Small prune tarts, puff pastry, with a cross of two pastry strips going from side to side.

PRUNEAUX FOURRÉS DE TOURS

I gave a recipe for this sweetmeat made and sold – at some expense – at

Maison Sabat in the main street of Tours, in *Good Things*. Here is a slightly different version. The two other pastes are made from prunes flavoured with rum (same method as the following apricot purée), and an almond paste coloured light green.

Bring 500 g (1 lb/2 cups) sugar to the boil with a litre (1¾ pt/scant 4½ cups) water, simmer three minutes, then pour on to 1 kg (2 lb) prunes. Leave for 24 hours.

Drain off the syrup into a pan, add 150 g (5 oz/⅔ cup) sugar, and bring again to the boil. Pour back on to the prunes. Leave 48 hours. Drain the prunes and slit them lengthways to remove the stones.

Stone and cook in as little water as possible 1½ kg (3 lb) apricots. Boil off liquid, and reduce apricots to a purée. Cook with 1 kg (2 lb/4 cups) sugar until the mass comes away from the side of the pan. Stir nearly all the time, or it will burn. If you like, to make a stiffer paste still, add 50 g (scant 2 oz) pectin to the mixture, but it is not really necessary. At the end of cooking time, add kirsch to taste, starting with a liqueur-glassful.

Let this apricot paste cool, then cut it into small rectangles and roll them into 'stone' shapes. Stuff them into the prunes. The stuffing should show in an oval, through the slit of the prune. Put each fruit in a little paper case, and range in a small basket if you have one, otherwise a box.

POMEGRANATE

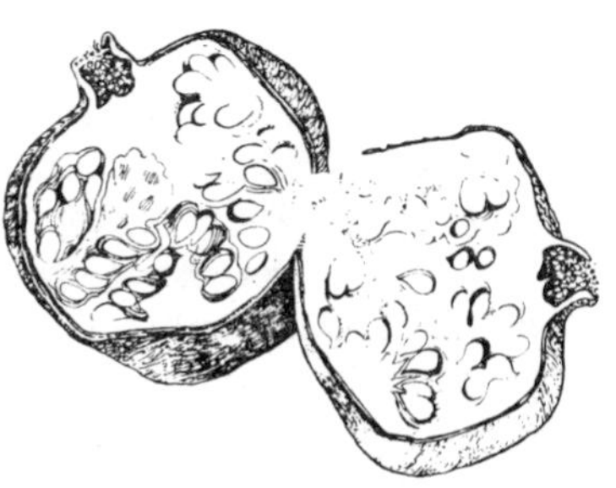

In *Les Nourritures Terrestres*, André Gide wrote a long poem in praise of pleasure, the pleasure being mainly fruit. He dedicated it to the pomegranate, calling it 'Ronde de la Grenade':

> A little sour is the juice of the pomegranate
>   like the juice of unripe raspberries.
> Waxlike is the flower
> Coloured as the fruit is coloured.
>
> Close-guarded this item of treasure, beehive partitioned,
> Richness of savour,
> Architecture of pentagons.
> The rind splits; out tumble the seeds,
> In cups of azure some seeds are blood;
> On plates of enamelled bronze, others are drops of gold.

But why the pomegranate? It is far less wonderful than the peach, less golden to eat than the mango. In any practical way, it is an unrewarding fruit. I have often wondered why so many are put out for sale in this country around Christmas. Every small town greengrocery has them, even ones that are shy of avocados and peppers. Of course, pomegranates are easy – the hard red and yellowish skins remain undamaged by the odd bump and jolt of transport. But what in the end do we get from them?

I know what Persians, for instance, get out of them. A purple, pleasantly sharp rich juice, that gives an edge to soups or stews, or provides a long drink. The seeds decorate a few dishes. But such things have never been part of our cookery, or of French cookery either. Persian practice does not explain pomegranates in the shops of Burford or Wootton Bassett.

Nineteenth-century *bons viveurs* felt that a cut pomegranate was essential to the fruit centrepiece on a shining mahogany dinner table. A decorous artistry and an inoffensive opulence. I suppose somebody spooned out the seeds when the display became dusty and needed refurbishing. For the pomegranate is no more than a closet of juicy seeds, each one gold in a deep pink jelly, the sections held firmly in a yellow astringent pith.

And yet this unrewarding fruit was respected by the grand and princely. Kings and priests had it embroidered on their robes. Henri IV took it as his device with the motto, 'Sour but sweet'. Anne of Austria, his daughter-in-law, mother of Louis XIV, adopted it, too, but with a different motto: 'My worth is not in my crown.' A modest reference to the Greek legend of the pomegranate – a nymph who had been told she would one day wear a crown, was seduced by Bacchus who promised her one. Instead of keeping his promise in the way the girl had expected, he turned her into a pomegranate, with a little crown-like calyx. Catherine of Aragon was another Pomegranate Princess. She must have reflected ruefully on this symbol of fertility as she went through pregnancy after pregnancy with only a girl and a divorce to show for it.

An encounter we had with the pomegranate was in Cyprus, with Persephone's seeds: they kept her underground with her dark husband, yet were also a sign that she would return to the earth and bring an end to winter. They were a promise of fruitfulness when everything seemed black. We had gone one day to the easternmost tip and promontory of the island to stand on the high site of Aphrodite's shrine, and look across to Syria where she had come from. The road went through a monastery courtyard. In the morning, it was empty. In the late afternoon, it was full of families, spending the day in the little cells. People rushed about, women carried their best bowls carefully and we saw they were full of wheat, decorated with silver bits and pieces and pomegranate seeds. Later on, I discovered that this special dish is called *Kolyva* and its ingredients have a special meaning. It is made for funerals, and I think that the Sunday we were there was the Greek Orthodox Easter, when the hopelessness of Lent turns into joy, the darkness of Good Friday becomes the Resurrection. In this dish, said to be a pre-Christian ceremonial dish, wheat stands for everlasting life, raisins and nuts for joy and sweetness, pomegranate seeds for plenty and fertility. The day ended in another church, with candles, everyone giving us candles, everyone lighting their candles from their neighbours'. As we drove home in the twilight, we passed other churches with their doors open and more candles burning inside. A greeting for Persephone on her return?

SOME GOOD ADVICE ON POMEGRANATES

Good eating varieties have a superb flavour and make a fine dessert fruit if time is taken to remove the masses of glassy red or pink seeds and to separate them out from the yellowish membranes. The whole mass, ice-cold and rather like a pile of red crystals, looks wonderful served in a silver dish. . . . A long drink made of freshly-squeezed pomegranate juice, chilled and mixed with Dutch gin (a 'Stobart Special') is fantastic for anyone looking for something bracing and new, but the problem with squeezing the juice from pomegranates is that too much tannin exudes from the membranes and makes the juice astringent. To remove some tannin, a little dissolved gelatine can be stirred in. The gelatine reacts with the tannin to form an insoluble compound – a cloud which can be settled, filtered or removed by fining. If the fruit is allowed to shrivel before being crushed, the tannin is less.

Tom Stobart: *The Cook's Encyclopaedia*

I suggest the gelatine treatment before you undertake the following recipes, except for the ones in which pomegranate syrup is used, which can be bought from Greek and Middle Eastern stores. In *Middle Eastern Food*, Claudia Roden remarks that the pomegranates on sale here are bland and tasteless by comparison with the sharp, acid fruit of the Middle East. She recommends lemon juice, for instance, as an alternative to pomegranate syrup in *faisinjan* (*see* over), but I think a mixture of pomegranate juice and lemon gives you something of the wonderful dark colour, with the sharpness.

POMEGRANATE SOUP

A soup that may seem strange at first to our western taste, but that grows on one rapidly. Sometimes it has little meat balls in it, but as I am not a lover of meat balls I leave them out.

1 large onion, chopped
butter
125 g (4 oz/½ cup) rice, or mixed lentils, split yellow peas and rice
1 bunch parsley, chopped
1 bunch spring onions, chopped, or a large leek
250 g (8 oz) fresh spinach, chopped
250 ml (8 fl oz/1 cup) pomegranate juice or
3–4 tablespoons (¼–⅓ cup) pomegranate syrup
250 ml (8 fl oz/1 cup) pomegranate seeds
sugar
1 tablespoon lime or lemon juice
1 tablespoon dried mint or 8 sprigs fresh mint

½ teaspoon cinnamon
pepper, salt

Soften the onion in a little butter until golden and melting. Stir in the grain(s) and add 2 litres of water (3½ pt/8¾ cups). Cook for 15 minutes, then add the greenery and cook for another 15 minutes. Put in the pomegranate juice or syrup, add seeds, and simmer another 15 minutes. Add sugar to taste, and the lime or lemon juice towards the end of cooking time, with salt and pepper.

Just before serving, fry the mint briefly in a little butter in a small pan. Stir it around to bring out the flavour, and add the cinnamon with the same amount of pepper. Stir in, check the seasoning and serve.

FAISINJAN

The aristocrat of the Persian stew-sauces, made with joints of duck or chicken, or lamb cut into 2½-cm (1-in) cubes from the shoulder. The defining ingredients of a *faisinjan*, as opposed to the other sauces that are also served with rice, are ground walnuts and pomegranate juice or syrup. Use good quality walnuts or the sauce can be bitter. The point of the lemon juice in this English version is to give a little acidity to our pomegranate juice which is on the mild side.

1 duck, jointed
or 1 large chicken, jointed
or 1–1½ kg (2–3 lb) boned cubed lamb
salt, pepper
butter
2 large onions, chopped
200–250 g (6½–8 oz/1½–2 cups) walnuts, ground
200–250 ml (6½–8 fl oz/¾–1 cup) pomegranate juice plus juice of a lemon
or 4–5 tablespoons (5–6 tbs) pomegranate syrup
sugar
beef stock
¼ teaspoon each cinnamon, nutmeg, black pepper
generous pinch saffron

Season the meat, and fry it in butter slowly until it is almost cooked – about 30 minutes.

In a separate pan, soften the onion in butter, and when it is golden, add the walnuts, stirring them thoroughly with the onion. Pour in the pomegranate and lemon juice, or the syrup, and 2 level tablespoons (3 tbs) sugar. Add enough stock to make the liquid quantity approximately 750 ml (1¼ pt/generous 3 cups) – this means you will need less stock with pomegranate juice, more with pomegranate syrup. Stew for 15 minutes, stirring often.

Bone the meat and add it to the sauce. Complete the cooking. The meat should be tender and bathed in a dark, sweet and sour sauce. It can be reheated successfully. Adjust seasonings to taste.

Serve with basmati rice, plainly boiled or cooked in the Persian *chilau* style, i.e. soaked, boiled, then steamed with butter or oil to give a crispy-based rice.

PHEASANT GRENADINE

As the pheasant came to us from Georgia – or, as some would say, was brought back to Greece from Medea's Colchis, now part of Georgia, by the Argonauts – I cook it sometimes with ingredients from that part of the world. You can buy *burghul*, i.e. cracked wheat, from health food shops or Greek delicatessens: brown rice is a good alternative. Be wary about pine kernels. They are expensive and sometimes rancid; taste a couple before you buy them, and switch to almonds if you are in any doubt. Quantities below for eight:

7 pomegranates
1 brace of pheasants
butter
1 medium onion, chopped fine
250 g (8 oz) burghul
light game or chicken stock or water
50–60 g (1½–2 oz/about ¼ cup) pine kernels or almonds
2 tablespoons (3 tbs) meat glaze, or
150 ml (5 fl oz/⅔ cups) strong beef stock
salt, pepper

Halve the pomegranates across, and squeeze six of them, to give you approximately 250 ml (8 fl oz/1 cup) juice. Remove the seeds from the seventh pomegranate, separate them with a fork, and set them aside.

Brown the pheasants in a little butter, then put them breast down into a closely fitting pot. Pour in the juice – no seasoning – and cover. Put into the oven preheated to gas 4, 180°C (350°F) and leave until cooked, turning the birds breast upwards after half an hour: they will need another 15–25 minutes.

Meanwhile cook the *burghul*. Soften the onion in a generous tablespoon (2 tbs) of butter in a heavy pan over a low heat. Stir in the *burghul*, then add enough stock or water and salt to cover it by 2 cm (¾ in). Put on the lid and simmer until the liquid is absorbed and the *burghul* tender. If the liquid disappears too fast, add a little more. Fry the pine kernels or almonds in butter and add to the cooked *burghul*. Spread it round a warm serving dish, leaving a hole in the middle (to take the pheasant), scatter with pomegranate seeds, and put into the oven when you take out the cooked pheasants. Either switch off the heat or leave the door ajar, so that the *burghul* keeps warm without drying up. If there is likely to be a longish wait, cover it with butter papers.

Strain the pheasant juices into a pan, then put the covered pot of pheasants back into the oven to keep warm while you complete the sauce.

Taste the juices and add meat glaze gradually to taste. Its purpose is to enrich and strengthen the flavour, not to dominate it. If you are using stock, boil it down by half and add it to the pomegranate juice in the same way. Bring to boiling point, cut up 60 g (2 oz/¼ cup) butter into rough dice, and whisk the dice into the sauce, off the heat. When the sauce is smooth and well amalgamated, taste it again and add any seasoning required. It can be kept warm over hot but not simmering water – but it is better to add the butter at the last minute which removes any risk of the sauce separating.

Carve the pheasants and season the pieces. Arrange them in the centre of the *burghul* and pour over the sauce.

ACCOMPANIMENTS To me this dish needs nothing more than a salad to follow, although I have served it with two purées, celery and pumpkin. They were made in advance, using the processor, and stored in small pudding basins in the freezer: I allowed plenty of time for them to reheat in a *bain marie*.

Chicory goes well with pheasant, too. Boil in salted water, plus a tablespoon of lemon juice and a teaspoon of sugar, until almost cooked. Drain and finish in butter, turning the heads over until they are very lightly browned.

VARIATIONS Use guineafowl instead of pheasant. Or the drumstick and thigh joints of chicken, and cook the dish as a sauté on top of the stove.

KOLYVA

This recipe comes, appropriately, from *The Art of Greek Cookery*, published by the women of St Paul's Greek Orthodox Church at Hempstead on Long Island. Or at least the recipe minus the one ingredient that makes it important from the point of view of fruit – the pomegranate seeds. And I would have thought it important, too, from the point of view of the symbolism. It is a funeral and a Lenten dish, as I explained in the introduction to this section. It put me in mind of those last meals of grain that are found in the stomachs of bog corpses in northern Europe, the victims sacrificed to the god of spring growth. You might also think it has a muesli air to it as well, although it is eaten as a dessert rather than a breakfast dish.

½ kg (1 lb) whole wheat, soaked overnight
½ teaspoon salt
300 g (10 oz/2½ cups) white flour
125 g (4 oz/⅔ cup) sultanas
1 tablespoon (2 tbs) cumin seeds
1 tablespoon (2 tbs) ground cinnamon
60–75 g (2–3 oz/¼–⅓ cup) sugar
250 g (8 oz/2 cups) walnuts, finely chopped
1 tablespoon (2 tbs) chopped parsley (optional)
icing sugar
blanched almonds
silver dragees and balls
seeds from one or two pomegranates

Drain the wheat, then put in a pan and cover generously with water. Simmer until wheat is soft, about two hours, add the salt and simmer for another minute or two. Spread on a cloth to dry. Brown the flour carefully and slowly in a heavy pan over a low heat until it is golden. Take care not to burn it. Mix half the flour with the wheat, add sultanas, spices and sugar. Spread into a large shallow bowl, making the top flat. Mix walnuts and parsley and spread evenly on top, then cover with the remaining flour.

Sift icing sugar over the top, pressing it down with a spatula or waxed paper. Decorate with almonds, silver dragees and balls, and the pomegranate seeds. At a funeral, there would be a silver cross in the centre, the initials in almonds on either side, and decorative borders of pomegranate seeds and the almonds and silver things.

POMELO, see GRAPEFRUIT

PRICKLY PEAR, INDIAN FIG, BARBARY FIG or TUNA

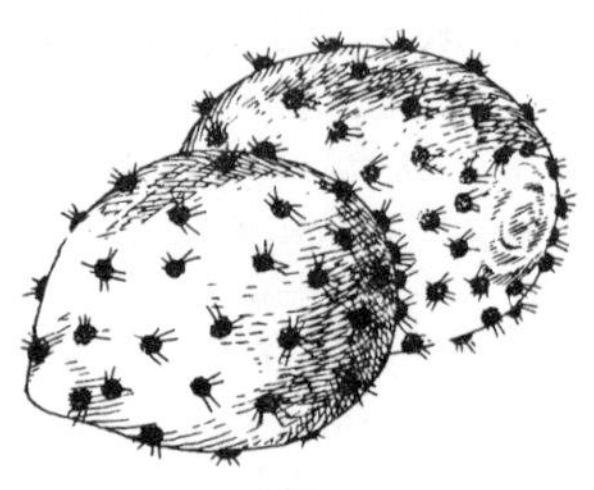

If you walk into the covered market in Kalamata in the Peloponnese, you will see the things you might expect to see – olives above all, and raisins – and then, if it is high summer, basket after basket of greenish-orange to red barrel-shaped objects. Behind each basket, a man or woman in black clothes, squatting. As you stop to look, someone will – if he likes your face – take up one of the objects carefully in a rubber-gloved hand and slice off the end parts. A few more quick movements, and he will hold out to you the opened outer peel, with a glowing coral-coloured fruit lying on it. This you take and eat, this prickly pear exposed in its sweetness – refreshing, juicy, full of edible seeds, but much in need of a little acidity. A squeeze of lemon or lime, and it would be perfect.

Once you catch on to the prickly pear in Greece, southern Italy and France, or in its native home of America, you see it everywhere (in some countries, Australia, India, South Africa, it has run wild and become a pest). As you come out of Pompeii on to the dusty road, and push past the ice cream sellers, you may find refreshment a little further on from the man with his cart of watermelon and prickly pears – *fichi indiani* (from its botanical name, *Opuntia ficus-indica*, meaning the American Indian fig, as it comes from Central America). As you go further south and into Sicily, such carts are on every street, or so it seems. From the train, you begin to realize that fields are defined by the prickly pear. You see the brown grass, dried in the heat, then the round-leafed green cactus striking out against the sun, dotted with orange and orange-red fruit. A stern beauty.

From a Corsican cookery book, I learnt the best way to pick the pears for myself. If you are staying in the south, you might as well profit from this source of free fruit: blackberry-pickers will understand that a little discomfort is well repaid. You need a bamboo stick, split at one end, with a cork pushed into the split to keep it apart. With a stick in the other hand to pull the fruit towards you, you can now hook off fruit quite easily with the split stick without needing to touch it. Rubber gloves are still advisable, in case you forget and pick up a fruit that has rolled away.

I should point out that you do not have to go to the Mediterranean or the southern States of America to find prickly pears. They sell them in many good greengroceries in the north of France these days, and I have bought them in London, principally at Selfridge's, and in Paris at Fauchon's.

In fact, the first lot I ever dealt with, I bought in Paris (appreciating, as I paid for them, how many opportunities we had missed to eat them free in the past). When we returned home, I thought that the shop must have singed the fruit clean of their spines, they looked so smooth and innocent. One fruit later came knowledge, and rescue from our daughter who knew exactly how they should be prepared. (She did not know how to remove my hundreds of invisible, determined spines unfortunately; much rinsing helped.) You steady the fruit with a fork, then slice off each end. Next you make a cut from end to end, trying not to go too deeply into the fruit. Slip the knife under one side of the cut, so that you can raise and push back the skin. Do the same to the other side, and you will end up with a flattened skin and the nicely-shaped orange fruit in the middle, waiting to be picked up and put into a bowl. Finish preparing all the fruit you have before you start slicing any of the edible part. Then you can clear away the debris, getting rid of the nasty spines.

Prickly pears are not one of the grand fruits. If you want to make a dessert of them, slice them and sprinkle them with the finely-grated zest of a lemon or – preferably – a lime, then squeeze the juice all over them.

Slices can be added to a fruit salad (the seeds that run through the whole flesh are quite pleasant to eat, adding a texture to the blandness). Or they can be infused, barely cooked, in a light lemon-flavoured syrup, and then sieved to make a brilliant orange-red juice. This can provide the liquid for a fruit salad, if you have not pushed through much of the debris; or a sauce for creams, soufflés and ices if you thicken the consistency by sieving through the flesh. Add the extra sharpness of citrus juice, to taste.

PRICKLY PEAR JAM

A recipe from Corsica that is simplicity itself. Note that you need no sugar at all.

Slice the prepared fruit into a pan, putting in a little water to prevent sticking. Cook gently, with a lid on the pan, until the fruit is very tender. Sieve it into a wide, heavy pan, leaving nothing behind but the pips. Bring to the boil and cook fast, stirring, until the purée becomes jammy and thick. Towards the end of the cooking time, add slivered almonds and lemon juice to taste.

Pot in the usual way. Eat with cream cheese mixtures of various kinds, as well as with bread and butter, etc. It makes a pleasant filling for little jam tarts, especially when they are baked with an almond crust. Put in the jam when the pastry is cold.

BARBARY JELLY

Jelly needs reinstating. Those garish blocks that make up into synthetic-tasting jelly have put us all off the real thing; they belong to fish-paste sandwiches and bought ice cream, the disgusting food that children are supposed to like. Proper jelly, made with real fruit (or with sweetened black coffee) and topped with a thin layer of farm cream, is a pleasure to delight the most demanding guest.

Do not worry that this jelly is not clear. You want to catch every scrap of flavour. In the end, the jelly has an exquisite pinkish-orange colour, and smells and tastes like paradise.

8 prickly pears
3 passion fruit
1 packet (½ oz/2 envelopes) gelatine
caster sugar
juice of an orange
lemon juice (see *recipe)*
150 ml (5 fl oz/⅔ cup) double cream

Carefully remove the skin of the prickly pears. Cut the fruit in slices and put into a heavy pan. Cut the passion fruit in half, and scoop out their pulp and put it with the prickly pear slices. Cover with water by 1 cm (generous ¼ inch) and put on the lid. Simmer gently until the fruit is tender and parts easily from the pips – the fruit should infuse in the water, rather than cook. Boiling spoils the flavour.

In a measuring jug, dissolve the gelatine in 6 tablespoons (½ cup) hot water. Sieve the fruit purée into the jug leaving only the seeds behind – keep back a few black passion fruit seeds for decoration, and throw the rest away. Add orange juice, and water to bring the quantity up to 500 ml (¾ pt/scant 2 cups). Stir in sugar to taste. Consider the flavour, and add a little lemon juice if you think the jelly should be heightened. Bring the final quantity up to 600 ml (1 pt/2½ cups) with water. Pour into 6 flute-shaped wine glasses, or other glasses, that hold 125 ml (4 fl oz/½ cup). Put into the refrigerator to set.

Just before serving, pour a thin layer of cream on to each jelly, and scatter three or four passion seeds in the centre of each one. Serve with almond biscuits or thin fingers of shortbread.

PURPLE GRANADILLA, see PASSION FRUIT

QUINCE

The largest quinces we have ever seen – at the market of Chania in Crete – were the size and shape of the marble breasts of Michelangelo's sleeping Night. When I first went to look at her, as a student, I felt he might not have been entirely happy with the female figure. Faced with those quinces, I wondered if they, or quinces like them, had not served as his model, another and easier symbol of love.

Of course, the Chania quinces had to be the best: they had been grown on exactly the right ground. They were eponymous quinces, one might say, the apples of ancient Kydonia – now Chania. The name of the apples of Kydonia changed as the quince travelled, rather as the plums of Damascus were transformed into damsons. In Provence, it became *cydonea* and *cotonea*, which the French turned into *coing* and *cotignac* (the paste made from *coings*), and so to the English quince. The name is really a mistake, as quinces come from much farther away, from Central Asia. They arrived in Europe from the east, with Aphrodite and her roses.

People say that for the Greeks and Romans, quinces were the golden apples of the Hesperides, the golden apples that prevented Atalanta winning the race. And a quince was the golden apple that Paris awarded to Aphrodite – quite rightly, for it was after all her fruit. The fruit of love, marriage, fertility. In spring on a warm day, if you sit in the lee of flowering quinces, you become quietly aware of a narcissus scent on the puffs of breeze. Very much the scent of the beginnings of love. The furling twist of the bud, pink and white, opens into a globe of pale pink, ruffed with leaves – its mildness goes unnoticed if you walk by without stopping.

Quinces and love: the association endured. Long after the time of Aphrodite or Venus an Andalusian-Arabic poem about the quince and love was written by Shafer ben Utman al-Mushafi, who died in 982 and was vizir to Al Hakam the Second of Cordova, a poem of this Mediterranean world, this quince world:

> It is yellow in colour, as if it wore a daffodil
> tunic, and it smells like musk, a penetrating smell.

It has the perfume of a loved woman and the same
hardness of heart, but it has the colour of the
impassioned and scrawny lover.

Its pallor is borrowed from my pallor; its smell
is my sweetheart's breath.

When it stood fragrant on the bough and the leaves
had woven for it a covering of brocade,

I gently put up my hand to pluck it and to set it
like a censer in the middle of my room.

It had a cloak of ash-coloured down hovering over
its smooth golden body,

and when it lay naked in my hand, with nothing more than
its daffodil-coloured shift,

it made me thing of her I cannot mention, and I feared
the ardour of my breath would shrivel it in my fingers.

Isn't that Aphrodite's apple?*

HOW TO CHOOSE AND PREPARE QUINCES

Not much good my giving advice on choosing quinces. You have to buy what you can find, and be thankful. Again, like the mulberry, the quince is a tree to look out for in other people's gardens in case they do not appreciate it, or are willing to share its fruit. A good tree for a wedding present, and faster growing than the mulberry. By a stream is the classic place, but in France we know quinces on the dryest of upland fields, and in the hottest of small village gardens. In the autumn as you go south, the sun shines on the fruit, which stands out as the leaves fall, like magic apples, gold and dazzling against the blue sky.

The trouble in the north is that our quinces tend to rot from the centre; this means that you may have some disappointments from time to time that are in no way the fault of the brave greengrocer who sells such an unusual line. A Greek greengrocer in London, near Charlotte Street, told a friend of ours that at home his family hang up the best quinces – come February, they taste soft and sweet enough to eat without cooking, 'like melon they are'. I wonder if they are a different variety? But perhaps

* from *Aphrodite* by Geoffrey Grigson, poem translated by A. L. Lloyd.

it is the climate. I hung up our best quince, and it rotted away in its string bag although it had been given the dryest steadiest corner of the kitchen, out of any steam.

With your own quinces, pick them when they are greenish gold if you fear the weather. Otherwise, leave them until they are yellow all over. Bring the best into the sitting-room or bedroom, which they will scent with the most heavenly smell.

In the damp and dirty past, quinces were put into boxes and stored on top of high cupboards, Some were put with the clothes or in chests of household linen. Stack them away, if you have a good crop and the space, on slatted wooden trays, making sure that they do not touch each other. In my experience, it makes little difference whether or not you wrap them in newspaper.

When you prepare quinces, you need to wash off the grey down that so often covers them in patches, if not all over. As you remove the peel and core – which are all worth keeping for the flavour that can be boiled out of them – drop the pieces into lightly salted water, or lemon water, as they will discolour like apples. With some dishes and preserves, discoloration does not matter as the quinces will be cooked until they are orange or orangey-red and any traces of brown will disappear.

Quince is about the best flavouring for apple or pear pies and tarts. You do not need much – half a quince peeled, cored and chopped small, in the bottom of the dish, with the debris cooked with the apple or pear debris to make the liquid. Later in the winter, use quince purée or syrup or jelly instead of the fresh fruit.

STORING QUINCES IN THE 16TH CENTURY

Pick ripe quinces without a blemish, in fine weather and at the waning of the moon. Put them gently into a large new bottle with a wide neck, having first rubbed off all the cottony down. Put a mat or ring of willow on top to press them down slightly, and cover with the best liquid honey.

This way of storing not only keeps the quinces well but also gives you a liquor with the taste of honey and of quince which can safely be given to those with *la sieure*.

Some people put quinces in little boxes on top of cupboards, where it's dry and cool, no smoke and no bad smells, with their heads up, stems down, not touching, in poplar or pine sawdust.

Antoine Mizauld: *Le Jardinage*, 1578

STUFFED QUINCES

DOLMEH BEH

Another recipe for fruit stuffed with meat from Maideh Mazda's *In a Persian Kitchen*. It is almost identical with her stuffed apple recipe on p. 8, but adjustments are made to cope with quinces which need longer cooking to soften.

Make the split pea, beef and cinnamon stuffing that you will find under the apple recipe.

Hollow out six perfect large quinces (save the pulp for jelly making). Stuff them and place them in a saucepan. Pour in water to come 2–2½ cm (¾–1 inch) up the fruit. Cover the pan, bring it to boiling point and leave it to simmer until the quinces are almost cooked.

Mix 6 tablespoons (½ cup) wine vinegar with 125 g (4 oz/½ cup) sugar and 6 tablespoons (½ cup) water in a small pan. Boil for 3 minutes.

When the quinces are almost cooked, pour over the boiling sweet-sour mixture and simmer a further 15 minutes. Remove the quinces to a serving dish and keep them warm. If the cooking liquor is on the lavish side, boil it down, then beat in a tablespoon (2 tbs) of butter and pour over the quinces.

RUSSIAN BEEF BRAISED WITH FRUIT

Fillet, rump or round of beef is sometimes braised with fruit in Russia, rather in the Georgian style. Scottish readers are lucky enough to be able to make the dish with a piece of shoulder fillet, a cut unknown in England except as sliced 'feather' steaks.

Lard and brown the meat. Pour off the fat, or transfer to a casserole. Add fortified wine and stock or water and fresh grape juice to come about two-thirds of the way up the joint. Add a decent-sized quince, peeled, cored and sliced. Put a butter paper on top of the meat and simmer in the covered pan until tender, either on top of the stove or in a moderate oven, gas 3–4, 160–180°C (325–350°F).

Transfer meat and quince pieces, if they are not too disintegrated, to a serving dish. Strain the juices into a wide pan. Degrease them, then boil down to make a concentrated sauce. If you used grape juice as the liquid, put in a handful of raisins, sultanas and currants. Red currant jelly can be used to sharpen the taste.

QUAIL WITH SLICES OF QUINCE
CAILLES AUX COINGS

Or should it be 'Quail Edward Lear'? The title looks like an alphabetical joke and perhaps it started that way, though I doubt it. The joke is mainly on the English side, and this is a French dish. At least, the happy idea of bedding the birds with quince, so that they become suffused with the scent, is French. After that I follow my own path.

Wild quail are not intended. They are protected here, and rare on French tables. Farm quail are the thing to use. They are not as easy to find as they are in France, but they are becoming more common. If you have any difficulty, substitute a brace of guineafowl or pheasant and follow the variations given in the next recipe. Once you get beyond a party of six people, which means a dozen quail, you will in any case find it easier to deal with two big birds rather than sixteen or twenty small ones.

2 large ripe quinces
12 quail
6 tablespoons (½ cup) Cognac or Calvados
quince or quince and apple jelly (p. 392)
butter
6 slices of white bread
150 ml (¼ pt/⅔ cup) poultry or game stock
salt, pepper

Wash the quinces free of grey fluff, then peel, core and slice them. Put the slices into a casserole with the quail fitted closely together and pour the alcohol over them. Cover and leave in the refrigerator for two days, turning the birds about from time to time.

To cook, remove the quince slices to a pan, add water to come halfway up, then add half a pot of jelly – 200–250 g or 6–8 oz. Simmer until the quince is tender.

Arrange half the slices in a buttered gratin dish. Place the quail on top with their alcohol. Brush them with melted butter, sprinkle with salt and roast at gas 6, 200°C (400°F) for 20 minutes until nicely browned on top and cooked. Baste with the liquor from cooking the quince slices from time to time.

Meanwhile fry 6 pieces of bread in butter, each one large enough to accommodate two birds. Place them on a serving dish and put the rest of the quince slices on each. When the birds are done, arrange them on the bread and keep the whole thing warm. Add salt and plenty of pepper.

Stir the stock into the roasting dish and strain the liquid into a little pan. Taste it, and if necessary boil down so that you have a generous teaspoon of concentrated liquor to pour over each pair of birds.

The success of the dish depends on getting a good savoury quince flavour whilst not overwhelming the birds. You might consider, for instance, that instead of putting quince slices on to the bread it would be better to keep them for another dish and just spread the slices with quince jelly, or leave them plain and put quince jelly on the table for those who like an extra emphasis. Quinces vary in strength from year to year, I find; it is wise to keep a check on the recipe as you go along, with this in view, and make adjustments.

PHEASANT OR GUINEAFOWL WITH QUINCE
FAISANS OU PINTADES AUX COINGS

Shut up a brace of birds with the quince slices, as in the recipe above, with 8 tablespoons (⅔ cup) alcohol.

Cook the slices and put a few into the birds, the rest into a well-buttered terrine. Brown the birds and place them breast side down on to the quince. Pour in the alcohol marinade and 150 ml (¼ pt/⅔ cup) stock. Put a tablespoon (2 tbs) of quince jelly (p. 392) on each bird, then a butter paper. Close the terrine, and put into the oven at gas 4, 180°C (350°F) for 45–60 minutes, until cooked. Turn the birds at half time. Carve the birds and put them on a dish, salt them and pepper them well. Mash the quince slices with pan juices and a little extra jelly if you like. The flavour can be softened by stirring in double cream, and a squeeze of lemon refreshes the flavour. Reheat the purée and serve round the birds, in spoonfuls, alternating with triangles of fried bread.

Note If you happen to have a bottle of *eau de vie de coings*, substitute that for brandy or Calvados.

RED QUINCE COMPOTE

A recipe from Audiger's *La Maison Reglée* of 1692, simple to do and very striking. It should be served with plenty of cream, or vanilla ice cream, or a cream cheese mixture, as the quinces are strong.

Cut up 3 or 4 quinces – use the windfalls for this – and put them, peel, core and all into a pan. Cover generously with water. Also peel neatly 5 or 6 perfect quinces. Add the peelings to the pan, and then the cores as you cut them into quarters. Prevent the pieces from discolouring by dropping them into a bowl of lightly salted water.

Boil up the panful of debris, and stew until it begins to turn a good rich red. Then strain off the juice, add sugar to taste and bring slowly to simmering point, stirring. In this syrup, cook the quince pieces until they are tender. A vanillla pod makes a good addition, and should be left in the compote as its blackness makes a good contrast to the red.

QUINCES BAKED IN THE FRENCH STYLE

COINGS AU FOUR

Allow one for each person. Peel and hollow out the cores of six to eight quinces, being careful not to pierce through the bottom of the fruit. Sprinkle each one with lemon juice as you go. Stand the quinces in a buttered gratir dish.

Mix together to a cream 150 g (5 oz/⅔ cup) caster sugar, 100 g (3½ oz/scant ½ cup) lightly-salted or unsalted butter, and 3 generous tablespoons (¼ cup) of whipping or double cream. Stuff the quinces with this mixture – if there is some left, add halfway through the cooking. Top each quince with a level tablespoon of sugar and bake at gas 5, 190°C (375°F) until the quinces are tender. Serve with cream and sugar.

Note Baked quince was Sir Isaac Newton's favourite pudding.

QUINCE AND APPLE SOUFFLÉ

We gave our daughter the two finest quinces from our tree in France: three days later she rang me up with this recipe.

250 g (8 oz) quince
250 g (8 oz) firm aromatic eating apples
2 cloves
1 tablespoon (2 tbs) butter
caster sugar
3 large eggs, separated
extra butter

Peel, core and cut the quince into small dice. Put into a heavy saucepan with a thin layer of water in the bottom. Peel, core and cut the apples into slightly larger pieces and put them on top of the quince, with the cloves. Cover and put over a low to moderate heat until the juices run and the fruit begins to soften. Raise the heat, and cook to a soft unwatery mash (remove the lid if necessary).

Sieve the fruit, add the butter and some sugar to taste. The purée should be on the sweet side. Mix in the egg yolks while it is still warm. This can be done in advance: reheat the mixture gently before continuing with the recipe.

Turn the oven to gas 6, 200°C (400°F). Whisk the egg whites until they stand in peaks – they should not be too dry – then stir a tablespoon into the fruit to slacken it, before folding in the rest of the whites with a metal spoon or spatula. Use a careful unhurried movement going right to the bottom of the bowl, so that the whites remain as airy as possible.

Put a baking sheet into the oven to heat through while you complete the final stage.

Butter a soufflé dish generously, of approximately $1\frac{3}{4}$-litre (3-pt/$7\frac{1}{2}$-cup) capacity. Put in a spoonful of sugar and turn the dish so that the sugar coats it evenly: tip out any surplus, or add more if needed. Put in the mixture. Slide the dish on to the baking sheet, and leave for 20 minutes. Open the door and carefully draw out the baking sheet just enough for you to sprinkle the top with caster sugar. Slide the tray back, close the door smoothly and leave for another 5 minutes.

Ovens are not always accurate, so it is wise to know what a soufflé looks like when it is ready. It should be nicely risen and a beautiful appetizing brown. When you give the dish a little push, it should wobble slightly. The ideal soufflé – for me at any rate – is creamy in the centre so that it needs no sauce.

A soufflé is one of the easiest things to make, and it is far more long-suffering than you would suppose. Anton Mosimann of the Dorchester says that you can stand the soufflé dish, complete with mixture, for a long time in a barely simmering *bain marie*, then finish it off in a hot oven. Prue Leith passes on a tip from the Café Royal: make the entire soufflé in the morning, including the egg whites and all, and put the whole thing into the freezer: bake it as usual, giving it about a quarter as long again.

OLD-FASHIONED QUINCE TART

These days we keep fruit purée and custard mixtures for ice creams and fools: it seems odd to put them in a pie crust and bake them. The idea, once so popular, has survived at least in America, e.g. in pumpkin pie.

Another old tradition we have dropped is using puff pastry for fruit tarts and pies. If it is light and made at home with butter, it does make a good contrast especially to thick fillings. Do not be tempted to use deep-frozen puff pastry, the lack of butter becomes very obvious, shown up by the fresh true flavour of fruit. Shortcrust is a better choice, if you are not a puff pastry maker.

Whichever kind you choose, line a 23-cm (9-in) tart tin and prick the base. Other ingredients:

150 ml (5 fl oz/⅔ cup) cooked quince purée
sugar
cinnamon or ginger
300 ml (10 fl oz/1¼ cups) whipping cream
2 large eggs

SAUCE
375 ml (12 fl oz/1½ cups) liquid from cooking quinces
sugar
1 level tablespoon arrowroot
quince eau de vie, *brandy or gin*

Sweeten the purée to taste, adding spice. Beat the cream with the 2 egg yolks and sweeten – it is always wise to season mixtures separately, final adjustments are then tiny and easily made. Mix smoothly with the purée. Whisk egg whites until stiff and fold in. Put into pastry case.

Preheat oven to gas 8, 230°C (450°F) for puff pastry, gas 7, 220°C (425°F) for shortcrust. Slide a baking sheet on to the centre shelf a couple of minutes before you put the tart on to it. After 7 minutes, check that the pastry is not rising through the filling (prick with a fork if it is). After 15 minutes, lower the heat to gas 4, 180°C (350°F) and leave for a further 20 minutes or longer. The filling can either be set firm or be slightly wobbly in the centre under the crust – whichever you prefer. Serve warm with the following sauce:

Heat the quince liquid and sweeten it to taste. Mix the arrowroot with a little cold water and stir into the quince syrup. Simmer until the sauce is clear and lightly thickened: if it seems gluey, add a little water or more quince liquid. Off the heat add the alcohol to taste.

QUINCE JELLY
GELÉE DE COINGS

A French recipe that is practical for people who grow quinces, and end up with the usual mixed bag of fruit, some perfect, some blemished, some a bit twisted, some very small. If it is a bad year, add windfall apples or cooking apples up to half the weight of quinces.

Set aside three or four fine quinces, after washing them all and rubbing away the grey fluff. Cut up the rest roughly into chunks and put them into a preserving pan, half full of water.

Peel your best quinces, then slice and core them neatly. Throw peel and core into the pan. Lay the slices on top. If the water does not cover the fruit, add some more. Bring to the boil and simmer until the fruit on top is completely tender – remove the slices with a slotted spoon. Boil the panful hard to get the most out of the debris. Crush it down occasionally with a potato masher. When the fruit is completely pulpy, strain off the juice through a jelly bag.

Measure the quantity: to each 600 ml allow 500 g sugar (pound to a pint 2 cups sugar to 2½ cups water). Put the reserved slices into the juice, bring to the boil, stirring, then boil hard until setting point is reached. Concentrate the slices in two or three sterilized jam jars. Top them up with jelly, then pour the rest into more jars. Cover and store in a cool dark place.

The quince slices come in handy, when stored like this, for adding to apple pies throughout the winter.

Spices can be added to quince jelly if you like, cinnamon stick and a few cloves for instance: put them in with the sugar.

QUINCE PASTE
COTIGNAC (FR.), MEMBRILO (SP.)

As you go further south in France, the quince trees which we treasure one by one north of the Loire and in this country, become far more common. Their fruit is larger, though not as large as the quinces I saw in Chania market in Crete, and as golden as any apple of the Hesperides can have been against the blue autumn sky. In her book on the customs and cookery of the Périgord and Quercy, Anne Penton remarks that 'in the Périgord Agenais, quince trees were formerly used instead of fences to mark off the limits of each property in the same manner as laburnum trees were used further north in the Dordogne towards the Charente'.

Then she gives a recipe for quince preserve. Cored and quartered quinces are cooked with an equal weight of sugar, plus water to cover: the cores are tied in a little bag to dangle in the liquid as orange pips do in marmalade. They take seven hours apparently to cook to tenderness, a state which is held back, I suppose, by the hardening presence of sugar. The fruit is put into pots, the syrup boiled to setting point and poured over them. This kind of preserve is excellent with mixtures of *fromage frais* and curd cheese, such as *coeur à la crème*.

We find quince paste more useful, being disinclined to jam and runny preserves. It can be served as a sweetmeat with coffee, or laid on grilled pork chops, or mixed into apple pies and sauces as a flavouring. If kept dry, it lasts for at least two years, probably longer, but ours has always disappeared by that time – however well I hide it.

This kind of cookery is very much a family enterprise, one of the first of the preparations for Christmas. The stirring can be laborious unless there is someone to help. And the pleasure is doubled if people come in and out, sniffing the wonderful smell that reaches even the furthest rooms of the house by the time you have finished.

If you have plenty of quinces, weigh out a couple of kilos (4 lb). If not, make up the weight with apples, or apples and no more than 225 g (7 oz) peeled, sliced oranges. Windfall cookers in reasonably good condition are perfectly acceptable.

Wash and cut up the quinces (and apples). Put them into a heavy pan with 300 ml (½ pt/1¼ cups) water. Bring to the boil and simmer until the fruit is tender. Towards the end, mash it down with a potato masher or beetle. The purée should be very soft and easy to push through a drum sieve. Weigh the purée, and put it back into the rinsed out pan with an equal weight of sugar. Stir over a low heat until the sugar has dissolved, then raise the heat and boil until the mixture thickens and candies, leaving the sides of the pan and turning dark red. It will explode and pop with an occasional fat burp. Only by stirring and constant attention can you prevent the paste burning once it begins to start this thickening process. Wrap your hand in a cloth and use a wooden spoon. Eventually you will barely be able to push the spoon through the paste.

Now you can remove it and spread it out in centimetre- (small ½-inch) layers in baking trays lined with Bakewell or greaseproof paper. Push it out as evenly as you can; then, when it cools down slightly, wet your hands and press it smooth, as if you were making *gnocchi*. Put the trays in the airing cupboard, on the rack over an oil or solid fuel cooker or in its plate-warming oven. You can also put it into a gas or electric oven switched on at its lowest point. The time required to dry the paste is about 12 hours, but try cutting a square occasionally to see how the paste is getting on. When it can be cut with a hot knife into firm squares, it is ready.

Dip the squares in granulated sugar (mixed with cinnamon if you like). Pack them in layers – with rice paper in between and two or three bay leaves – in a box or an air-tight container if there is any risk of humidity.

Another way is to keep them in a great deal of granulated sugar in a plastic box. As the quince paste is eaten, the flavoured sugar can be used for apple or pear dishes, as you might use sugar from the vanilla pod jar.

QUINCE BUTTER

Fruit butters come between fruit jelly and fruit paste. They are soft enough to spread, yet they look firm and thick like a fruit paste. Quince butter can be eaten with bread and butter, or with pork.

The process is simple. Quarter and cut up the quinces into fairly small pieces, no point in removing skin or core. Add a little water and cook until the fruit is soft enough to go through a sieve.

Measure this purée by volume, return it to the pan and stir in three quarters of its volume of sugar, and a knob of butter. Set over a low heat until thick – stir all the time, as with the quince paste, because the mixture burns easily. If you are not used to this kind of preserving, tackle a small quantity first.

With butter added to the fruit and sugar, there is unlikely to be any scum. But if there is, remove it before potting in small sterilized jars.

Note This recipe can be used for other fruits. Remove large stones before you start cooking, if this is convenient, and crush the soft fruits in the pan before you start cooking them.

QUINCE VODKA

A long time ago I wrote that quinces made anything delicious. 'Yes,' replied one reader, 'but first catch your quinces.' And he enclosed a good recipe for quince vodka, for those who can only bag a couple.

Allow the quinces to become really ripe and yellow. Wash them well, rinsing away any grey fluff that might remain. Then grate them – peel, core and all – and put them into a litre bottling jar (1¾ pt/scant 4½ cups). Add 60 g (2 oz/¼ cup) caster sugar. Fill jar with vodka (or rum, gin or brandy for that matter). The jar need not be full, but the fruit must be covered. Close tightly. Leave in a dark place for at least two months. Taste, and decide whether to leave for another two months or longer – it improves with time, and much depends on the quality of the quinces in the first place which can vary from year to year. Add extra sugar if you prefer a liqueur sweetness; strain off the liquor into a clean bottle.

RAMBUTAN, see LYCHEE

RASPBERRY and LOGANBERRY

The raspberry is that rare thing – a delicious improvement of a wild northern fruit. It is nice to think of Eskimos eating raspberries in Alaska, even if they're the wild, unimproved variety. And of children picking wild raspberries in the woods of Scotland. When we were young, a particular pleasure of ours was being let loose amongst a neighbour's raspberry canes on a sunny afternoon – I understand what Parkinson meant when he said that raspberries were an 'afternoon dish' in summertime. Buzzing and warmth and crushing raspberries with your tongue against the roof of your mouth and everyone pleased that the crop should not be finished by the birds.

It is children, in fact, who come to the aid of the cultivated raspberry crop, commercial raspberries. School holidays are timed to coincide with the six-week harvest. Young heads bob up and down among the raspberry canes of Perth in the Carse of Gowrie and of Angus, all 7947¼ acres of them, as they pick what amounts to ninety per cent of all the raspberries we eat in Britain. Scottish raspberries are the best in the world, the long summer day in the north, not too hot, suits them exactly. Machines of the sort they have in America to pick raspberries have not been widely used. Our raspberries are juicier and too delicate for mechanical treatment on the whole; though adaptations are being studied, it seems that raspberries are often destined to rot by the ton on the canes of Scotland.

Many gardeners claim that for flavour, the yellow raspberry is superior to the more common red one. This is so, I agree. I have a book written in 1920 about the good things of France, taken month by month, and I read that when the raspberry crop of the outskirts of Paris is exhausted in September, 'the fruit growing valley between Metz and Thionville is covered with the golden grains of the yellow raspberry, which is mild and sweet'. I wonder if that is still true, sixty years later? It must be a sight, the first of the golden displays of autumn. We had a corner of yellow raspberries once, but the canes were exhausted after a number of years and never replaced. Raspberries of blessed memory.

Buy them, if you see them; if you intend to grow raspberries, ask your nurseryman for a yellow-fruiting variety. You will not be disappointed. In the shops, though, to keep life easy, I imagine that raspberry-coloured raspberries will always be preferred. They are a difficult fruit to sell in any case, red or yellow, turning so easily to dampness and mildew. Some growers have the pleasant habit of picking them into punnets lined with soft green leaves which cradle the berries more gently than the white cardboard punnet on its own.

Raspberries are above all the fruit for eating raw, preferably on their own with Jersey cream and sugar. They are companionable with peaches and melon, if there are just a few of them, or with cream cheese mixtures. These remarks do not of course apply to people who grow so many raspberries that they are desperate to use them up, freezers overflowing, children stuffed and red-mouthed beyond their desires. They will not mind a loss of quality in exchange for variety: it is for them that the recipes following are mainly intended.

The cooking berry *par excellence* is the loganberry, which can be harsh when raw if not absolutely ripe. It tastes like a cross between raspberry and blackberry but is, like the boysenberry, a cultivar of the Pacific blackberry. Any of the raspberry and blackberry recipes can be adapted to these two berries, one named after Judge J. H. Logan in whose garden at Santa Cruz it was raised, and the other after Rudolf Boysen who helped to introduce it, in the Twenties.

HOW TO CHOOSE AND PREPARE RASPBERRIES AND LOGANBERRIES

Buying raspberries is a hazardous business, as I have said. Look out for stains at the bottom of the punnet. You will do much better to locate a local field where you can pick raspberries for yourself, and where you can go regularly through the season. Dealing with a large quantity, unless you have a lot of help, can be very wasteful or exhausting. In fact, raspberries freeze very well, far better than strawberries. Of course you lose the finest aroma, the finest velvety touch of the fresh fruit, but in mid-winter they can provide a welcome treat.

Avoid washing raspberries – as they grow high off the ground, they should not be dirty. Just pick them over gently to pull away any bits of stalk.

When making raspberry summer pudding, raspberry jelly for dessert or jam, remember that a quarter of the weight in red currants makes a good addition. In Richard Bradley's *Country Housewife and Lady's Director* (1732 edition), he observes very justly that 'where you put any of the raspberry syrup, the flavour of raspberry will prevail . . . unlike mulberry syrup which . . . carries no flavour that will be predominant over that of another fruit'. This applies to mixtures of the fruit as well.

If you want to make a raspberry sauce to go with puddings or ices, crush damaged fruit and boil with an equal weight of sugar, plus a little water to get them going. Put the best raspberries in a dish. When the syrup is sweet and thick enough for your taste, strain it boiling over the uncooked fruit and leave the whole thing to cool slowly. To make a concentrated sauce Melba – Escoffier's *pêche Melba* was a cooked peach half on vanilla ice, with raspberry sauce – process raspberries with icing sugar to taste, then push through a sieve.

GUINEAFOWL WITH RASPBERRIES
PINTADEAU AUX FRAMBOISES

In La Varenne's *Le cuisinier françois*, he has a section headed 'Cooking with the army'. The first recipe, which seems unsuited to the battlefield, is for turkey with raspberries. Having no fresh turkey in the raspberry season, I decided to try a guineafowl with raspberries instead.

My large bird was enough for 5 people, after a good first course. For 6 or 8, buy two of ¾–1 kg (1½–2 lb) each, and increase the other quantities slightly.

1 large guineafowl
250 g (8 oz) raspberries
butter
1 litre (1¾ pt/scant 4½ cups) poultry or giblet stock
6 tablespoons (½ cup) fortified wine, e.g. Madeira, port
1 tablespoon (2 tbs) raspberry jelly or jam
salt, pepper, cayenne
250–300 g (8–10 oz/about 2 cups) brown or wild rice, cooked
60–90 g (2–3 oz/¼–⅓ cup) unsalted butter

Stuff the guineafowl with two-thirds of the raspberries. Butter it generously and put it to roast in a hot oven, gas 7, 220°C (425°F), reducing the temperature to gas 4–5, 180–190°C (350–375°F) after 20 minutes. According to size, the bird will take 45 minutes or a little more. Put the rice on to boil.

Meanwhile reduce the stock to 250 ml (8 fl oz/1 cup) adding the wine about half way through. You should end up with a rich-tasting juice that is slightly syrupy. Skim it well, add the jelly at the end.

When the bird is cooked, scrape the contents of its belly into the sauce. Pour fat from the pan juices and add them to the sauce as well. Simmer gently, while you carve the guineafowl, and put it in the middle of a ring of boiled brown rice, or wild rice if you can find and afford it. Season well and scatter the fresh raspberries over the top. Cover with a butter paper and keep warm.

To finish the sauce, strain it into a clean pan and taste it for seasoning; a pinch of cayenne is sometimes a good idea, as well as the usual pepper. It may well be salty enough from the reduction – that depends on the stock you started with. Bring to boil.

Whisk in the butter in small bits, raising the pan from the heat, then putting it back, as you do so. This keeps it hot, but under boiling point. Taste when you've put in two-thirds of the butter: it may need no more.

Pour over the guineafowl, and serve. A plain green salad is all that is needed afterwards.

GROUSE WITH WILD RASPBERRIES

If you are lucky, both inhabit the same ground, at the same time, when you are there (possibly with a gun), on or soon after 12 August. I am quite happy to wait for a while, not being so entirely devoted to grouse as perhaps I should be, and to use domestic raspberries. Other wild berries can also be used, whortleberries, bilberries and rowan berries for instance.

3 young grouse
salt, pepper
3 or 4 heaped tablespoons (4 or 5 tbs) berries
butter
9 rashers fat streaky bacon
3 thick slices crustless bread

Season the grouse inside and out. Mash fruit roughly with 3 good tablespoons ($\frac{1}{3}$ cup) of butter, and divide between the cavities of the birds. Wrap them in the bacon.

Butter the bread lavishly and place in a buttered roasting pan. Put grouse on top. Put into the oven at gas 5, 190°C (375°F). After 20 minutes, remove the bacon and baste with the melted butter. Baste again 5 and 10 minutes later. After 35 minutes, they are likely to be done, pinkish, rather than rare. Serve on the bread with any juices, surrounded with watercress and straw potatoes. Cut each grouse in half to serve.

SUMMER FOOL

An idea I have taken from Hannah Wolley, from her *Accomplisht Lady's Delight* of 1675, is to flavour raspberry creams and fools with rose-water.

For six to eight people, use ½ kg (1 lb) of raspberries. Remove three tablespoons and crush them on a plate with a fork; push the purée through a sieve into a pudding basin. Sugar the rest of the raspberries.

To the raspberry purée, add a tablespoon (2 tbs) of icing sugar, 250 ml (8 fl oz/1 cup) whipping cream and 125–150 ml (4–5 fl oz/⅔ cup) double cream. Whisk until fluffy. Fold in the sugared raspberries with any juice.

Mix in gently a tablespoon of triple-strength distilled rose-water. Try it for strength and add more, gradually, until you just catch a mysterious and harmonizing flavour at the end of the taste.

Cover the bowl with film and store in the refrigerator. Before the meal, bring out into the kitchen and divide between individual glasses. Do not put back into the refrigerator, as it is best eaten cool rather than chilled.

Sponge finger biscuits or crisp vanilla thins should be served as well.

WINTER FOOL

An inexpensive way of using raspberries you have put by in the freezer, the less glorious ones. Raspberries straight from the canes deserve Loseley cream or some other fine farm cream, but this butter and milk substitute does well for frozen fruit which has lost its freshness but still retains a vigorous flavour.

250 g (8 oz/1 cup) unsalted French or lightly salted Danish butter
250 ml (8 fl oz/1 cup) milk
1 rounded teaspoon gelatine
½ kg (1 lb) frozen raspberries
caster sugar
pinch salt
3–4 tablespoons kirsch, framboise or gin (optional)

Dice butter and melt it with the milk and gelatine over a low heat. The mixture should come nowhere near boiling point. Tip into the processor and whizz for 30 seconds, then put in the frozen raspberries and continue to blend until you have a lurid pink cream. Taste and add sugar – quantity depending on how much sugar was frozen with the raspberries. Push through a sieve.

A pinch of salt may help with the flavour, and the alcohol gives it an extra subtlety.

Put into 6 or 8 flute glasses or custard glasses. Serve with a sponge finger biscuit stuck in at an angle, or with almond or hazelnut biscuits served separately.

You can pour a thin layer of whipping cream over the top of each fool if you like, or raspberry syrup (raspberries cooked in syrup until it takes on their colour and flavour – cool and pour through a sieve without pushing the debris through). You can scatter a toasted almond or hazelnut or two over the pink cream. But none of these things is necessary.

HAZELNUT AND RASPBERRY MERINGUE CAKE

A classic Austrian recipe that has become popular in the last few years for birthdays and special occasions. Raspberries make the ideal sharp fruit filling, to contrast with the hazelnut meringue, but strawberries, peaches, apricots can be used instead. Packets of toasted chopped hazelnuts can be bought, which saves a little time, but if you have a nut-mill or processor, you will find that freshly toasted nuts give more flavour.

125 g (4 oz/¾ cup) shelled hazelnuts
5 large egg whites
300 g (10 oz/1¼ cups) vanilla caster sugar
½ teaspoon wine vinegar
300 g (10 fl oz/1¼ cups) whipping cream
250 g (8 oz) prepared fruit
icing sugar

Put the nuts on a baking sheet and put in the top of the oven set at gas 2, 150°C (300°F) for 10 minutes or until pale brown all through. Cool, then reduce to a coarse powder in a nut-mill or processor.

Whisk the egg whites until stiff. Then add the sugar a tablespoon at a time, still whisking, then the vinegar. Fold the nuts carefully into this bulky, satin-sheened mass.

Line two 20–23-cm (8–9-in) sandwich tins or flan rings with vegetable parchment, and spread the mixture evenly between the two, whirling the surface of one with a fork. Bake 35–40 minutes at gas 4, 180°C (350°F) and then leave to cool.

Just before the meal, whip the cream and fold in fruit and icing sugar to taste, and use to sandwich the two meringue discs together. Or whip the cream with some sugar and spread over the bottom disc, arrange the fruit on top, bedding it into the cream, and then put on the top disc – the one with the whirled surface. Sprinkle with icing sugar.

Note I imagine that this cake is the origin of the Australian pavlova, another soft chewy meringue.

SUMMER PUDDING IN THE FLORENTINE STYLE

If some national opinion poll decided to find out our favourite pudding, many people would, I think, reply, 'Oh, summer pudding – with plenty of cream.' And whether filled with black currants, or with raspberries and red currants, it is indeed one of the most delectable things imaginable – so long as it's made with proper bread. If sliced bread happens to be your doom, use sponge cake instead.

I chose this pudding in the Otello restaurant in Florence, because I was intrigued by the tracery of brown lines, which made it look like a phrenologist's model head.

1 large sponge cake of the Victoria or Genoese type (p. 465)
375 g (12 oz) raspberries
100–125 g (3½–4 oz/scant ½ cup) icing sugar
2–3 tablespoons (3 tbs–¼ cup) orange liqueur or kirsch
250 ml (8 fl oz/1 cup) double cream
4 tablespoons (⅓ cup) single cream

Slice the top from the sponge cake fairly thinly, and set it aside. Cut the rest of the cake into three layers, and use them to line a large pudding basin, cutting the pieces to fit as irregularly as you like, and pushing them closely together. The brown edges of the cake will form the irregular pattern of brown lines I was talking about.

Sprinkle the raspberries with the sugar. Leave for an hour or two until they are bathed in a certain amount of juice – about 150 ml (5 fl oz/⅔ cup). Strain off the juice, mix it with the liqueur or kirsch, and use it to moisten the sponge lining: stop before it becomes too wet.

Whip the creams until thick and standing in peaks. Fold in the raspberries and fill the sponge-lined basin. Put the top of the cake on as a lid, trimming it to fit if necessary. Place a plate inside the rim of the basin, and leave overnight. Next day, remove the plate, and invert a serving dish on top. Turn the whole thing upside down to serve. The sponge cake will have acquired a pinkish look. No extra cream is required for serving; if some of the filling is left over, spoon it round the base of the pudding.

OLD-FASHIONED RASPBERRY PIE

It's a good rule, I agree, never to stew raspberries – who, after all, wants to eat raspberries cooked when they can eat them fresh? – but raspberries baked in a pie is another matter. This version of one of our best national dishes comes from the eighteenth century. Cooks in those days used to pour mixtures of cream (or wine) and eggs into a pie at the end of cooking time: this 'caudle', as it was called, took the place of an accompanying jug of custard or cream. It thickened the juices slightly and, in the case of a fruit pie, it softened the acidity of the fruit which is emphasized by cooking. Cooks also used puff pastry for most pies, but a flaky or shortcrust is more popular nowadays. The thing to avoid is *pâte sucrée* – English pies are best without it.

For a round, 2½-cm (1-in) deep glass pie plate, you will need:

pastry made with 375 g (12 oz/3 cups) flour
about 500 g (1 lb) raspberries
125 g (4 oz/½ cup) caster sugar
1 egg white, lightly beaten
extra sugar
300 ml (10 fl oz/1¼ cups) whipping cream
2–3 egg yolks

Line the plate with pastry, setting aside enough for the lid. Pile in the raspberries, interspersing them with sugar: there should be enough to mound up above the level of the rim of the plate. Cover with pastry, making a central hole large enough to take a small kitchen funnel. Decorate discreetly and brush over with egg white. Sprinkle with sugar.

Bake at gas 5, 190°C (375°F) until the pastry is cooked, about 40 minutes. Check from time to time to make sure the pastry is not browning: a fruit pie is best with the pale, crunchy finish that comes from the egg white and sugar. Warm the cream to boiling point, whisk into the yolks, and pour slowly through the central hole down the funnel. Be careful it doesn't overflow. Return to the oven for 10 minutes. Any cream and egg left over can be sweetened and gently heated to form a little extra custard.

RASPBERRY JELLIES

Fruit jellies, set in wine glasses, are simple to make, and elegant as the last course of a dinner party. The point to watch is the set, which should not be too sloppy nor too firm.

Put 500 g (1 lb) raspberries into a heavy stoneware jug. Cover tightly with foil. Stand in a pan and add hot water to within 2½ cm (an inch) of the top. Heat gently either in a very low oven or plate-warming oven, or on top of the stove. From time to time pour out raspberry juice through a strainer into a litre measure. Return the debris to the jug, and continue as before. The raspberries shrink rapidly, and when they are reduced to a greyish-red debris, pour in a little water and heat them for the last time. Add to the liquid in the measure. Sweeten it to taste, and add water to bring it up to 600 ml (1 pt/2½ cups), if necessary.

Now dissolve a packet (½ oz/2 envelopes) gelatine in 6 tablespoons (½ cup) very hot water, using a large basin. Pour on some juice, stir well, then gradually add the rest. Put a little into a small glass and chill in the refrigerator, to test the set. Add more water if it is too firm, but it should be about right. Pour into 150-ml (5-fl oz/⅔-cup) wine glasses and chill. Just before serving, run a layer of cream, a thin one, over the top. Loseley or other farm cream is ideal, otherwise use double cream. Serve with oatmeal or hazelnut biscuits, p. 459.

Note Strawberries and currants can be turned into jellies the same way.

RASPBERRY VINEGAR

Put ½ kg (1 lb) raspberries into a bowl and pour over it a litre (1¾ pt/scant 4½ cups) of white wine vinegar. Cover and leave 24 hours. Next day, strain the liquid over another ½ kg (1 lb) raspberries. Next day, do the same thing again. Finally decant into a bottle in which there are some more fine raspberries – enough to quarter fill the bottle – and close tightly. Leave for a good month.

This kind of raspberry vinegar is a favourite ingredient of the young chef, Jean-Pierre Billoux, at the Hôtel de la Gare at Digoin. He uses it to deglaze the pan after cooking calf's liver, for instance. The juices are thickened and enriched by beating in little bits of butter, and then strained on to the plate. The liver is then put on top, with its discreet garnishing.

RED CURRANT, see BLACK CURRANT

RHUBARB

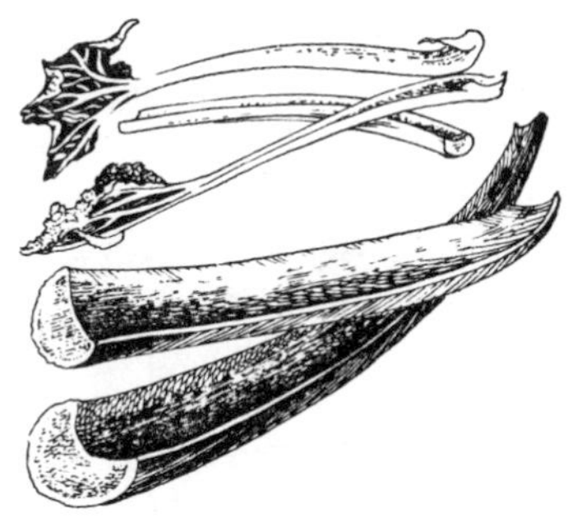

Time: Easter holidays. 'Good for you, dear. Good for you.' And there it lay, pink and slightly green. Livid. Bird's custard yellowing on one side of the plate, and pastry – admittedly the pastry was nice – a little pink and flavoured, but not too much, with rhubarb juice.

Nanny-food. Governess-food. School-meal-food (cold porridge with rhubarb for breakfast). And I haven't got over disliking rhubarb, and disliking it still more for being often not so young and a little stringy. Also rhubarb's country of origin is Siberia. Stewed rhubarb with frozen mammoth?

Of course, you may ask whether stems or stalks should be in this book, as fruit, but then our nomenclature does work out a little oddly (is tomato fruit, or vegetable?). Rhubarb edges in as pre-fruit, or first fruit or semi-fruit.

A word more on 'being good for you'. This is a very ancient tale, extending or dwindling from the powdered roots of a Chinese species of rhubarb, *Rheum officinalis*. The Greeks called this medicinal rhubarb *rha barbaron*, because it reached them via foreigners, barbarians, down the river we call the Volga and they called the Rha. Then in the sixteenth century, Europeans began growing a different kind, our *Rheum ponticum*, in their physic borders, again for the roots, but also for making rhubarb syrup. Nothing like it, they thought, for a gentle purge.

It is only about 1800 – and then in England – that rhubarb begins to be tarted under pastry and enjoyed, though – Good for you, dear – not by everyone. By the 1880s, at least the French had taken to the rhubarb tart, conscious of its English origin, and to stewed rhubarb, and rhubarb omelettes, rhubarb jam, etc. My French neighbour's children had their whack of stewed rhubarb – *ça te fait du bien, chérie.*

All the same, this once rhubarb-sufferer does have to admit there are some delectable, or not too undelectable, ways of employing rhubarb. I do have one good memory, of sitting with my sister on a doorstep, each with a stick of rhubarb, and a saucer of sugar between us. We dipped and chewed, dipped and chewed in the warm sun, with clucking hens stepping round us.

HOW TO CHOOSE AND PREPARE RHUBARB

Only young pink rhubarb is worth eating. Once the stems become green and thick, it tastes coarse, and the acidity dominates the flavour. If you grow rhubarb and cannot use the stalks up when they are young, pick, blanch and freeze them for later use. Much more satisfactory than making people suffer as the stalks grow thicker.

Be careful to remove all the leaf, and chop off any light brownish bits at the opposite end. Cut the rhubarb into pieces, removing any stringiness as you go. Young rhubarb does not need peeling and should not have any strings to speak of.

Orange and ginger improve rhubarb and, with citrus fruits, it makes a delightful jam. If you must have custard with rhubarb, purée the cooked rhubarb and add it to cream and eggs, then sieve or process and bake in little pots in a *bain marie* – an egg to 150 ml (5 fl oz/⅔ cup) rhubarb and cream combined is the minimum to get a set. Two yolks will give a softer texture, or one egg and one yolk.

SPRING SOUP

The first course of Mrs Lowinsky's dinner to impress your publisher (p. 182), the alternative to *moules marinière.* A successful soup with a spring sharpness, not unlike sorrel's in acidity. The cayenne pepper is important, so are little *croûtons.* The recipe also needs sugar, though it is not listed. Enough for 8–10 people.

48 sticks of young rhubarb (approx ½ kg or 1 lb)
2 sliced onions
1 carrot, chopped
30 g (1 oz) chopped, uncooked ham or gammon
60 g (2 oz/¼ cup) butter
2¼ litres (4 pt/10 cups) chicken stock
90 g (3 oz/1½ cups) fresh white breadcrumbs
salt, cayenne, sugar
little bread cubes fried in butter

Peel the rhubarb and cut the stems in half. Blanch them in boiling water for 3–4 minutes. If you use frozen rhubarb, it will not need further blanching. Drain it well, then put into a large pan with the onion, ham and butter. Stew gently until the rhubarb is tender. Pour in the stock and breadcrumbs. Boil for 15 minutes, not too fast. Skim off fat. Season with salt and cayenne and liquidize, process or sieve.

Reheat, and check for seasoning. Dilute a little more if the consistency is on the thick side for your liking. Serve with the croûtons in a separate bowl.

Note The rest of the menu was roast chicken with the grape salad on p. 182, followed by Milan soufflé similar to the lemon soufflé on p. 216. The savoury was either cucumber and mushrooms or fried Camembert (dip pieces twice in egg and breadcrumbs first). The table decoration, by Mr Lowinsky, was 'Flowers in a chromium-plated vase made in the form of a crate.' Chromium-plate, like the refrigerator, was an important item of smart living in the Thirties.

LAMB KHORESH WITH RHUBARB

The idea of using rhubarb to give a piquant edge to a rich stew had never occurred to me until I read *Persian Cooking* by Nesta Ramazani. It would be delicious with duck, as well, though this I have not tried. You need not admit to the rhubarb if you fear a domestic riot. Nobody would guess, though the interesting nature of the sourness might puzzle them. I have made one or two alterations to Mrs Ramazani's recipe, though nothing I think that would spoil its authenticity.

Boned shoulder of lamb is the right cut to use; although if your butcher or supermarket sells those little knuckles of lamb, they can be used instead. Allow one for each person, with perhaps a couple extra for the larger appetites, and leave the meat on the bone.

1½ kg (3 lb) boned shoulder of lamb, or 6–8 lamb knuckles
2 large onions, chopped
butter
generous pinch saffron
600 ml (1 pt/2½ cups) beef stock
4 tablespoons (⅓ cup) lemon juice
salt, pepper
2 bunches fresh parsley
6 large sprigs fresh mint
5 stalks rhubarb

Trim excess fat from the lamb and cut it into 2.5 cm or inch cubes. Brown the onions in a heaped tablespoon (2 tbs) of butter, cooking them slowly at first, then raising the heat to colour them. Push to one side and brown the meat. Put in the saffron, and stir for a minute. Then add stock, juice and seasoning. Cover and simmer gently for an hour.

Wash and chop the parsley leaves finely, with the mint. Fry them for a minute or two in another heaped tablespoon (2 tbs) of butter and add to the stew. Simmer another half hour. Cut up the rhubarb, stringing it if necessary, into 3–5-cm (1–2-in) lengths. Add them to the stew for the last quarter of an hour.

Using a slotted spoon, transfer the meat etc to a serving dish. Then skim fat from the liquid in the pan, and boil it down hard to concentrate the flavour; pour over the meat. Persians like their stews to have a lot of sauce which they eat with a lot of rice. In the west, we prefer to have less liquid and less rice, more meat.

Serve with basmati rice in a ring round the lamb, either boiled or boiled and then steamed with butter in the bottom of the pan so that some of the grains become crisp and make a light brown crust.

BAKED RHUBARB COMPOTE

The only time I have ever eaten plain rhubarb with pleasure was in the Scandinavian style, with pork, as we serve apple sauce.

Choose young, pink rhubarb. If it is really tender, do not peel it. Or just peel the coarse parts. Cut it into 3–4 cm (1½ in) lengths.

Layer the pieces into an ovenproof dish with 125 g (4 oz/½ cup) sugar for each ½ kg (1 lb) fruit. Cover closely and bake at any convenient oven temperature – in the bottom of the oven, for instance, below the meat – from moderately hot down to low. Allow 20 minutes before checking. Time will vary according to the temperature, of course.

Add no water. The slow cooking enables the rhubarb to stew in its own juice.

STEWED RHUBARB

Orange seems to be the most popular amelioration for rhubarb at the moment. If you do not have the oven on and only wish to cook a small quantity, this is a good way to do it.

Squeeze the juice of a large orange into a pan. Stir in 2 heaped tablespoons (¼ cup) of sugar (or of honey, if you like it in cooking) and some grated nutmeg or allspice. Cut up ½ kg (1 lb) of young pink rhubarb and put it in with the other ingredients. Cover closely after bringing the panful to simmering point, and leave for 8 minutes. Remove from the heat and leave to cool down. By this time, the rhubarb will be tender.

If the rhubarb is beyond its first young pink stage, be prepared to cook it for a further 2 to 4 minutes. But take the pan from the heat before the rhubarb really softens, as it continues to cook in the dying heat.

RHUBARB WITH GINGER AND ORANGE

A friend first put me on to the idea of poaching fruit in an electric deep-frier. An ideal method if you are preparing the fruit in quantity, or if you need to take special care as you do with rhubarb which collapses so easily.

Here is his way of dealing with rhubarb. Cut it into 3–4-cm (1½-in) lengths, after cutting away the leaves and brown bits. Cook by the method given on p. 437, adding a slice or two of peeled fresh root ginger to the syrup. Taste occasionally and remove the ginger when the flavour is right, not too strong.

After the fruit is all dealt with, reduce some or all of the syrup until it is very thick. Add the finely-grated zest of an orange, and orange juice to taste. Pour over the rhubarb, being careful not to drown it. Cool and chill before serving.

RHUBARB PLATE PIE

One French writer noted, in 1890, that 'the French eat as much rhubarb pie as the English'. His recipe for *gâteau de rhubarbe* was simple, identical with ours. Pieces of young rhubarb with sugar, between a double crust of pastry, sugared on top (brush it with egg white first to hold the sugar). Some *pâtissiers* also made open tarts of rhubarb, glazing the fruit with syrup.

Line a deep pie plate with shortcrust pastry, and fill it with chunks of young rhubarb. To every 250 g (8 oz), you will need about 90 g (3 oz/good ⅓ cup) sugar. Moisten the pastry rim and cover the pie, pressing the edges together in the usual way and making a central hole. Brush with water or egg white and sprinkle with sugar. Bake for 15 minutes at mark 7, 220 °C (425°F), then lower the heat to mark 4, 180°C (350 °F) for between 20 and 30 minutes.

Finely-grated orange peel can be sprinkled in with the sugar and rhubarb.

I reflect that the French domestic passion for rhubarb seems to be as strong nearly a century later. Only at Trôo do friends present me with bundles of thick greenish rhubarb stalks.

RHUBARB AND GRAPEFRUIT JAM

A friend who was amused at my intense dislike of rhubarb told me about this jam of Pamela Westland's from *A Taste of the Country*. She was right, it is good, and I put it in the rhubarb section although I feel that virtue comes from the grapefruit.

¾ kg (1½ lb) trimmed, washed rhubarb
2 grapefruit
¾ kg (1½ lb/3 cups) sugar

Put the rhubarb into a bowl, and grate the grapefruit rind over it, or remove it with a zester which makes fine shreds. Peel the fruit and cut it in half. Press it through a sieve with your hands or a pestle into the bowl. Pour the sugar on top, cover and leave overnight. Next day, bring slowly to the boil in a heavy pan, stirring often. When the sugar has completely disappeared, raise the heat and boil hard for 15 minutes. Test to see if setting point has been reached. Pot as usual. Enough for four ½ kg (1 lb) jars.

ROWAN BERRY, see SORB APPLE

SAPODILLA

If you chew gum, you will appreciate the existence of the sapodilla tree which provides chicle gum. The fruit is not chewy at all, but meltingly soft and brown, something of the size of a Chinese gooseberry, round or oval, with a brown skin. The flesh is honeyed, brownish-fawn, the taste reminds people of brown sugar. The seeds set off this restrained harmony by being black and shiny, svelte and long, neatly finished with a little twist at the top and a white stripe down one edge. As you would guess from the gum, it is native to America, tropical Central America that is, and the West Indies.

I suspect the sapodilla loses more than most fruit on its journey to Europe. Days of travel, however carefully the temperature is regulated, must blunt its special virtue. I hope they manage to get it right one day, because the sapodilla has the reputation of being among the best fruits of that part of the world, and the ones I have eaten were disappointing. I am sure that Descourtilz must have known what he was talking about when he wrote, in Flore des Antilles, that 'an over-ripe sapodilla is melting, and has the sweet perfumes of honey, jasmin, and lily of the valley'.

To get at the fruit, slice it down or across, according to whether you wish to cut it in wedges or to scoop out the inside with a spoon. Be careful to remove the pips. The softness of flavour needs lime juice, or lemon, but preferably lime juice. Sprinkle the slices and serve them as they are, or with some coconut cream sauce (p. 61), or in a fruit salad. The softest unsliceable fruit should be mashed and used in creams, fools and other puddings.

SHARON PERSIMMON, see PERSIMMON

SORB APPLE, WILD SERVICE BERRY and ROWAN BERRY

We were sitting in a large garden, our feet in cool grass, watching Russian dancers. A country occasion, with an early Renaissance gatehouse and manor for background, plane trees to keep off the worst of the showers. Across the small damp stage glided two rows of pale embroidered girls, the lines moving through each other in that even motionless movement that catches your breath with its beauty. Each one carried a branch of sorb leaves; it was called the dance of the sorb tree, something to do with love and fertility. For the Russians, perhaps; to West European doctors, the sorb was good for the digestion, for fevers accompanied by diarrhoea and for haemorrhages. To French countrymen in the past, the sorb was good for a kind of cider. People tell us it was drinkable, but I notice they do not make it any more. What they still make, what we learnt to make from them, is the most delicious liqueur.

We had come across sorbs before we went to France, many years ago in a Piedmont village shop. These bright red and yellow apples still on the branch, hung from a hook in the ceiling. The woman explained, but our Italian was suited to art rather than the intricacies of horticulture and we did not understand. We bit into the apples at our picnic – what a disappointment! The sharp astringent juice set our mouths on edge. We tried cooking them, even worse. We threw them away.

Coming to France and slowly learning the ways of the countryside, we came to understand about sorb apples, *sorbes* or *cormes* as they are called. We plotted their position, noted the hamlet and farm names that included *corme* in one or another form. The wild service tree, the *alisier*, was more difficult to find. Then one day we walked down a great alley of trees from a favourite little château, and there they were, tall beyond our reach, magnificent trees, which needed long sticks to bring the clusters of berries to a tall person's picking level.

The sorb tree is also a magnificent sight, a noble sight, conspicuous across a landscape of fields, often growing alone by the side of the track. It grows slowly and looks like it, having a most deliberate trunk with a grey, tight-patterned bark. The leaves are fine and long, making a sturdy lace pattern over the great height. The hard, fine-grained wood was once much in demand for screws and by wood-engravers. In October, or late September, the fruit turns this beautiful yellow and red among the long oval leaves, against a blue sky. But you should leave the fruit – 'Leave upon the tree,' wrote John Evelyn, 'till you perceive by their falling in great numbers, they admonish you to gather them.'

This is the give-away sign, at least for sorbs. The ground will gradually be covered with the fruit which turns brown, and as you might think rotten. In fact, this dark softness means that you can eat them, all astringency gone. It's a process that medlars have to go through as well, a process called bletting. Once the fruit has fallen, or even if you pick it from the tree, you can bring it into a warm house and it will gradually soften. You can eat the sorbs directly, but they are not very rewarding, less so than medlars. Their best use is for liqueurs.

Wild service berries, being reddish brown and in clusters like rowan berries, are more difficult to see whether on the tree or on the ground beneath. It is best to hook the branches down and pick the berries from the tree, then bring them indoors to blet. Although the berries resemble the rowan's they are unmistakeably different in colour, and the trees are quite different. I have not tried rowan berries in a liqueur, but keep them for a tart bright red jelly to eat with game or lamb (cook them with apples, up to fifty per cent, or with crab apples: see the recipe for cornel cherry jelly). When I read in books on spirits and liqueurs that one of the Alsatian *eaux de vie de fruits* is made from rowan berries, I wonder if it is a mistranslation or misunderstanding for the fruits of the *alisier*, the wild service. For they give the finest flavour of all to a liqueur, with an almond hint of noyau.

The wild service (*Sorbus torminalis*), like the rowan (*Sorbus aucuparia*), was introduced into this country in the sixteenth century, but has not had the same success. Rowans take a precarious foothold by the side of Cumbrian becks, or sit in tamed crimplene comfort along suburban avenues. Their brilliant berries catch the eye, and wise people make jelly from them to serve with game, venison in particular. Wild service trees are not so happy, and grow sparsely in the south. I know of none in our region.

The sorb (*Sorbus domestica*) one would have expected to catch on in large gardens, on account of its fine presence. There was one in the famous Tradescant garden in the seventeenth century – it was a later introduction than the wild service – which grew to nearly 40 feet and had a large crop of big sorbs every year. It inspired market gardeners round

London to grow them. Sales continued every autumn with wild fruit coming in to London from Cornwall, Worcestershire and Hertfordshire at least until the middle of the nineteenth century. In his *Encyclopaedia of Gardening* (1848), J. C. Loudun remarks that one of the Dukes of Richmond brought in seeds from trees around Aubigny in France and raised a great many sorbs at Goodwood. I wonder if they or rather their descendants are still there?

SORB OR WILD SERVICE LIQUEUR

As the fruit blets to a soft squashy brown, put it into bottling jars. Make a syrup by simmering together two mugs of sugar and one mug of water. Give it four or five minutes, then cool and pour enough into the jar to come just under half, or just over one-third, of the way up. Top up with brandy or vodka or gin. Close tightly and put into a cool dark place at least until Christmas, better still until Easter.

You can strain off the liqueur into clean bottles, say in February, and then cover the fruit with a fresh batch of syrup and alcohol.

You can of course eat the fruit when it is bletted and soft, or after it has been in the liqueur bottle. It is not exciting. That it should flavour a liqueur so well is surprising.

RUSSIAN SORB JAM

Soak 1 kg (2 lb) of firm sorb apples in plenty of water for two days, changing the water twice a day. Drain and cook the fruit in more plain water. When it is tender, put it on to a hair sieve, or flat draining plate and pour cold water over it.

Boil together to make a syrup 1 kg (2 lb/4 cups) and 500 ml water (¾ pt/scant 2 cups). Put in the fruit and continue boiling until the syrup begins to thicken. Skim and cool before putting into preserving jars. Serve in little dishes with glasses of clear tea.

From *The Russian Cook Book*, compiled and translated by Princess Alexandre Gagarine (1924).

SOUR-SOP, see CHERIMOYA

STRAWBERRY

Do you remember the kind and beautiful girl in Grimm's fairy tales, who is driven out by her stepmother to find strawberries in the snow? How she comes to the dwarves' house, and shares her crust of bread with them? And how, as she sweeps the snow aside with their broom, she finds growing there – strawberries? That vivid image of delight, of fruit and snow against forest darkness, is never forgotten. It's our northern winter longing for summer, a joy of the mind. And yet, in the sudden snow of winter a couple of years ago, I went to sweep our doorway – and found strawberries.

Strawberries on a card from Paris, where *Le panier de fraises des bois* – the basket of wild strawberries – was a star of the great Chardin exhibition. Against muted browns of table and wall, a cone-shaped mound of these small fruit, piled into a wicker basket. I counted over fifty on the half-surface that the painting shows.

When Chardin lived, wild strawberries were the only ones. His contemporary, Diderot, described them as being like 'the tip of wet-nurses' breasts'. Our plump round large strawberry did not exist – anywhere. For centuries, people had brought the wild roots of wood and verge into their gardens. But all they could do was encourage them to grow a little larger. Royal gardeners bought these wild plants by the thousand. Poor people with a cottage patch went after them, too:

> Wife into a garden and set me a plot
> With strawberry roots, the best to be got.
> Such growing abroad amongst thorns in a wood,
> Well chosen and picked prove excellent good.

In mainland Europe they were – and are – lucky enough to have a second wild strawberry: its large drops of bright red berry taste – and feel – more luscious than the wild strawberry that we know, which can be tiny and dry, a little seedy on the tongue. This larger fruit from the high woods is appropriately called the Hautbois or Hautboy and, a little less appropriately, the Alpine strawberry. It was taken into gardens, sent to England and reckoned to be the best for flavour. This is Chardin's strawberry. And you may sometimes find it on the menu of good French and Italian restaurants, in high-class fruit shops (at a price) and you can buy it from the best nurserymen to grow in your own garden.

The wood strawberry has been taken as a bright relic of our time, in a humorous and ironic poem by Jacques Prévert. He wrote as an archaeologist of the twenty-fifth century, living at a time when nothing grew any more and making what he could of the twentieth century from the few bits and pieces dug up in the Champ de Mars (the site of the Eiffel Tower):

> Behind a triple wall of glass
> in the great museum of machines
> in a little block of ice
> a wood strawberry is shown
> the whole world crowds in to look at that strawberry
> charts show the height of the tree that gave
> such fruit
> nine hundred feet

He tries to fit everything in, picturing the Eiffel Tower tree hung with dentist's drills – *fraises* in French – and wood strawberries, and interprets it as a relic of the time when they were only just beginning to tar over the great oases of Africa.

By comparison, the modern strawberry may seem brash. But this is not fair. It's a different fruit with different virtues. And if you enjoy the thought of arduous travel and the richness of continents reduced to one small delight, you will like its extraordinary tale.

It begins in Virginia. Early settlers liked the wild scarlet strawberry they found there. John Tradescant the elder, great seventeenth-century gardener and introducer of plants, brought it to England. Others had taken it to France. It tasted better than our own wild strawberry, but not so good as the garden Hautboy. It could be crossed with neither, as its genetic make-up was different. So it was modestly enjoyed and – in a sense – it waited, like Sleeping Beauty, since we are remembering fairy stories. Then, in 1712, a French naval officer called Frézier – which sounds just right, though it is spelled differently – managed to bring back to Brest five plants of a large fruiting, yellowish pale strawberry, tasting a little of pineapple, that he had found on the west coast of South America: the Chilean pine or sand strawberry. The plants were given to Monsieur Jessieu, Professor of Botany to the Royal Garden at Paris.

He sent plants all round Europe, the fruit they produced was as large as a walnut, sometimes as big as a hen's egg, sometimes the size of a small apple. Philip Miller, at the Chelsea Physick Garden, who tells the story in his *Gardener's Dictionary*, acquired plants of it in Holland in 1727. In England, in Chelsea, they thrived, but produced little fruit. And, anyway, it did not taste as good as our wood strawberry, so what point was there in its size?

What this large South American strawberry needed was the small strawberry from Virginia. I had the notion that these two, separated by mountain and desert on their own continent, met by chance in the Vilmorin acclimitization garden in Brittany.

Reality was less sensational. The botanist Duchesne realized that the two American strawberries were compatible scientifically, and produced hybrids with the size and pine flavour of the Chilean strawberry, the vivid scarlet and stronger flavour of the Virginian. The French Revolution held things up, but eventually growers here put the first large modern strawberry on the English – and European – market in 1821, when Michael Keens, a market gardener at Isleworth, showed his Keens' Seedling. Nothing like it had yet been seen for size and flavour. The Royal Horticultural Society awarded him a silver cup. It was a sensation. The next major strawberry was Downton, raised by Thomas Andrew Knight, and then in 1827 came the Elton. And so it went on. The culmination came in 1891 with Scarlet Queen and 1892 with Royal Sovereign, raised by Laxton's at Bedford.

I have a booklet on cultivated strawberries published in Le Mans in mid-century, claiming to list all the strawberries grown in France, Holland and England. Among the large strawberries, 67 out of the 153 named varieties are obviously of English origin. Indeed to the French, the large-fruiting varieties were known as '*fraises anglaises*'.

The English could never have too many strawberries, and were passionate to have them as early as possible. The most sheltered growing place was at Plougastel, across the sound from Brest where Frézier had first arrived with his strawberries from Chile. By 1850, buyers crowded into that gentle backwater during the early season, after the delicious white and pale rose fruit. The opening of the Paris-Brest railway in 1865 meant, too, that these marvellous early strawberries could be sent rapidly to Paris as well. Speed being the thing. Eventually the growers of Plougastel resented the middlemen who came to cash in on their hard work, and formed their own companies. Their solution was to send the fruit by steamboat to Plymouth, for despatch to Manchester, Birmingham and London. So fast and efficient was the system that the strawberries were on the market thirty-six hours after being picked.

I remember when we first went to Trôo in the early sixties, one high pleasure was buying canned strawberry jam from Plougastel. I have never tasted a commercial strawberry jam to touch it. Then one year it came no longer, the last factory had closed. Now the countryside from Plougastel-Daoulas to the sea grows many other things than strawberries. Their early advantage was not early enough once Israel began sending fruit to Europe in February.

We learnt a lesson in wily horticulture when we were invited to Israel to see the fruit and vegetables being grown for export. The thing that brought home to me the whole position of Israel was the small strawberry gardens enclosed in cages of plastic, and cherished. The rows of plants grow up through holes in black plastic sheets which keep in the humidity and prevent weeds. Along each row runs a tube, pierced every so often, so that each strawberry plant receives exactly the moisture necessary, mixed with the right nutrients for its best growth. That precious water, measured out drop by drop, comes down from Galilee through the one waterway, through the narrow width of the country, to give us strawberries in the snow of a European winter.

HOW TO CHOOSE AND PREPARE STRAWBERRIES

It is surprising how much flavour the early imported strawberries have, whether they come from Israel or America, but one should not expect too much, too sharp or scented a flavour. Early English strawberries can be disappointing – we just do not have enough sun. I remember an enormous bowl of them being brought to our table in a Bath restaurant. They were beautiful, the first Cheddar strawberries of the year, and absolutely tasteless.

Once you get to the season, you will soon find out which of your local growers provides the best. Or the ones that you like best. Going to the fields with the family has become quite an occupation during the summer months. It makes an outing, and you get all the strawberries you could ever want at a reasonable price. What do you do with them next?

Jam-making for the smaller ones. Freezing for the larger perfect fruit. I never find whole strawberries very successful, whether they are open-frozen, or packed in loose sugar or syrup. People say, 'Add them to fruit salads while still frozen', but to me they are always a disappointment; you look at the charming shape, but the savour has all gone. I find it much better to purée them with a little sugar, in the processor or liquidizer. This gives you strawberry sauce, or the purée base needed for chilled soufflés, ices and Bavarian creams: this seems to me much more of a winter treat than trying to pretend that the whole strawberry can keep its soul in the freezer.

Try to avoid washing strawberries. If you have to do this on account of muddiness, rinse them fast before hulling them. Once the little plug of leaves is out, water can get in and spoil the fruit.

If you have to leave strawberries overnight, sprinkle them with sugar. And cover them with plastic film, or else their smell will dominate the refrigerator for days. But avoid punnets wrapped in plastic film, it squashes the strawberries down on top of each other, and prevents you seeing what the ones underneath are like. It takes a brave person to tip out every punnet in a shop before choosing the best one, and I do not think it is a fair thing to do as each time the punnet is tipped out and rejected, the strawberries suffer slight bruising. It is much better to stick to the greengrocer with a rapid turnover, who you know from experience has good fruit. And it is not his fault if the strawberrries are not quite up to scratch in a wet summer. In tricky seasons, you are much better off making the slight effort required to pick your own.

Everyone knows about strawberries and cream. Many people know about strawberries in wine, and strawberries sprinkled Italian style with a little wine or lemon or orange juice. I have eaten strawberries with a sprinkling of wine vinegar under a friend's persuasion, and found that this works every bit as well as wine if you do not overdo it.

Cream can be improved – unless you can buy unpasteurized farm cream of Loseley quality. We are lucky in having a supply near Hungerford at Prospect Farm a mile or two further on from a good strawberry field. The bland double cream of the big dairies can be lifted by mixing it with soured cream, half and half. This approximates to the slightly sharp *crème fraîche* that you find in France and which sets off fruit so well. Another French trick is to add a stiffly beaten egg white or two, to whipped cream. You end up with a light and foamy bulk, which is not as bad for your waistline as it looks. Cream cheese drained or set by gelatine (p. 470) in little heart-shaped moulds turns a simple enough dish into something pretty and special. For me, this is the way strawberries should be. Very direct and garden-like – strawberries piled on to a crisp lettuce leaf or two look extraordinarily beautiful without affectation. You will not find many recipes in this section because the strawberry themes are few and well known. You do not need instruction in making little sweet shortcrust boats, filling them with strawberries and brushing them over with a red currant glaze. The one new and old idea of value I have come across for strawberries, while writing this book, is strawberry fritters: they sound shocking but taste delicious. The snag is that you have to eat them the moment they are cooked, which means either martyrdom or a meal in the kitchen, or everyone cooking their own which is not always the best way of conducting a dinner party – an electric thermostat frver makes it easier.

PRUE LEITH'S AVOCADO WITH STRAWBERRY SAUCE

Prue Leith told me about this dish that they serve in her restaurant. She sounded apologetic – 'I know it seems bizarre and looks far too pretty to be good, but try it.' And I did, the moment I got home, working out quantities from her brief description and adding a few toasted almonds. As she explained, the sauce is really a vinaigrette, the acidity coming from strawberries rather than wine vinegar.

The method is quick and simple. Reduce 250 g (8 oz) hulled strawberries to a smooth purée in a blender or processor. Keep the machine going and add 100 ml (3 fl oz/scant $\frac{1}{4}$ cup) each of olive and sunflower oil mixed together; do this slowly, tasting from time to time, and stop when the balance of oil to strawberries seems agreeable to you. Add salt and pepper, and a little sugar to bring out the flavour.

Halve and peel three ripe avocados, throwing away the stones. Rub the cut sides with half a lemon, then place on a board, cut side down. Slice each one across a dozen times or so, keeping the shape. Squeeze lemon juice over to prevent discoloration. Slip a palette knife under the first avocado half and transfer it to a side plate, of a size which holds it comfortably, not too spaciously. Fan back the slices to spread them out a little. Repeat with the remaining halves.

Spoon the sauce round them, and scatter a few toasted slivered almonds on top of the avocados. Serve at room temperature rather than chilled.

VARIATION You can turn this into a main summer dish, by adding smoked chicken. Arrange the sliced avocado halves on a large serving dish, pour the sauce round, and put pieces of chicken, nicely cut, between the avocados. Almonds on top.

STRAWBERRIES ROMANOV

FRAISES ROMANOFF

Marinade strawberries in orange juice and one of the curaçao orange liqueurs, in a closed bowl, for several hours. Serve with whipped cream, that has been beaten over ice to get the maximum volume and lightness, and sweetened with icing sugar.

STRAWBERRY FOOLS

I find that people prefer the slightly lumpy effect achieved by mashing the strawberries with sugar, and then folding them into whipped cream.

Or into whipped cream, lightened with whisked white of egg.

Or into whipped cream folded into a light fresh cheese, e.g. *fromage frais*.

Serve with thin shortbread biscuits or sponge finger biscuits.

STRAWBERRY AND RASPBERRY FOOL

Equal weight of strawberries and raspberries, mashed and sugared. Folded into whipped cream.

The quantity of fruit to cream can be equal in volume. Or less cream to fruit, according to your tastes and pocket.

Icing sugar is easiest to flavour these dishes with, as it dissolves so easily. But caster can be used, too.

STRAWBERRIES WITH ALMOND CREAM

Almonds have an affinity with fruit – almonds and raisins, almonds in apricot tarts, almonds and plums, Elizabeth David's almond and mulberry pudding on p. 248 – but above all with strawberries. Almond biscuits are an obvious partner to strawberry fools and ices for instance. And this almond cream to serve with strawberries makes a good change from double cream. It goes back to the almond milk of the Middle Ages, as far as the method goes. Try it with other fruit as well, apricots, peaches and pears, lightly poached in syrup.

5 heaped tablespoons (10 tbs) almonds
½ litre (15 fl oz/scant 2 cups) whipping cream
caster sugar
German bitter almond essence, or Langdale's natural almond essence
generous ½ kg (1–1¼ lb) hulled strawberries

Blanch and sliver a heaped tablespoon (2 tbs) almonds. Spread them out on a baking sheet, and toast under a grill or in the oven. Set aside to cool.

Rinse and dry remaining almonds. If you have a blender or processor, reduce them to a sludge, skin and all, with 250 ml (8 fl oz/1 cup) water. Otherwise grind them in a nut or electric mill.

Pour the cream into a wide shallow pan. Note the level. Stir in either the almond sludge or the ground almonds, plus the same quantity of water. Bring to boiling point and cook steadily – but do not let it boil over – until the level is the same or very slightly more than the original level of the cream. Put through a hair sieve or squeeze through a double muslin. Taste and add sugar. If the almond flavour needs bringing up, add a drop or two of essence, but do not overdo it.

Divide the strawberries between 6 glasses or small bowls. Sprinkle on a little sugar (this is better than oversweetening the cream). Let them stand awhile, until you are ready to serve the dish, then pour over the almond cream and scatter with the browned almonds. Do not chill.

Note The almond debris can be incorporated into bread dough most successfully.

STRAWBERRY FRITTERS

BEIGNETS AUX FRAISES

One evening in Paris, we went to the Escargot restaurant in the rue Montorgeuil, in search of snails of course, and of Montreuil peaches (*see* p. 297). I did not expect to find anything special about the strawberry fritters which I chose out of curiosity for dessert. It had been a good meal, in black and red surroundings of past elegance, with a close friend. Those fritters were to be the fault, the way Chinese and Japanese potters deliberately put a fault into their vases to make them human. But I was not allowed my fault after all. The fritters were perfection, the batter crisp, the strawberry inside firm but meltingly delicious. Even the *crème anglaise* was right.

What an original idea, strawberry fritters! But is there ever an original idea in cookery? Eight months later, I came across a recipe for them in an 1810 translation into French of an English cookery book, *Le Cuisinier anglais universel ou le Nec plus ultra de la gourmandise*, by F. Collingwood and J. Woollams.

The Escargot fritters, half a dozen, were served in the centre of a large plate with a thin custard covering the base. A good combination, so long as you make the custard of eggs and milk, or milk and cream. Pour it on to each plate and leave in a cool place.

Sort out and hull six fine strawberries for each person. Make a batter with:

125 g (4 oz/1 cup) flour
¼ teaspoon salt

grated peel of half a lemon
1 tablespoon white wine
2 egg yolks

and enough water to make a single cream consistency. Just before cooking the fritters, stir in a tablespoon of oil and 2 egg whites stiffly beaten.

Spear the hulled strawberries, one by one, with a skewer, or two-pronged fork, dip into the batter and then into deep-frying clean oil, preheated to 195°C (385°F). Cook about six at a time. You can keep them warm in the oven, but it is better to serve people straight from the pan even if they do have to eat seriatim. That way, the batter stays crisp.

OLD-FASHIONED STRAWBERRY SHORTCAKE

The idea of putting strawberries with some sort of cake, to make them go further and to enhance their flavour by softening it, is not particularly original. Our solution is the sponge cake, a peaceful unhurried solution as the cake can be made in advance and filled at leisure, and can be glorified by pastrycooks. It's rather a soft solution. I am not sure I don't prefer the American shortcake, which is a grand form of scone for its crispness and butteriness. The snag is that it tails off the longer it waits.

In other words, the shortcake is family food, pudding-from-the-garden like our summer pudding. It's quickly made and quickly baked, an impromptu affair for unexpected friends, for unexpected sunny days in the cool of the trees.

Other fruit can stand in for strawberries. I have the feeling that the stronger they taste, the more successful the result. Bilberries or blueberries, hardly cooked at all. A mixture of different currants, or some gently heated red currants and sugar into which a lot of raspberries have been stirred as the pan is taken from the heat.

½–¾ kg (1–1½ lb) strawberries
sugar
500 ml (15 fl oz/good 1¾ cups) whipping cream

CAKE
½ kg (1 lb/4 cups) flour
3 rounded tablespoons (⅓ cup) sugar
1 level teaspoon salt
5 level teaspoons baking powder
300 g (10 oz/1¼ cups) butter
up to 300 ml (10 fl oz/1¼ cups) chilled whipping cream

Prepare the fruit. Set aside the best berries for putting on top of the shortcake, halve the rest and sprinkle them with sugar. Whip the cream until it is very thick, but pourable, and put it in a jug.

To make the cake, sift together the dry ingredients. Rub in 200 g (6½ oz/good ¾ cup) butter as if you were making pastry. Mix to a soft dough with the cream. Knead the dough for a minute, then roll it out thickly just enough to be able to cut out twelve 7½-cm (3-in) rounds with a plain scone-cutter.

Rub a baking sheet with butter paper. Place half the rounds on it. Melt a little of the remaining butter and brush it on the rounds. Put the rest of the rounds on top. Bake at gas 8, 230°C (450°F) for about 15 minutes – check after 10 minutes – and remove when they are golden brown.

Pull the cakes apart. Spread the inner sides with the last of the butter. Put the sugared strawberries on half the rounds, the inner sides, then put the remaining rounds on top, inner side up, and top with the best strawberries.

Serve at once with the cream.

Note You can also bake the dough as one large double scone in a 20–23-cm (8–9-in) baking pan, at gas 7, 220°C (425°F) for 20 minutes or a little longer. Pull apart, butter and fill as above.

STRAWBERRY AND ORANGE ICE

Frozen strawberries can quite well be used for this recipe, which makes a refreshing winter dessert.

250 g (8 oz) strawberries
60 g (2 oz/¼ cup) icing sugar
juice of a large orange
juice of a medium lemon
thinly-cut rind of a medium lemon
125 g (4 oz/½ cup) granulated sugar
2–3 tablespoons (3 tbs/¼ cup) orange liqueur or gin

Liquidize or process strawberries and icing sugar, then sieve to remove the seeds. Add citrus juices. Simmer rind, sugar and 300 ml (½ pt/1¼ cups) water for 5 minutes. Cool and strain gradually into the strawberry purée, tasting every so often to see if you have added enough (quantity will depend on flavour and sweetness of strawberries).

Freeze in the usual way, turning edges to middle as they solidify. When the ice is consistent but not hard, beat it well and add the alcohol. Make any adjustments of flavour. Complete freezing. Serve with almond biscuits.

STRAWBERRY SOUP
SOUPE AUX FRAISES

Soup for the wrong end of the meal, and soup of a deliciously light yet thick consistency.

From 1 kg (2 lb) strawberries, remove 6 or 8 of the best. Hull the rest – they need not be the finest-shaped strawberries or the biggest, but they should taste good – and cut up roughly into a pan. Add 125 g (4 oz/½ cup) sugar, and a very little water, just enough to get things going. Bring to simmering point, and as the strawberries begin to give up their juice, stir in a small wineglassful – about 100 ml (3½ fl oz/½ cup) – red wine mixed with a rounded teaspoon of cornflour. Stir and continue to cook until the strawberries have had 10 minutes simmering in all. Liquidize or process and sieve, or just sieve, into a serving bowl, beating the mixture as it cools from time to time. Taste when cold, adding more sugar or wine as you think fit (strawberries vary and so do people's tastes).

Chill and serve with the best strawberries halved on top and blobs of whipped cream: the consistency should be enough to hold the strawberries and cream so that they are not entirely submerged from view.

BACHELOR'S JAM
RUMTOPF

This preserve which starts with strawberries is destined for Christmas and New Year parties. You need a 5-litre or gallon (20-cup) stoneware jar, or a special *rumtopf* from Austria or Germany which can sometimes be bought in this country.

Before you start, a word of warning. The method is easy, foolproof. The tricky part is the quality of the fruit. It must be of the finest, and preferably from a garden that you know has not been much subjected to sprays: in the opinion of one expert, whom I consulted after a batch had developed mould for no reason that I could see, it is difficult these days to make a *rumtopf* that you can rely on, and he is convinced that the reason is the chemical treatment that commercially grown fruit undergoes.

This poses a real dilemma with the required soft fruit. You can scrub an orange or an apple without harming it: a strawberry that becomes acquainted with water loses its virtue.

Another point to watch is the alcohol. Cheap rum and brandy sometimes means weaker rum and brandy.

If after these warnings, you decide to try your luck, this is what you do:

Prepare a 1 kg (2 lb) of strawberries, removing their hulls. Sprinkle them with a ½ kg (1 lb/2 cups) sugar and leave overnight. Next day, tip the whole thing, juice included, into a well-washed and dried *rumtopf*. Pour on 1 litre (1¾ pt/scant 4½ cups) of rum (as in Austria and Germany) or brandy (as in France).

Put a clean plate directly on to the fruit to make sure it stays below the surface. Cover the jar with plastic film and the lid. Keep in a cool dark place.

Add more soft fruit and sugar – half quantities are fine – as the summer progresses. Add more alcohol, too, from time to time to keep the liquid level up. It should clear the fruit comfortably. Remove and replace the plate each time.

Suitable fruits include:

sweet and sour cherries (including stones), raspberries, loganberries, boysenberries, mulberries, peaches, apricots, greengages, mirabelles (include an occasional stone), melon, pineapple (cubed), one or two apples and pears.

Although gooseberries, currants and blackberries can go in, they do tend to go hard and uncomfortable.

When the pot is just about full, top it up with a final dose of alcohol. Cover with fresh cling film, then the lid, and leave until Christmas, at least a month.

TO USE Serve in wine glasses with cream floated on top.

Serve as a sauce with ice creams, vanilla or honey for instance.

Mix with champagne, mostly the liquid with not too much fruit, for a cocktail.

Make a triangular build up of boudoir biscuits, dipped in black coffee, glued together and covered with whipped cream that includes sugar and finely ground coffee – but not too much – and serve the rumtopf as a sauce.

STRAWBERRY GIN OR VODKA

Fill a large bottling jar with hulled strawberries. Choose clean ones that do not need washing. Tip in caster sugar to come just over a third of the way up the bottle. Add gin or vodka until the fruit is covered. Close the bottle, and store in a dark cool place. Turn it over occasionally, and leave for a month. The sugar gradually dissolves, and the strawberries lose their colour, becoming flabby and pallid.

You can now strain off the liquid through a double muslin into a clean bottle. Or you can leave the whole thing untouched for months longer, until it is required.

This cordial can be drunk as a liqueur, or used to flavour strawberry fools, charlottes and mousses, etc.

STRAWBERRY SODA

Buy ½ kg (1 lb) of jam strawberries for this refreshing drink. Set a few of the best ones aside, then wash and hull the rest. Bring 175 g (6 oz/¾ cup) sugar and ½ litre (¾ pt/scant 2 cups) water to the boil, stirring to make sure the sugar dissolves before boiling point is reached. When the syrup is bubbling hard, tip in the strawberries and remove the pan immediately from the stove. Put on the lid and leave to infuse until cold.

Blend or process, then put through a fine sieve or double muslin to remove the last of the debris. Half fill a glass, add soda water and ice, and put the strawberries you save on top, halving or slicing them first according to size. A sprig of mint can be added too.

For children, you can turn the drink into a milk shake by adding ice cream, vanilla or strawberry. To do this, return the sieved strawberry liquid to the rinsed out blender or processor, put in the ice cream and whizz for a couple of seconds. Serve straightaway.

STRAWBERRY FREEZER JAM

Mash 500 g (1 lb) strawberries with 1 kg (2 lb) sugar and a teaspoon lemon juice. Leave until sugar is dissolved. Mix in 125 ml (4 fl oz) bottled pectin. Put in small containers. Leave at room temperature 8 hours. Put in refrigerator for 2 days, until set. Freeze. Thaw overnight in refrigerator, stir well and use up fairly rapidly.

TANGERINE, see ORANGE

TUNA, see PRICKLY PEAR

UGLI, see GRAPEFRUIT

WATER-MELON

Water-melons are a new pleasure. Even if we never leave the United Kingdom, we can delight in their coolness since Israel now sends us a conveniently small variety every summer, smaller than the Mexican fruit of José Juan Tablada's haiku

> Summer's loud laugh
> Of scarlet ice
> A melon
> slice

The high season and place for water-melons in Europe is on 10 August, in Florence. It's the feast day of San Lorenzo, the patron saint of cooks, the name-day of Lorenzo de' Medici and, by repute, the hottest day in the year. Outside the saint's grey church next to the market, people saunter down the wide steps after Mass into a world of water-melons, refreshment for an overheated saint. 'Turn me over,' said San Lorenzo to his tormentors, as they grilled him to death over charcoal. 'The other side isn't cooked.'

Florentine water-melons are huge, more than an armful, and heavy. They are piled up by the stands like green cannonballs, though some are halved and cut up to show the fiery red flesh, spotted with coal-black seeds. The grinning slices rock on their smooth rind. As you pick one up and bite into its delicious ice-coldness, the juice dribbles stickily down your chin.

At one time botanists thought that the water-melon's original home might have been southern Italy. But no wild plants were found. Asia and India were other ideas, but again no wild plants.

So the matter rested until David Livingstone found in the 1850s, 'great tracts of water-melon (called kenwe or keme) growing wild in the Kalahari desert'. Imagine the slow ups and down of sand, the heat by day, then – a mirage – stretch after stretch of water-melons. In fact no mirage, but drops of liquid enough to cool Dives' tongue in hell, let alone refresh a Glasgow missionary explorer and his band.

In view of this happy, if occasional grace of nature, it seems odd that no word – which means no reality – for water-melon existed in Greek or

Latin. Water-melon leaves, water-melon seeds have been found in New Kingdom tombs in Egypt, yet European botanists did not describe it until the thirteenth century. Perhaps, like the South-East Asian aubergine, it came to us via the Arabs and their gardens in Spain.

We were introduced to it in 1597, though growing it was not much of a success in our climate. For most of us, until lately, water-melons have been one of the magic items of an Italian holiday – along with the green and purple sea, shining monuments, the midday smell of parmesan and oleanders.

There was also the grinning water-melon, chocolate-coloured-coon image from the Southern States of America. Indeed the slaves from Africa probably brought it with them, as they did okra, and the first written record of its cultivation is in Massachusetts in 1629. Two of the modern American varieties are Charleston Gray and Crimson Sweet, but to us the best known is Sugar Baby, a small, early melon that we import from Israel in June and July. Its four to five kilos are nothing compared to the great water-melons of Italy, but it's easier to carry home, easier to eat and to store, but every bit as cooling and almost as sweet. Should a whole melon, even in this size, be too much, shopkeepers are willing to serve them in halves and quarters.

BUYING AND STORING

It is difficult to judge a water-melon. As a general guide, it should be firm and nicely shaped, with a good even colour except on the side where it has lain on the ground and become slightly yellowish-green. The rind should have a soft waxy shine. Whole water-melons should be stored in the refrigerator, wrapped in film: they last a week in good condition. The average Sugar Baby should be enough for 8 people, for more if it is mixed with other fruit, or used for soup.

It is easier to pick out a good piece of cut water-melon. The red should be vivid and firm, not woolly or dry-looking. If the pieces are not sold in plastic film wrapping, wrap them when you get home before putting them in the refrigerator. Again, they keep well.

With large, thick-skinned water-melons such as you see in Mediterranean countries, it's possible to make a pickle from the white rind between the pink flesh and green skin (*see* melon section, p. 244). Spaniards and French candy it and sell it in boxes of mixed glacé fruit, for baking.

The seeds can be toasted and skinned. From Israel east, melon and other seeds occupy the place of chewing gum in the States. Something to nibble at boring moments. By bus stops, you will see a trail of husks going back to the seed stall a few yards away, and people chewing with blank faces.

WATER-MELON SOUP

Follow the recipe for melon soup on p. 239, using extra water-melon as the taste is so delicate. Serve in a hollowed out water-melon shell (*see* the following fruit salad recipe).

WATER-MELON SALADS

For a simple salad to go with cold chicken or ham, cut the flesh into chunks or into balls with a cup-shaped potato cutter (keep the bits and pieces for an ice or cold drink). Remove and discard the black seeds. Arrange the melon on a bed of lettuce – it must be crisp lettuce to match the melon, Webb's Wonder is an obvious choice. Or surround it with watercress. Pour over a little French dressing and scatter some chopped mint and a few mint sprigs on top, or rings of green pepper finely cut.

If you use a nut oil instead of olive oil for the dressing, with cider vinegar, scatter a few toasted chopped hazelnuts on top or a little broken walnut.

For a mixed salad, add a roughly equal quantity of water-melon pieces to some diced cooked chicken. A little chopped smoked ham adds relish. If you use *lachschinken* or Danish hamburgerryg (smoked pork loin from Danish food centres), or Parma, Bayonne or Westphalian ham, the salad will be even better.

WATER-MELON FRUIT SALADS

In Tuscany, especially in Florence, in the high summer, cafés put huge water-melon halves in their windows, packed with fruit. As well as the pink-fleshed melon, the salads contain strawberries, blackberries, grapes, the fruit in season. They tempt customers in from the pavement and the dust that swirls round their knees in the afternoon heat.

For extra refreshment, the fruit is layered with crushed ice like a Persian *paludeh*, or the sherbet described by Sei Shonagon in tenth-century Japan, one of her Elegant Things:

> A white coat worn over a violet waistcoat.
> Duck eggs.
> Shaved ice mixed with liana syrup and put
> in a new silver bowl.

A rosary of rock crystal.
Snow on wistaria or plum blossoms.
A pretty child eating strawberries.

Incidentally the syrup that comes from sugaring the fruit is sparked with alcohol, sometimes with gin which comes as a surprise.

You will find our smaller water-melons adequate for a party of ten to twelve people. Slice off a top lid. Or, if you are good at this kind of thing, cut away a lid in Vandyke points. Some people go so far as to draw a scalloped line round the melon with the aid of a tenpenny piece, to make a guide for the knife. I do not. I do take the precaution of cutting a small nick in fruit and lid so that they can be fitted together exactly for serving.

Scoop out the flesh, remove the seeds and trim the pieces into reasonable cubes. Or use a potato-ball cutter, and keep trimmings for a sorbet or milk sherbet (recipes follow).

Put the pieces into a large bowl, and add
either the Florentine mixture as above
or a diced fresh pineapple and 1 kg (2 lb) strawberries, garden or wood strawberries, or the two mixed
or the cubed, seeded pulp of a green- and an orange-fleshed melon, plus seeded or seedless grapes.

Sprinkle the fruit with caster sugar and leave for an hour or two, until the juice is drawn. Add alcohol – gin, vodka, kirsch, maraschino, fruit brandy – to taste.

Pack the fruit into the water-melon shell, and pour on the syrup – fortified if need be, with extra sugar and alcohol. Replace the lid. Tie into a plastic carrier to prevent the smell pervading the refrigerator, and chill until required.

Note The shell can be stored in the freezer and re-used. Wrap it in cling film, and place it in a box so that it does not get damaged by other packages.

WATER-MELON MILK SHERBET

An easy American ice cream – just put the following ingredients into a liquidizer or processor, to reduce them to the smoothest possible purée:

grated rind of half a lemon
juice of a lemon

125 g (4 oz/½ cup) caster sugar
1 litre (1¾ pt/scant 4½ cups) water-melon pieces, minus seeds
300 ml (10 fl oz/1¼ cups) evaporated milk

Taste and adjust seasonings of lemon and sugar if necessary. Add a pinch of salt. Pour into a container and freeze at the lowest possible temperature (or according to the instructions if you use an electrical machine). Turn hard sides to middle after an hour or more, when most of the mixture is set, but not really hard. Add two tablespoons (3 tbs) of sherry, or more to taste – dry or medium sherry is best, but sweet can also be used.

Have ready two egg whites stiffly beaten. Keep the beater going and add the frozen mixture in spoonfuls. Return to the container or put into new containers and complete the freezing.

The ice can be served from a water-melon shell, but it is best frozen in a refrigerator box, then transferred to the shell not long before the meal.

WATER-MELON SORBET

¾ kg (1½ lb) water-melon pieces, minus seeds
300 g (10 oz/1¼ cups) sugar
2 sticks cinnamon
juice of a large lemon

Mash or liquidize or sieve the water-melon to make a smooth pulp. Simmer sugar with ½ litre (¾ pt/scant 2 cups) water and the cinnamon for 5 minutes. When cool, remove cinnamon, and add the syrup gradually to the water-melon, stopping when the balance of fruit and sweetness seems right. Use the lemon juice to bring out the flavour.

Freeze in the usual way, at the lowest possible temperature. If you can stir the sorbet from time to time, to keep the texture even, you can serve it at the granita stage, when it is a thick grainy sludge. Very cooling.

As they say in Egypt, 'Fill your stomach with a summer water-melon' – in other words, don't worry, relax, enjoy yourself.

APPENDIX

THE BEST FRUIT SHOP?

Does anyone know of a shop that can better these two lists of fruit that I made in Fauchon's in 1979? Fauchon's is in the Place Madeleine, in Paris. and I recommend anyone who is on holiday there to go and study what a food shop can be. The asterisked fruits in the list of 7 August were also in the shop on 3 September.

7 August

apples, akene
golden delicious
Granny Smith
reine de reinette
apricots
bananas, red
yellow
cherries
Chinese gooseberries (kiwi)
*figs
gooseberries
*grapes, Belgian hot house
muscat
Belgian hot house royal
nouveau
Chasselas
Muscat de Hambourg
*greengages
grenadillas (yellow passion fruit)
*guavas
kumquats
lemons
limes
*mangos, Egyptian and Mexican
*melons
nectarines, *yellow
white
oranges
*passion fruit
peaches, white
yellow
*peach-apricots
pears, Guyot
*Williams'
*pineapple
plums, Monsieur
rambutans
*raspberries
red currants
sour sop
*strawberries
strawberries, wood
*water-melon

3 September (including those asterisked in the list above)

combavas (a special lemon from Malaysia)
dates, fresh
grapes, perle
mandarins
mangosteen
mirabelles
papaya
physalis or Cape gooseberry
plum, Quetsch
pomegranate
prickly pear
quince

EDWARD BUNYARD'S MARRIAGES OF FRUIT AND WINE

Here is a summary, from *The Anatomy of Dessert*, 1933 edition, of Edward Bunyard's conclusions:

APPLES: dry port or sherry. The masculine group of Orléans Reinette, Blenheim orange – tawny port. The more highly scented Cox, a dry sherry. Summer apples – Montrachet.

APRICOTS: Sauternes wines, or – André Simon's suggestion – Monbazillac.

CHERRIES: 'have so far defied any matrimonial efforts. I must leave them in unwedded bliss.'

FIGS: sherry or Marsala – 'a few drops of either inserted with an eye-dropper into the fruit before serving will win much applause for the gardener'.

GOOSEBERRIES: a Moselle.

MELONS: almost any wine – 'it would seem that an unsuitable wedding is hardly possible'.

NECTARINES AND PEACHES: a sturdy red wine, e.g. a Burgundy or Châteauneuf du Pape. André Simon advises champagne.

ORANGES: ruin all wines – 'let them remain in the nursery with the viscid Banana'.

PEAR: Pomerol or St-Émilion, or a Burgundy 'set forth the fruit's lighter graces'. Sauternes are too close in character.

STRAWBERRIES: claret or Beaujolais.

MIXED FRUIT RECIPES

POACHING FRUIT

The safest way of poaching fruit, especially if any quantity is being cooked, is to use an electric deep-frier.

Fill it up to the maximum mark with water and sugar in the proportion of 600 ml (1 pt/2½ cups) to 300 g (3½ oz/scant ½ cup) sugar. Add some curls of lemon peel, or cinnamon sticks, or vanilla pods. Switch on to 110°C (225°F), stir until the sugar has dissolved, then leave until the red light goes off.

Meanwhile prepare the fruit, and divide it into batches. It is always best to cook a small quantity at a time, starting with the most delicately-flavoured fruit.

Put the first batch into the basket. Lower it into the syrup and allow it to come back to temperature if necessary, immediately switch it down to 90°C (200°F). Some fruit, e.g. rhubarb, needs only 2 minutes. Whole peaches may need 10 minutes, but test after 5 with the point of a sharp knife.

When all the fruit is cooked, some or all of the poaching syrup can be boiled down by at least half – do this in a wide shallow pan with sides that slope out – and poured over the fruit. Do not attempt to reduce syrup in an electric fryer, the minimum liquid level must *always* be maintained.

Left-over syrup can be stored in the refrigerator to be used on other occasions. Or for making fruit salads, ices and jellies.

LE GRAND DESSERT

The nearest to a grand spectacular that the austere and skilful simplicity of the *nouvelle cuisine* allows. A 'tasting plate' of fruits and ices, served with little delicate biscuits or a few large almond *tuiles*. The arrangement, a chef's palette of colour, is made in front of the diners, showing the skill, and occasionally the lack of skill, of the waiter as well.

Le grand dessert is today's successor to the macedoine or salad of fruit that glorified the sumptuous tables of the nineteenth century. Then such dishes, in huge bowls or set into towering *millefeuilles* of jellies, were the perfect response to large dinner parties and the supper tables of the Victorian ball. Once the Edwardian age was over, we lost the enthusiasm for feeding a crowd of people well, it was no longer a challenge but a hazard to be overcome as cheaply as possible. Fruit salad became disgusting, which was unfair. It needed and deserved recreating. *Le grand dessert*, though much of its preparation is the same, is far more appropriate to the small parties of six and eight people that we prefer – and can afford. The zeal for the perfection of each fruit, something that was once taken for granted, is a defiance to the daily 'convenience' of tinned, dehydrated or frozen food.

As with fruit salad, you need discrimination in the first stages of *le grand dessert*. The choice of fruit, the choice of ices – six different fruit, well contrasted in colour, taste and texture, two ices is the minimum requirement. Ten fruit, three ices are the maximum.

Discrimination is also required when you decide how to treat the fruit you have chosen. Divide them:

Soft, pulpy fruit to be kept for ices, e.g.

- banana
- bilberry
- cherimoya
- custard apple
- mangos, ripe and soft
- passion fruit
- persimmons, ripe and soft
- prickly pear
- sour sop

Soft fruit to be eaten raw, after minimal preparation, e.g.

- berries of most kinds, apart from gooseberries that are sharp
- Chinese gooseberries
- grapes, peeled, halved, pipped
- guavas, ripe
- lychees
- mangosteen

Fruit to be macerated in sugar and liqueur, or sugar alone, e.g.

- citrus fruits, sections cut free of enclosing skin
- melon, all kinds
- pineapple
- papaya
- red currants, if very ripe
- Sharon persimmon

Fruit to be lightly poached in syrup, e.g.

- pears
- peaches, cut in wedges
- eating apples, rich-flavoured
- nectarines, cut in wedges
- greengages, halved and stoned
- mirabelles, stoned
- apricots, stoned
- plums, halved and stoned
- dried fruits, prunes, apricots etc, pre-soaked
- gooseberries that are sharp
- red, white and black currants

Fruits of the last group are arranged in ascending order of strength of flavour. If you have an electric deep-fryer, now is the time to use it. If not, use a large saucepan and a blanching basket, with a thermometer.

ASSEMBLY: in a restaurant, a trolley holds all the fruit, with special insulated, covered tubs for the ices. There will be a dish of tiny biscuits or wafers etc, to be put on the table before serving starts. At a convenient corner of the trolley, you may see a large bowl of half-whipped cream, with large forks in it.

The waiter will arrange tablespoonfuls of the ices on each plate, in a petal sort of arrangement, either in the centre or to one side. Then he will put little mounds or fans of fruit all round, with regard to colour and shape. Finally, he will lift one of the forks from the cream bowl and rapidly take it to and fro across the plate – needs practice, or the rim gets messy – to leave a light veil across the whole thing. And there you are.

OBSERVATIONS: this grand dessert is more complicated to describe than it is to do. People who have less access to fruit trees – and supermarkets – can make their ices of other things than fruit: vanilla ice, chestnut ice with glacé fruit, coconut or almond or toasted hazelnut, or crushed praline ice. The sorbets can be made with dessert wine, liqueurs, or plain lemon juice.

Gardeners and those who live further south will see in this paradisal assembly some reflection of what Milton's Eve served to the angels in the Garden, where the divine trees are –

> Delectable both to behold and taste;
> And freely all their pleasant fruit for food
> Gave thee, all sorts are here that all th' Earth yields,
> Variety without end.

What a descent to canned fruit cocktail!

MACÉDOINE OF FRUIT OR FRUIT SALAD
MACÉDOINE DE FRUITS

A *macédoine* – referring to the Macedonian Empire that Alexander the Great assembled into harmony, from a muddle of small warring states – was first applied in cookery, probably by Carême, to a mixture of vegetables. That was round about 1814, in France. By 1830, it began to be used for mixtures of fruit in syrup, fresh fruit and cooked, dried or glacé (the last two were revived in water, then gently heated through until tender). The syrup would often be flavoured with a liqueur, and the whole thing would be served very cold. This chill was a luxury, too, in a time of few ice houses and no domestic refrigerators. Sometimes the different fruits were set, layer by layer, into a Babel tower of jelly.

The impulse to this magnificent dish came from the increasing importation of new exotic fruits and the development of better and better varieties of old favourites. The commercial ice trade grew fast in mid-century: caterers, at least, could supply iced puddings for parties and special dinners. Parties and special dinners were more within the reach of the middle classes who were rising fast due to the prosperity of industry.

It surprises me that macédoine of fruit took so long to cross the Channel. Although Mrs Beeton gave a recipe, rather a tame one, in her *Book of Household Management* of 1859–61, it was unfamiliar to such epicures as Sir William Hardman and his wife, and their close friend George Meredith the novelist. In Paris in the autumn of 1863, the Hardmans dined at the Trois Frères restaurant. They ate a 'novelty' of fresh fruit 'embodied in a thin half-set jelly with Maraschino', and on their return gave a dinner to which Meredith was invited, with this dish as a centrepiece.

They brought back other novelties, too, such as special fish knives, the idea of serving a separate vegetable course, and of placing melon at the start of a meal. The dinner was a great success.

TO MAKE A MACEDOINE OR FRUIT SALAD: choose and divide your fruit into three categories – raw, raw and macerated, and poached in syrup. Proceed in the same manner as for *le grand dessert*, but mix the fruits together in a large bowl, once they have finished macerating or are completely cold.

Reduce the syrup hard, taking into account the juices that will have oozed out of the fruit in its bowl. Flavour the syrup when no more than tepid with liqueur, maraschino or kirsch are standard, but orange and other liqueurs can be used when they are appropriate to the mixture of fruit. When the syrup is cold, strain enough over the salad to come just below the top of the fruit: avoid a swill of liquid. Cover with plastic film and chill thoroughly. If you like, you can serve the bowl set in another bowl of cracked ice.

By cooking strongly-coloured fruit – black cherries, black currants, quinces – last of all, you can end up with a glorious coloured syrup. People with mulberry trees might sacrifice some of their fruit to this end (mulberries included in a fruit salad, should be raw). I sometimes make a collection of golden fruit, plus lychees and black cherries, all floating in a royal purple syrup – Emperor salad. Tinned lychees, golden berries (physalis) or peaches are permissible if they are drained of the heavy syrup and quickly rinsed, so is unsweetened pineapple, but keep such things to a minimum. Avoid canned red cherries or canned pears and mandarins which unfortunately acquire a synthetic taste and texture from the process.

TO MAKE A JELLIED MACEDOINE: after preparing, macerating and poaching the fruit, keep them in separate bowls. Reduce the syrup a little, if the flavour needs concentrating, but keep it silky. Measure out rather more than you will need, and add gelatine in appropriate quantity (dissolve it first in some of the measured syrup, reheated). Flavour with a little liqueur or fruit brandy, if you like.

Choose a simple charlotte mould if this is your first effort, and you aren't experienced in jellied puddings. Pour a layer of jelly into the base, and put it into the freezer to set quickly. Arrange the prettiest, most delicate fruit on top of the jelly, remembering that the whole thing will eventually be turned upside down and that the first layer will be the one most seen. Run jelly round the fruit – it should be set almost to the egg-white stage, but still runny enough to go round the fruit. Put back in the freezer. Pour on a thin layer of jelly to enclose the fruit completely, and make a smooth bed for the next fruit. You cannot do this in one stage or the fruit will float.

The snag, of course, is catching each layer of jelly when it is *just* set, before it hardens completely. If the layers set very firm, they may separate slightly when you turn the whole thing out. Should your jug of jelly with which you are working set firm, stand it in some warm water, and mix it together as it melts again.

I am no enthusiast for this kind of pudding, as you will have gathered. A plain fruit jelly is a delight, an elegant or a riotous fruit salad can give a lot of pleasure and be fun to make. But a towering combination of the two, which has taken you all day to make, does not taste good enough to justify the effort.

KISSEL

A Russian fruit pudding originally set with soured cereal water, *keesel*. Nowadays, potato starch is used and you will find arrowroot even better. *Kissel* is served well chilled with a big jug of cream or milk.

STRAWBERRIES/RASPBERRIES: hull, sieve or liquidize and sieve a generous ¼ litre (8 fl oz/1 cup) fruit. Dissolve 125 g (4 oz/½ cup) sugar in 350 ml (12 fl oz/1½ cups) water, and heat slowly. Meanwhile dissolve 2 level tablespoons (scant ½ cup) potato starch or arrowroot in very little water. When the syrup boils, stir in starch or arrowroot, and cook until thickened. Off the heat, add the fruit purée, taste to see if more sugar is needed, then pour into bowls and chill in the refrigerator.

BILBERRIES/CHERRIES etc: sieve or liquidize and sieve enough fruit to measure ¼ litre (8 fl oz/1 cup). Put the debris, e.g. cherry stones and skins, into the pan with sugar and water as above. Then strain and finish in the same way. This method gets all the flavour possible from the fruit.

APPLES, BLACK, RED OR WHITE CURRANTS, SOAKED DRIED FRUIT: measure a generous ¼ litre (8 fl oz/1 cup) of prepared fruit. Put into a pan and cover with cold water or soaking water. Simmer until fruit is cooked and hardly any liquid remains. Sieve. Make the thickened syrup as above and add to fruit purée.

Note With such overall instructions, you need to taste and reflect on small adjustments that will improve any particular *kissel*. Fruits vary so much in strength of flavour.

FRUIT WATERS

With our processors, liquidizers, freezers, I think that fruit waters could do with a revival. They are much more refreshing and much pleasanter to drink than commercial 'squashes'.

With the basic recipe below, you can make up any kind of fruit water. They come from Audiger's *La Maison Reglée* of 1692. We find the red currant water particularly reviving, especially with a little vodka or gin.

STRAWBERRY WATER: process or liquidize ½ kg (1 lb) of strawberries with 600 ml (1 pt/2½ cups) water, 125–150 g (4–5 oz/½–⅔ cup) sugar and the juice of a lemon. Leave to infuse for 15 minutes – longer if you like – then strain into a jug and chill. Serve with ice cubes and soda water, plus spirits if you need picking up as well as refreshing.

RED OR WHITE CURRANT OR RASPBERRY WATER: as above, omitting the lemon.

LEMON AND OTHER CITRUS WATERS: bring 600 ml (1 pt/2½ cups) water to the boil with the peeled zest of three lemons. Leave to infuse until cool. Strain on to 125–150 g (4–5 oz/½–⅔ cup) sugar. Add the juice. Chill. With sweeter citrus fruits, use two of them with one lemon.

PEAR, PEACH AND APRICOT WATERS: cut up 4–8 fruits fairly small – peel, core and all with pears, no stones for the other two. Put them with 600 ml (1 pt/2½ cups) water and bring slowly to boiling point. Give one proper boil. Remove the pan. Stir in 125–150 g (4–5 oz/½–⅔ cup) sugar. Leave to cool. Strain and chill.

POMEGRANATE WATER: squeeze 2 or 3 pomegranates and add to water and sugar as above. Liquidize with a spoonful of red currant jelly, if the colour is too pale.

FRUIT PRESERVES

PRESERVING FRUIT

In this book, which is devoted to the use and enjoyment of fresh fruit, I have only given preserves of the simplest and most foolproof kind, preserves which carry no risk of being damaged by a mistake or misunderstanding in their preparation, preserves which carry no risk of damaging you.

The whole business of preparing food for storage is complex. General principles for each process can be set out, but these often need small variations when you come to the different items – the explanations of these variations is what takes up space. My own two bibles are *The Preserving Book*, devised and edited by Caroline Mackinlay and Mike Ricketts, available in Pan paperback; and *Putting Food By*, an American publication by Ruth Herzberg, Beatrice Vaughan and Janet Greene, brought out in this country by Dent's.

I find two ways of preserving fruit simple and useful. This is partly because we spend the summer working in France, where the fruit is so much more abundant, so much cheaper than it is here, and where I do not happen to have a deep-freeze or much space. Anyone who takes a house or flat in a hot climate, might at least think of devoting the occasional 20 minutes to making a few bottles of fruit liqueur to bring home for Christmas. Or they might put out fruit on sheets in the sun to dry (bring it in at night for fear of moisture, or when rain threatens). These are the best souvenirs of a holiday, and give pleasure far beyond the work involved. It is something the whole family can share.

FRUIT SPIRITS AND LIQUEURS

Fruit liqueurs are the simplest preserve of all. So many people make sloe gin, yet never think of applying the principle to other fruit as well. Strawberry gin and raspberry vodka are just as delightful. Best of all are small hard apples of the sorb tree and the clustered berries of the related *alisier* (wild service), left until they are bletted, i.e. soft and almost rotten, *see* p. 414.

DRY FRUIT SPIRITS: for this you need soft fruit – strawberries, raspberries, cherries – and a good vodka, or Dutch gin. Avoid washing the fruit, and pick out the best specimens. Put 250 g (8 oz) into a litre (1¾ pt/scant 4½ cup) bottling jar, and pour over the spirit. Fasten tightly and leave in the dark for at least a fortnight. The longer the better. When you want to use it, strain off the liquor into a clean bottle.

If you find it too dry, add a very little caster sugar to taste.

FRUIT LIQUEURS: the general principle is to infuse fruit with two-thirds brandy or other spirit and one-third sugar. At the end, you often have the bonus of a delicious fruit to eat as well as the liqueur. Serve together in small glasses.

The difference in the two following methods is determined by the nature of the fruit. With firm fruit, quinces, peaches, apricots, you need to make a syrup. With soft juicy fruits, you can get away with tipping sugar directly into the bottling jar, before filling up with spirit. The sugar gradually dissolves, and you can help it along by turning the bottle upside down, or giving it an occasional shake.

The simple way of making the syrup is to dissolve two mugfuls of sugar with one mugful of water in a pan over a low heat, stirring it. Once the liquid boils, stop stirring and simmer until you have an oily-looking syrup – about 5 minutes. Cool before adding to the jar. Spices, orange peel, a vanilla pod can be cooked with the syrup, or put straight into the jar with the fruit.

Firmer Fruit: in this category I include peaches, nectarines and apricots, as well as pears, sorbs and wild service berries. Pineapples, peeled and sliced, can be done either way.

You will get more fruit into the bottle if you halve it, when the fruit is as large as a peach or a pear. Leave the skins and the core of pears: crack peach, nectarine, apricot stones and add the kernels to the bottle. Or just put in the stones, if you are storing the liqueur for several months, some of them cracked.

Make a syrup as above, and pour in to come one third of the way up the fruit. Top up with alcohol, and close.

Soft Fruit: with cherries, sloes, mirabelle plums, which should like all other fruit for preserving be ripe but firm and unblemished, prick them with a darning needle before packing them into the jar, right to the neck. Pour in caster sugar to come one third of the way up the fruit: fill up with brandy, gin or vodka. Strawberries, raspberries, mulberries, black currants can be treated in the same way – a few raspberry or black currant leaves may be put in as well.

DRIED FRUIT LIQUEURS: to improve the dreadful country distillations of grape and apple marc (the debris left after wine and cider making), people put a few prunes into the bottle. A desperate remedy, but from it the French have developed a more agreeable preserve. Fill the jars two-thirds full of partially soaked prunes or apricots. Push in a cinnamon stick (prunes) or vanilla pod (apricots), and a curl or two of orange peel. cooled syrup (above) to come one third of the way up the jar, then finish with brandy for prunes or for apricots gin or vodka or brandy. The fruit will gradually soak up the liquid. Should it rise above the liquid level, top it up if there is room, or remove a premature apricot or prune as a taster and mark the progress. With this preserve, you end up with fruit for dessert, rather than a liqueur. Serve it on its own, or with vanilla ice cream, or cream and soft cheese mixed, or just cream on its own, and small thin biscuits.

TO FINISH THE LIQUEURS: drain them off into clean bottles and label them. An alternative is to drain off a little of the liqueur as you need it, and to top up the level with spirit on its own, or a mixture of one third syrup and two thirds spirit. Choose the method according to whether you think the liqueur is too sweet, or just about right. The fun of making these fruit liqueurs is the tasting and adjusting.

As a general rule – I have never known an exception, but there may be one – leave all these liqueurs for at least a month. Often the interval between the season of the fruit and Christmas is exactly right. With firmer fruit, especially quince vodka, *see* p. 395, a year is not too long. If the contents of the jar begin to look revolting, which can happen with sorb and *alisier* liqueurs, filter off the liquid through a double muslin so that it regains its clarity.

DRYING FRUIT

Drying fruit is fun, and not difficult. After all, you know what you are aiming at from experience with bought prunes and dried apricots. The point of the process is to remove the moisture of the fruit steadily but slowly so that the fruit is not cooked. It follows that the essential thing is a source of low, constant heat. It need not be continuous heat: if you need to use the oven for a meal, which means higher temperatures, just remove the trays of fruit for the time being. The only snags are that the whole business takes longer, and the kitchen is cluttered up with the wooden trays.

The particular advantage of drying fruit, apart from the obvious one of not wasting a glut is that you provide yourself with an extra ingredient. If you freeze fruit, you end up with much the same result as fresh fruit. Dried fruit is very different in taste and texture, being a concentration of sweetness and flavour even after it has been soaked and cooked.

The important thing is to have the right fruit. In the first instance, it must be ripe but firm and unblemished (not always easy in this climate, if you grow your own fruit). In the second instance, some varieties are much more successfully dried than others. Grapes from you own vine will not be as good ultimately as imported Muscat grapes which have grown in a hotter country. The Agen variety of plum, the Perdrigon varieties, Ickworth Imperatrice, are all recommended: the old favourite was the Sainte-Catherine. Bramley's Seedling and Russet are good apples to choose, and the old Moorpark apricot dries well. The Williams' pear should be picked when firm but not quite ripe, then ripened in the house before drying.

Putting Food By, recommended in the previous section, gives detailed information for each fruit of the desirable extra attention it should have. The summary below is rough and unsubtle, for the occasions when you have unexpected extra fruit and have to deal with it quickly (though I do think that any kitchen bookshelf should have a good preserving book on it).

APPLES: peel, core, slice and put immediately into brine to prevent discolouring. Dissolve a tablespoon (2 tbs) of untreated sea salt in a litre (1¾ pt/scant 4½ cups) water for the brine.

APRICOTS, NECTARINES, PEACHES: halve and remove stones.

BANANAS: peel, slice down lengthwise.

FIGS: dry whole unless they are enormous, in which case halve them down.

PEARS: peel, halve and scoop out the main core with a potato-baller. Pull out the thready part.

PLUMS, CHERRIES, GRAPES: dry whole. Separate the grapes from the bunch.

Note that all cut fruit should be dried cut side up at first, and that pieces should not touch each other. As the pieces dry and contract, you can rearrange the trays.

For trays, use wooden fruit boxes with slatted bottoms. Scrub them well. Put in a double layer of washed butter muslin (old nappies, for instance) or stockinette or sheeting. The bottom rack will have to be raised from the base of the oven so that the air can circulate freely. Apart from this, the trays may be stacked one above the other.

TO DRY: you need an airy steady temperature between 50° and 60°C (120° and 150°F). The plate-warming oven of a solid fuel or oil-fired stove is often a good place, so is a warm airing cupboard. Gas and electric ovens can be used: you will need to keep the door open, as even the lowest thermostat setting is too hot. Whatever method you use, buy a thermometer so that you can be sure the heat is within the right range, since the only thing that is really important is to keep the temperature absolutely right.

You know when the fruit is ready to cool and pack away by its appearance. Apple rings should be leathery. Soft fruit should yield no juice when you cut a piece of it. Bananas will have a wavy edge and be drier than the kind you buy: they will be less sweet, too, and we much prefer them, whether as sweets with toasted nuts as a homely kind of *petits fours* with coffee, or as sustenance on a long walk when we need to travel light and want to picnic. If you grow pears, it is certainly worth drying them as nothing can take their place in, for instance, Hamburg eel soup.

Cool the fruit thoroughly before packing it into paper bags or cardboard cartons. Avoid plastic as any slightly moist fruit may go mouldy when trapped in so airless a situation. Store in a cool, dry place; hanging the bags from a ceiling hook is a good idea, but all too few kitchens these days are provided with such sensible aids to storage.

TO RECONSTITUTE: pour boiling water over the dried fruit, enough to cover it properly plus a little extra. If the fruit absorbs it all and needs more, add it gradually. Cook the fruit in its soaking water (that is why you do not want too much of it). Any sugar needed should be added when the fruit is tender.

Dried fruit, soaked and cooked, can be a good basis for a simple form of candying (*see* the apricot recipe on p. 34). It can also be soaked in syrup and alcohol, and eaten without cooking, *see* previous section on fruit liqueurs.

SOUR-SWEET FRUIT PRESERVES

Pieces of fruit and whole fruit in spiced and sweetened vinegar, fruit chutneys, are all good store-cupboard items. As the method alters from fruit to fruit, you will find recipes under the fruit concerned.

For pickling, some fruits are cooked in water first or, in the case of prunes, tea. Others are simmered in a mixture of sugar and vinegar with spices, without previous cooking. And the quantity of sugar to vinegar can vary, according to the fruit. If you cannot afford wine vinegar, use cider vinegar rather than malt. And when cooking the fruit, be careful not to leave it until mushy: melon can be tricky in this respect, but it makes such a good pickle, it is worth persisting with it.

Chutneys I often find unsympathetic. They are too strong, and too mixed. Of course, there are exceptions, and a lot of people will disagree with me very strongly. The problem is that although chutneys look simple to make – cut up a lot of fruit, raisins and onions, and stew them with sugar and vinegar and a few spices, until the whole thing is pulpy – they are very difficult to make really well because of the balance and variation in quantities. I imagine this is why certain recipes are treasured in families, and shared only with close friends.

One pleasure, though, is to make your own mango chutney. This has become possible now that West Indian and Indian shops, and some greengroceries and supermarkets, begin to stock under-ripe mangos. I cannot claim that the result is cheap, but it is fun to make it occasionally and see how it compares with the brands on sale.

SPICED FRUIT

I have found the whole business of cooking soft fruit, whether for spicing it or for a fruit salad, much easier since a friend told me to use a blanching basket and a deep pan. I now use an electric deep-frier (*see* p. 437), heating the water, dissolving the sugar and boiling them together to make a syrup, in the case of a fruit salad or compote, and substituting vinegar for water if I am preparing spiced pickled fruit.

Of course, it means large quantities of syrup, and inevitably some is left over. It goes into the fridge and comes in useful in all sorts of ways, either on its own, or as the start of a new batch of syrup when the next load of soft fruit has to be dealt with.

I have the feeling that apricots, peaches and nectarines keep a better appearance and shape if they are pickled whole. The problem is that you need far more bottling jars than if you halved and stoned them first. The snag of the halved fruit is that the pieces rapidly become ragged; one minute they are not quite ready so you leave them, the next minute they have flopped.

Another decision to be made, or rather born in mind as you finish the process, is whether to leave the spices in the syrup. They look attractive, but they continue to add their flavour which can become too strong – cloves are the tricky spice, and they are best removed. Small cinnamon sticks or a slice or two of fresh ginger can be left, or a small red chilli. In any event, taste the spicing mixture occasionally.

To 1½ kg (3 lb) fruit ready for cooking:
½ kg (1 lb/2 cups) sugar
300 ml (½ pt/1¼ cups) wine vinegar
5 cm (2 inch) stick cinnamon
6 cloves
2½ cm (1 inch) fresh ginger, peeled, sliced
2 small red dried chillis

In a heavy pan or electric fryer, stir the pickle ingredients over a low heat until the sugar is dissolved, simmer 5 minutes, then lower in the fruit in a blanching basket. Cook until just tender. Remove, cool and pack into rinsed out jars that have been dried upside-down in a warm oven. Boil down the liquor until syrupy, then pour boiling over the fruit, making sure it is covered. Seal and leave in a cool dark place for at least a month.

If you want to use a deep-frier, follow the instructions on p. 437.

OTHER FRUIT PRESERVES

The general principles of jams, jellies, fruit curds, butters and cheeses, the candying of fruit and peel, are again matters for specialized books such as the ones mentioned on p. 444.

Such recipes occur throughout the book – lemon curd, passion fruit curd, candied grapefruit or pomelo or citron peel, cornel jelly, quince paste, uncooked raspberry jelly, fig jam. They are included because I find them especially popular and they are not in the usual books, or because my methods have some little quirk that makes them different from the standard versions.

Bottling I have left out altogether. It is not difficult, but it needs accuracy and adjustments according to the fruit being dealt with. You do not bottle an odd jar of fruit, as you might make a jar of lemon curd; it is a large-scale enterprise which a lot of people enjoy as an agreeable autumn ritual. I confess that I dislike bottling wholeheartedly. Something to do with the horrible bottled plums of wartime: all that labour for something so disagreeable (I seem to remember that they were put down without sugar). When I see expensive jars of bottled peaches and apricots and pears in grand French food shops, I realize the folly of my prejudice: but somehow all the fruit that comes my way goes into the freezer instead, which I find more convenient and more versatile.

PASTRY

Long explanations of the classic ways of making puff and choux pastry, *pâte sucrée* and brioche dough are out of place in a specialized book of this kind. Beginners should turn to *The Observer French Cookery School* (Macdonald, 1980) for the careful instructions devised by Anne Willan and her chefs at La Varenne cookery school in Paris. Or to *Leith's Cookery Course* by Prudence Leith and Caroline Waldegrave (Deutsch, 1980). If you are beyond the learning stage, and just need a reminder of quantities and temperatures, you will probably find enough assistance in this section or in any general cookery book (though I certainly recommend turning to Anne Willan or Prue Leith for a refresher course from time to time).

A point to look out for with pastries is the way cookery books express the measurement required. Our convention is to indicate weight by stating the quantity of flour required to make it. In other words

250 g (8 oz) shortcrust pastry

means shortcrust pastry made with 250 g (8 oz/2 cups) flour, plus other ingredients in the usual proportion. Which means that the total weight of the dough will be over half as much again, by the time you have added butter or butter and lard and the mixing liquid.

However, the convention falls down with puff pastry. Cookery books up until about 1950 follow it because in those days there was no frozen puff pastry. Nowadays it is recognized that most people do not have the time to make puff pastry, which is a prolonged business, and buy it frozen (or substitute another kind of pastry, of course). Some writers get round this by saying

250 g (8 oz) weight puff pastry

which means a small packet, weighing 250 g (8 oz) in all. An earlier book, saying the same, would mean a total weight of about 500 g (1 lb) as quantities of flour and butter are more or less equal. This is something to watch out for as you may well need double the quantity of pastry that you thought you needed at first glance.

Once you have been cooking for a while, you know by instinct how much pastry you need for any dish. And if you have a freezer, you might just as well make the required pastry in large quantity and freeze the surplus in convenient amounts. Pastry freezes well.

Here is the La Varenne summary of pastry quantities: useful to keep this in mind, but reflect that ultimately the number of people served will depend on the rest of the meal, and on their appetites.

You will note that the weight is expressed as the quantity of flour needed, other ingredients in proportion.

An 18–20-cm (7–8-in) flan tin needs 125 g (4 oz) pastry; and serves 3–4
A 24–26-cm (9–10-in) flan tin needs 200 g (6½ oz) pastry; and serves 5–6
A 28–30-cm (11–12-in) flan tin needs 250 g (8 oz) pastry; and serves 8

TO BAKE PASTRY BLIND

The more water you use in making pastry, the more it will shrink if you bake it blind. If you stretch it too much when rolling it out, it will also shrink back. Two ways of minimizing the risk, is to mix the dough with cream, egg or egg yolk, and to make sure that it is well chilled before you roll it out.

For tarts intended for a non-liquid filling, prick the pastry well. You do not need to put in foil and beans.

For tarts intended for a moist filling, you cannot prick the pastry or the filling will flow through it. Line the pastry with foil or greaseproof and weight it down with dried beans. When the pastry is set but not browned, lift out the foil or paper and beans, and return the tin to the oven for the base to lose its moist uncooked appearance and for the edges to colour lightly. Cool and store the beans for another occasion.

Elizabeth David has commented on the pallid underbaked pastry of English cookery: *pâte sucrée,* she finds, is rarely brown enough or crisp enough. When you bake a pastry case blind, the time and temperature will depend on whether or not there is to be a second cooking time, whether you want a firm but pale crust, whether you want a crust that is just set.

Gas 6, 200°C (400°F) for 10 minutes gives you a set crust, then you must judge whether to leave it at that heat, or to lower the temperature to gas 4, 180°C (350°F). Bear in mind that oven thermostats are not always accurate. Judge by the appearance of the pastry.

IMPROVED FROZEN PUFF PASTRY

Many of us find it difficult to get puff pastry right. We follow the instructions, and somehow the result is disappointing considering the time and effort involved. I once asked Anne Willan, when I was visiting La Varenne cookery school, about this problem. She agreed with me that it needs regular practice, and the *tour de main* only develops if you are making it more or less daily. Her pastry chef takes a week or two to regain his skill when he has been away on a month's holiday. If he finds it tricky, the rest of us may be forgiven for resorting to frozen puff pastry.

A suggestion – if you do make your own puff pastry, try using strong plain bread flour. It will rise far better. If you want to eat puff pastry cold say as *millefeuilles* or some other cake, you should make it yourself with butter. Or order it from a reputable pastrycook.

What is wrong with frozen puff pastry? Two things, the flavour of butter is missing as the manufacturers insist on using innominate fats, and it is never quite as light as puff pastry made by a skilful *pâtissier*. In daily life it is easy to forget standards of perfection. We went a couple of summers ago to a meal at the Troisgros brothers' restaurant at Roanne. One of the courses was a *feuilleté*, a puff pastry sandwich. The ungreasy richness, the light crispness of the many layers, the even rising, the golden brown colour, the way it disappeared in one's mouth, made me realize that I had never known before what puff pastry could be.

Ever since that meal, I've been reflecting on and off about puff pastry. First in anger. Why do manufacturers skimp on the fat? Intervention butter from the EEC is nowadays so cheap for the trade that there is no reason why they shouldn't use it. No doubt the problem comes from the way butter deteriorates in the normal deep freeze. But then why should they not use butter and date-stamp the packets? A little more work for supermarket managers? No, not work, just a fraction more thought and organization. And what a gain in quality for the consumer! But then consumers are only there to feed the Morlocks of commerce as far as this country is concerned.

Anger at the well-known shortcomings of the big food firms, irritation at my own lack of skill and the time to develop it, at least prodded me into thinking that surely some improvement could be made?

Eventually this is what I did. I left a large packet of puff pastry to thaw, then rolled it out into a thin oblong on a floury tray. The whole thing went into the refrigerator, while 100 g (3½ oz/scant ½ cup) butter was melted, then allowed to cool until opaque but still very soft.

One third of this butter was rapidly brushed over the pastry, which was folded and turned in the usual way. This was done twice more. Then the pastry was left to chill in the refrigerator until it was needed.

After baking, the pastry had a better flavour than before, and a better colour. It had a more home-made appearance. Certainly it was not that Troisgros perfection of pastry, but it was certainly worth eating and enjoyable.

QUICK FLAKY PASTRY

Weigh out the flour you require for the size tin (p. 453) and tip it into a bowl, with salt (level teaspoon to 250 g, 8 oz or 2 cups flour).

Take a block of well-chilled butter from the refrigerator, and fold back the paper leaving enough for a hand-hold at one end. Grate the butter coarsely on to the scales, weighing out two thirds of the quantity of flour. Mix quickly into the flour, using a knife, and bind with as little liquid as possible – a tablespoon of lemon juice plus iced water is recommended.

Chill the dough well. You will see that it has large flecks of butter in it.

Avoid mixing the dough electrically or in a processor.

DUTCH PROCESSOR PASTRY

Weigh out equal quantities of butter, plain flour and well-drained yoghurt or *fromage frais* or curd cheese. Break up the butter and put with remaining ingredients and a pinch of salt (or sugar if a sweet pastry is needed) into the processor. Switch on until mixed to a dough. Pat into shape with floured hands, and chill for an hour, in a plastic bag, or plastic film, or foil etc.

This makes a rich, short pastry verging on flaky pastry. Good for little fruit turnovers.

You can also make this pastry with an electric beater, or with a fork, if you soften the butter first. Avoid handling it.

CREAM SHORTCRUST PASTRY

Another version of rich, almost-flaky shortcrust. Again use a processor or electric beater if you like.

Crumble 150 g (5 oz/⅔ cup) butter, or butter and lard mixed, into 300 g (10 oz/2½ cups) plain flour and a level teaspoon of salt. Bind with single or whipping cream or *crème fraîche*. Double cream can be used, but it is really too thick. Chill, as above.

Add 125 g (4 oz/½ cup) caster sugar for a *pâte sucrée*.

PÂTE SUCRÉE

Classic *pâte sucrée* is bound with egg yolks. It can be made with a processor or electric beater perfectly well. Here is the correct hand method:

Tip 200 g (6½ oz/1⅔ cups) plain flour through a sieve on to a marble pastry slab. Make a well in the centre, put a pinch of salt in the middle, with 100 g (3½ oz/scant ½ cup) soft but cool butter, 75 g (2½ oz/⅓ cup) caster sugar and 3 egg yolks. Mix the centre ingredients together with your finger, gradually drawing in the flour. When you have a pile of large crumbly bits of dough, mix them into one mass using the heel of your hand. Work fast and do not be afraid to handle the dough. Chill for an hour, wrapped in a plastic bag or cling film.

CREAMED PÂTE SUCRÉE

This is a good method if you want to use an electric beater or processor, or if you prefer beating mixtures to working them with your hands.

If the butter is on the firm side, warm the beater bowl. With a processor, all you can do is cut the bits small after waiting for the butter to soften slightly.

125 g (4 oz/½ cup) soft butter
2 rounded tablespoons (¼ cup) vanilla sugar
1 large egg
pinch salt
250 g (8 oz) plain flour

Cream butter and sugar, add egg and salt. When the mixture is reasonably well amalgamated, add the flour. The dough should need no water. Chill in the refrigerator for at least 30 minutes.

ELIZABETH DAVID'S SWEET FLAN PASTRY

Easy to remember, and efficient in practice.

If you follow the metric system, remember 1:2:3. If you stick to Imperial, remember 3:6:9.

In other words, 100 g (3 oz/6 tbs) caster sugar, 200 g (6 oz/¾ cup) butter and 300 g (9 oz/2¼ cups) flour.

Mix with iced water, using as little as possible for the lightest pastry.

CRACKLING CRUST AND OTHER NUT PASTRIES OATMEAL PASTRY

Crackling crust, an eighteenth-century recipe, should be made with all ground almonds and no flour. But this means it is impossible to handle, and drops into too many bits. It works better if you use the following quantities:

Equal weights of flour, freshly ground nuts or medium or porridge oats, and butter. Mix with iced water, beaten egg or cream.

Even with this amount of flour, the pastry is still a little tricky to handle. If it cracks as you line the tin, don't worry. Just press the edges together gently but firmly, and patch holes with little bits of pastry. It can be stored in the refrigerator for at least a week, wrapped in cling film.

SUET CRUST AND STEAMED PUDDINGS

A suet crust is the easiest of all pastries to make. It can be baked, like shortcrust, but is mainly used for boiled and steamed puddings. Quality is at its best when you buy a piece of beef suet from the butcher, and remove the skin yourself, before chopping it evenly (a processor helps). This suet is fresher than the package kind; moreover, it has not been rolled in flour to prevent it sticking together during a long shelf life.

The usual amount, suitable for a 1-litre (2-pt/2½-cup) pudding basin, which is enough for 6:

250 g (8 oz/2 cups) self-raising flour
125 g (4 oz) prepared, chopped suet
¼–½ teaspoon salt, according to the filling

With your hands, mix these ingredients together thoroughly. Stir in ice cold water – 100–125 ml or about 8 tablespoons (⅔ cup) – to make a soft dough. It should not stick to your fingers. Add a little more flour if it does.

Form into a ball and roll out evenly on a floured board to large dinner plate size. Cut out approximately a quarter, and set this piece aside for the moment.

Butter the pudding basin generously, or use lard for meat puddings. Drop in the large piece of pastry, press the two cut edges together, and ease the whole thing into the basin to line it evenly. The pastry should lie slightly over the rim.

Form the pastry you set aside into a ball, and roll it out to make the lid.

Put the filling into the basin, and lay the lid on top. It should come 2–3 cm (about 1 inch) below the rim of the dish, to allow for rising. Flip over the edge of the underneath pastry, moisten the edges and press them together.

Cover the basin with a circle of greaseproof paper, box-pleated in the centre to allow extra rising space. Then cover with pleated foil, and tie.

TO BOIL: if you have a blanching basket large enough, put the basin into it. If not, make a string handle across the top of the pudding, so that you can lift it in and out of the saucepan.

Stand in a saucepan and pour in cold water to come a good halfway up the basin. Remove and bring the water to a rolling boil. Lower the basin into it and adjust the heat once the water returns to the boil, so that it simmers steadily. Put a lid on the pan and leave for 3 hours for fruit puddings, meat puddings need 4 hours.

Keep an eye on the water level, and replenish it with *boiling* water.

When the pudding is done, remove the foil and paper, cutting away the string. A meat pudding is usually served in the basin, with a white cloth tied round it. Fruit puddings and other sweet puddings are usually turned out. Run a knife round it, to ease it from the basin. Put a dish on top, bottom side up; protect your hands with an ovencloth and then turn the whole thing right over. Slip a knife under the basin edge, and carefully raise it.

If the pudding has to wait about in a warm oven, leave the basin in place to keep the pudding in good shape once you have made sure that it can easily be removed.

TO STEAM: half-fill the lower part of a steamer with water, bring to the boil, then set the upper part in place with the pudding inside. Cover and steam for 3–4 hours; check the water and top up with *boiling* water. These two methods can be used to cook all steamed and boiled puddings, whether or not suet pastry is used. A pressure-cooker speeds things up, follow the instruction booklet.

FRUIT DUMPLINGS

One of the best Austrian puddings which is made with small peaches, apricots or plums, or – more easily, with their jams. Try to stone the fruit without halving it, by using a stoner or pushing the stone through. Replace with sugar lumps.

Boil, peel and sieve 375 g (12 oz) floury potatoes. Cool slightly, then mix with 30 g (1 oz/2 tbs) butter, an egg yolk, pinch of salt and as much flour as the potato will absorb – about 150 g (5 oz/1 good cup) – to make a soft coherent dough. Rest it for an hour.

Divide into lumps the size of a small egg. Roll out and enclose fruit, or fold over teaspoon of jam. Slip into simmering water in batches. Half-cover and leave for 10 minutes to cook. Serve with 30 g (1 oz/1 cup) fine breadcrumbs fried in 30 g (1 oz/2 tbs) butter and icing sugar.

BISCUITS, BREAD ETC.

NUT OR OATMEAL BISCUITS

Make up one of the pastry doughs in the recipe on p. 456, using cream if possible, although egg or water will do. Form the dough into a thick roll, and then chill in the refrigerator for several hours, wrapped in cling film.

Cut thin slices from the roll. Put them on greased baking sheets and sprinkle them either with granulated sugar or coarse sea salt. Bake at gas 5, 190°C (375°F) until nicely browned. These very short biscuits are particularly good with fruit soups (and with pâtés of fish, or fish rillettes, and many soups of vegetables or meat).

CRISP WALNUT BISCUITS

Mix 6 tablespoons (½ cup) walnut oil and 125 g (4 oz/½ cup) butter – or 220 g (7 oz/⅔ cup) butter – with 150 g (5 oz/⅔ cup) sugar, ¼ level teaspoon salt and 250 g (8 oz/2 cups) self-raising flour. Bind with an egg, then add 90 g (3 oz/¾ cup) shelled, coarsely-chopped walnuts. Form into a roll. Chill well and store in the refrigerator. Slice off thin rounds and bake them at gas 5, 190°C (375°F) on a greased baking sheet, until crisp and lightly browned

Hazelnuts and almonds, with the appropriate oil, can be used instead of walnuts.

GREEK FESTIVAL SHORTBREADS
KOURAMBIETHES

If you make these plump half-moon shortbreads in a very small size, they go admirably with fools and ices and creams. Visitors to Greece will recognize them from the way they are hidden in a bran-tub of icing sugar, in baskets on the counter. The best I have ever eaten came from the bakery on the quay at Hydra.

250 g (8 oz/1 cup) unsalted or lightly salted butter, softened
60 g (2 oz/¼ cup) caster sugar
1 large egg yolk
1 teaspoon Ouzo, Pernod or brandy

125 g (4 oz/1 cup) blanched almonds, ground
60 g (2 oz/½ cup) cornflour
300 g (10 oz/2½ cups) plain flour
orange-flower water
icing sugar

Cream the butter until white (the mixture can be made in a processor or by an electric beater). Add sugar, egg yolk, alcohol and almonds (do not use ready-ground almonds, they are too fine). Sift cornflour and plain flour together and add enough to the mixture to make a soft but firm and unsticky dough.

Divide into about 40 pieces, and form into sausage shapes. Bend the sausages round a finger, to make a crescent. Place on vegetable parchment-lined baking sheets, allowing them room to spread a little. Bake at gas 4, 180°C (350°F) until slightly coloured and crisp, 25–30 minutes. It is better to underbake them and put them back when you have tried one, than to brown them too much.

When cool, dip them one by one in orange-flower water, then roll them in icing sugar. Store them in plenty of icing sugar in an air-tight tin or plastic box.

SHORTBREAD KNOBS

The browned butter and vanilla seeds make this Danish version the best of all shortbreads. They go well with fruit dishes and ice creams.

175 g (6 oz/¾ cup) slightly salted butter
vanilla pod
175 g (6 oz/¾ cup) caster sugar from vanilla jar
225 g (7½ oz/scant 2 cups) plain flour

Stir the butter, cut in lumps, over a moderate heat until it cooks to a beautiful golden brown. Allow it to cool, stirring occasionally. Meanwhile, split the vanilla pod, scrape all the black seeds and soft part into a bowl and return the pod to the vanilla sugar jar. Add caster sugar and flour to the seeds, mix well and tip into the butter. Stir to a smooth dough.

Form into tiny balls, using half a teaspoon of the mixture at a time. Put quite closely together – they spread a little but not much – on two baking trays lined with non-stick vegetable parchment. Bake 20–25 minutes at gas 2, 150°C (300°F), until cooked and lightly coloured. Swop the trays round at half time, keeping them as near the centre of the oven as possible.

SUGAR, LEMON OR GINGER THINS

Refrigerator biscuit dough is one of the blessings of modern life. I always keep a roll handy in the visiting seasons of the year, with another roll in the freezer. Thin slices can be shaved off and quickly baked, as you need them. Because you chill the mixture, it can be rich – and that means very thin, crisp biscuits, of a delicate flavour.

Electric mixers or processors are ideal for making the dough. If you do it by hand, cream the butter and sugar, then add the other ingredients down to and including the baking powder.

250 g (8 oz/1 cup) softened butter
250 g (8 oz/1 cup) caster sugar from vanilla jar
1 large egg
1 tablespoon double cream
300 g (10 oz/2½ cups) plain flour
½ teaspoon salt
1 level teaspoon baking powder
inside scraped from a vanilla pod,
or grated zest of a lemon and tablespoon juice,
or powdered ginger to taste
extra sugar

Mix ingredients down to the baking powder, then add whichever flavouring you prefer. Or divide dough into two or three, and flavour each one differently.

Put a piece of greaseproof on the table and spoon the dough on to it in a rough roll. Bend the paper up and over the dough. Using both hands, as if you were gently pushing a rolling pin, roll the paper-covered dough to even it out. Aim for a cylinder or roll about 5 cm (2 in) in diameter. Leave like this, or flatten gently to form a slab the shape of a binoculars case (which will slice to make squashed oval finger biscuits).

Chill overnight in the refrigerator. Next day, slice off thin pieces of dough, as many as you need. Put on to a baking tray lined with Bakewell paper. Sprinkle extra sugar over them and bake for about 10 minutes at gas 5, 190°C (375°F). They should not be browned.

Put remaining dough back into the refrigerator, or freeze it if this is more convenient.

WALNUT GRANARY BREAD

If you want bread for banana sandwiches, or for honey sandwiches to eat with lightly-cooked pears, this is a good one to make. As you will see, I believe in speeding up the business with Harvest Gold fermipan yeast which goes straight into the flour from the packet, and an electric dough hook or processor for mixing and kneading. Making bread by hand is fine if you do it occasionally, or have the time, but if you have a job and want to provide homemade bread as a normal part of the family diet, you need all the help you can get. With the aids I describe, daily bread-making is no chore and well within the competence of a seven-year-old.

500 g (1 lb/4 cups) granary flour
150 g (5 oz/1¼ cups) strong white bread flour
1 packet Harvest Gold yeast, or 2 level teaspoons fermipan dried yeast
2 level teaspoons fine sea salt
2 tablespoons (3 tbs) walnut oil or melted butter
about 300 ml (½ pt/1¼ cups) water at blood heat
125 g (4 oz/1 cup) shelled walnuts, coarsely chopped

Mix together the first four ingredients, add the oil or melted butter and enough water to make a coherent dough that is not too soft to handle. Use an electric dough hook first at the lowest speed of all, then gradually working up to 3. Or mix in the processor. Follow your instruction leaflet

Put the dough in a lightly-oiled bowl, or leave in the mixer bowl. Tie it into a plastic bag and leave in a warm place. Set the kitchen timer for 1 hour; if the dough has more or less doubled, it can be punched down and the walnuts mixed in. If it has risen less than that, leave it for another half hour or more according to need. If you kneaded the dough electrically, it will not need a second kneading – just punch it down.

Put into warm, greased tins, or put the roll of dough on to a warm, greased baking sheet. Leave for 30 minutes, then bake in a preheated oven at gas 8, 230°C (450°F) for half an hour. Remove from the tin; put back into the oven for 5 minutes if the lower crust is soft and pale. Brush over the top with milk, and leave to cool slowly, near the stove, either placed across the top of the loaf tin, or on a wire rack.

Note If you cannot get Harvest Gold or fermipan dried yeast, use 2 level teaspoons of ordinary dried yeast, and whisk it into half the warm water. Leave for 10 minutes or until it has frothed into a yeasty mass.

With fresh yeast, weigh out 15 g ($\frac{1}{2}$ oz/$\frac{1}{2}$ cake compressed yeast) and fork it to a mash in a bowl, adding half the warm water. Leave until it rises into a foamy head, 10–15 minutes.

Add either yeast to the dry ingredients with the oil, then add the rest of the warm water according to need.

BRIOCHE LOAF

PAIN BRIOCHÉ

In French bakeries, you can often buy *pain brioché*. It is made from a yeast dough enriched with eggs and fat, which should be butter, but in smaller quantities than for brioche proper. It's easy to recognize from its yellowish colour and enormous length: *pain de mie*, a close-textured white bread rather like ours but lighter, is baked in the same tins but the crumb looks whiter, and the crust less rich and brown. You do not have to buy the whole thing. The dough is divided into separate mounds, and they are put into the tin side by side, so that they rise and bake into sections which can be broken off and sold in ones or twos.

A brioche loaf can be used for apple charlotte, for *croûtes aux abricots* (and other fruits), for cutting into bread boxes and deep-frying as containers, say, for sliced apples cooked in butter and sugar.

If you want to flavour the dough with nuts or olives (125 g or 4 oz/1 cup nuts or $\frac{3}{4}$ cup olives), or with a slightly larger quantity of lightly-fried onions and walnuts, or onions and mushrooms, add the items after the first rising.

400–425 g (13–14 oz/3$\frac{1}{2}$ cups) strong white bread flour, or half bread flour/half plain flour
1 packet Harvest Gold yeast (see *previous recipe)*
1$\frac{1}{2}$ level teaspoons salt
2 eggs
5 tablespoons (6 tbs) melted butter or appropriate oil
up to 150 ml ($\frac{1}{4}$ pt/$\frac{2}{3}$ cup) water or milk at blood heat

Mix first 3 ingredients together, then add eggs, butter or oil and enough water or milk to make a soft but coherent dough. Knead electrically or by hand (*see* recipe above for walnut bread).

When dough is silky and leaves the sides of the bowl, oil the bowl and put back the dough, turning it over so that the top is lightly oiled too. Tie into a plastic bag and leave in a warm place until doubled in bulk, 1 hour or more. Punch down, put into warm, greased tins (I often use three tall round coffee tins) and return to the plastic bag for 30 minutes. Bake at gas 5, 190°C (375°F) until bread is well risen and browned, 35–45 minutes. Remove from tins, put back for 5 minutes if crust is pale, then brush over tops with milk and cool on a rack.

Note If you want to make an apple galette, p. 24, or a pizza, add 60 g (2 oz/½ cup) chopped toasted hazelnuts to half the brioche dough above, after its first rising. Leave it to rise a second time, then punch down again and put in a plastic bag in the refrigerator to chill thoroughly. You will then be able to roll it out to the large thin circle of dough required. Unchilled dough is impossible to roll out, you flatten it firmly – then it shrinks back. You try again and it shrinks back again. In the end, you lose your temper.

Of course you can make fruit galettes without adding nuts to the dough, but they do give an extra interest and bite.

PAPANAS FROM ROMANIA

A book of the world's pancakes would be immense. Certainly they are one of the best-loved foods of the human race; they all have some tiny adaptation of ingredient or manufacture which places them, although the basic idea is so simple that it might seem to exclude invention. In this recipe, which deserves to be more widely known, the use of a moist fresh cheese and soured cream proclaims its origin in Eastern Europe. There papanas make a savoury dish, but I find they go well with apple purée, or apple slices fried in butter with cinnamon. Dried apricots, lightly cooked, are also set off well by their crisp richness.

250 g (8 oz/1 cup) fromage frais *or* quark
or substitute half curd cheese/half yoghurt
2 large egg yolks
5 level tablespoons (1 cup) plain flour
1 stiffly beaten egg white
pinch salt

Put the fresh cheese (or curd cheese and yoghurt) into a bowl. Beat in the yolks, and then the flour, sprinkling it on to the mixture gradually. Finally fold in the beaten egg white and the salt just before cooking the papanas.

Heat up a frying pan with 100 ml (3½ fl oz/good ½ cup) oil and fry tablespoons of the mixture, flattening them slightly as they cook into round cakes. Serve immediately with soured cream, sugar and fruit.

Note In Romania, papanas are served with soured cream and chopped chives.

LACE PANCAKES
CRÊPES DENTELLES

These crisp pancakes with their lacy (*dentelle*) edges are a speciality of the Quimper district of Brittany. Their delicate texture is better suited to a sweet course than our usual flabbier pancakes which are more suitable for fish and cheese fillings. They can be served with orange or lemon or banana and honey butters, or with fruit fillings of many kinds.

125 g (4 oz/1 cup) flour
2 eggs
scant 125 g (4 oz/½ cup) caster sugar
60 g (2 oz/¼ cup) clarified butter, melted
3 tablespoons (scant ¼ cup) oil, e.g. walnut or hazelnut or sunflower oil
2 tablespoons (3 tbs) rum, gin, brandy or whisky
pinch salt

Mix the ingredients in the order given, plus 150 ml (¼ pt/½ cup) tepid water. Or put everything into the processor. Or into the liquidizer, apart from the flour, which should be added gradually at the end.

Allow to stand before cooking if this is convenient. There is no other reason for doing so.

Note It is the sugar and alcohol that gives the right texture. Alcohol burns off as the pancakes cook, and the taste is not perceptible, so *crêpes dentelles* are entirely suitable for children. Or for teetotallers.

GENOESE SPONGE
GÉNOISE

The most useful cake mixture of all, as far as fruit is concerned, lighter than a pound cake, mellower tasting than a true sponge. The mixture can be flavoured with grated lemon or orange rind (add to the flour), a teaspoon or two of orange-flower water, or very strong coffee (mix in with egg and sugar), or cocoa (substitute for a quarter of the flour).

If you have a heavy electric beater – not the kind that you hold – you can dispense with the need for whisking over hot water. Just beat until the mixture is bulky and the beaters leave a trail when they are taken out.

4 eggs
125 g (4 oz/½ cup) caster sugar
flavouring
125 g (4 oz/1 cup) flour
large pinch salt
60 g (2 oz/¼ cup) butter, melted and warm

Start by brushing out a 23-cm (9-in) cake tin with melted butter. Cut a circle of Bakewell or greaseproof paper, put it into the bottom of the tin, and brush with melted butter. Sprinkle a tablespoon of flour on to the base, then turn the tin about until base and sides are evenly coated: tip out surplus flour. If you want to bake a large single layer of genoese, use a rimmed baking sheet roughly 30 × 40 cm (12 × 16 in) and treat it in the same way. Butter and flour are extra to the quantities above.

Beat eggs, sugar and flavouring, if appropriate, over a pan of almost simmering water, until they are thick and foamy, much expanded, and the beater leaves a trail. Remove bowl from pan and beat until cool.

Sift flour with salt. Then sift half on to the egg mixture. Fold it in with a metal spoon, lifting gently. Sift over the rest and pour the clear butter round the side, leaving the white sediment behind. Fold flour and butter in together.

Pour into the round tin, or spread out and level gently on the baking sheet.

Bake in a preheated oven at gas 4, 180°C (350°F) for 35–40 minutes (round cake); or at gas 5, 190°C (375°F) for about 20 minutes (baking sheet).

The cake is done when the top is nicely browned, and when the top springs back if you press it lightly with your finger.

Cool in the tin or on the sheet for 5 minutes, turn out on to a wire rack and leave until quite cold. Chill the cake before you slice it across in half, or stamp it out to make little cakes.

CREAMS, SUGARS, ETC.

FRESH FRUIT AND CREAM FILLING

Suitable for pound cake, sponge cakes with and without butter, chocolate sponge cake, or American shortcake. It can also be used to fill a roulade, or pavlova, and meringues.

Prepare 250 g (8 oz) of one of the following fruit – strawberries, raspberries, apricots, peaches, Chinese gooseberries (kiwi), mulberries, blackberries. Avoid washing the fruit unless it is absolutely necessary.

Put it into a bowl, sprinkle with sugar to taste and add 2 tablespoons (3 tbs) of appropriate liqueur, fruit brandy, gin or genever. Leave for an hour, then drain the fruit, and sprinkle the juice over the cut sides of the halved cake, or the inner side of a roulade. Don't do this if you are filling a pavlova, or other meringue cake.

Whisk 250 ml (8 fl oz/1 cup) double cream with icing sugar to taste, or caster sugar from the vanilla pod jar. Strengthen the vanilla flavour if you like, with a drop or two of proper vanilla essence (not vanilla *flavouring*).

Set a little of the cream aside if you wish to decorate the top of the cake with piped rosettes, and some of the fruit. Either mix the rest of the fruit with the remaining cream and spread it on the cake. Or spread half the remaining cream on the cake, then all the remaining fruit, then the rest of the cream. Put the top of the cake in place, after sprinkling it with icing sugar, or decorating it with the cream and fruit set aside.

By using a plainly patterned doily, or cutting strips of paper and laying them on the cake, and then sifting on the icing sugar, you can produce an agreeable pattern without the fuss of piping cream. This I much prefer, but it is a matter of individual taste and skill.

With a pavlova, you mix some of the fruit with all the cream and pile it into the meringue nest, which is then decorated with the remaining fruit. One delightful way is to flavour the cream with sieved passion fruit pulp, and decorate the cream with slices of Chinese gooseberry peeled and cut thin.

With meringue discs, you divide the cream and fruit according to the number of layers.

FRUIT BUTTER CREAM

Use strawberries and raspberries and mangoes raw, reducing between

200 and 250 g (6–8 oz) prepared fruit to a purée in liquidizer or processor. Then strain through a sieve. Apricots are best cooked with a little butter and sugar, after stoning, then drained and puréed: dried apricots should be soaked and simmered in their soaking water first. Passion fruit gives a strong flavour of great delight – scoop out the pulp of three or four ripe passion fruit, and heat it very gently; then rub through a sieve so that the seeds are left behind. Currants need light cooking and draining: cranberries should be cooked in a little water until they begin to pop open – about four minutes.

For orange cream, use the grated rind of a couple of oranges, plus orange liqueur to taste. For lemon, use the rind of two lemons, and a little juice. If you use a processor, the peels can be put in with the yolks, in strips, and whizzed until they are quite broken up.

The quantities below make enough butter cream for filling a cake generously. If you want to cover the cake, double the quantities at least, or make two batches. Unless you are used to making this kind of butter cream, it is wise to work with small quantities at first. The important thing is to pour the syrup on to the eggs really hot, and either to have a helper who can keep whisking, or an electric beater.

2 large egg yolks
60 g (2 oz/¼ cup) sugar
4 tablespoons (⅓ cup) water
175 g (6 oz/¾ cup) softened or creamed slightly salted butter, or unsalted butter with a pinch of salt
150 ml (¼ pt/⅔ cup) fruit purée
appropriate liqueur or fruit brandy

Whisk the eggs. Dissolve sugar in the water over a low heat. Turn up the heat, bring to the boil and stop stirring. Boil until the soft ball stage, 115°C (239°F), i.e. when a little dropped into a mug of iced water turns to a soft ball. Pour on to yolks, whisking all the time. Whisk until smooth and thick and just tepid. Add butter, then the purée and alcohol to taste.

GLACÉ ICING

Sift 250 g (8 oz/2 cups) icing sugar into a basin. Mix to a paste with 5 tablespoons (6 tbs) liquid – water, or a fruit juice, or strong coffee, or alcohol. For chocolate icing, grate in 45 g (1½ oz/¼ cup) bitter chocolate and mix with watei oi coffee or rum and water.

Put over a pan of almost boiling water, and stir until the icing is very smooth and warm. It should, like custard, coat the back of the spoon. Do not prolong the heating or the icing will turn irrecoverably grainy.

For fruit-flavoured icings, a drop or two of colouring is sometimes a good idea, but keep the tone pale and understated. Nothing puts people off more than a bright orange or apple-green cake.

Stand the cake, free of all paper, on a wire rack over a tray. Pour on the warm icing, spreading it with the aid of a palette knife. Leave to cool, then scrape up the icing from underneath the cake, re-warm and use to coat again where necessary.

Note For large fruit cakes, Christmas cakes and so on, double the quantity of icing, and be prepared to add extra icing sugar to make the second layer of icing thicker.

SABAYON

ZABAGLIONE

A foamy Italian pudding or sauce that goes well with such fruit dishes as baked bananas, or lightly-cooked pears. It amounts to a light custard. The problem is that it should really be made between courses so that it can be served immediately. If you have visitors who can relax, or if you are eating in the kitchen anyway, this makes no difficulty.

Quantities for one person are usually as follows – but for a sauce you could reduce quantities by a third:

1 large egg yolk
1 rounded tablespoon (2 tbs) caster sugar
2 tablespoons (3 tbs) Marsala, or other sweet wine

Put ingredients into a large pudding basin, set it over a pan of barely simmering water which should not touch the bottom of the basin, and whisk steadily with a rotary beater. Or an electric beater, which will halve the time needed.

The mixture will increase bulkily up the bowl, turn pale and take on a rich but light consistency. Serve at this point for a warm sauce or pudding. Continue to whisk for another 5 minutes, then remove from the pan and whisk until cold, if you are serving the sabayon at a later stage.

As a pudding, it is usually served in glasses.

CUSTARD SAUCE
CRÈME ANGLAISE

Simple to make, so long as you do not overheat the egg and curdle it. Custards can begin to look a little grainy before simmering point is reached; if this happens, rush them into liquidizer or processor and whizz until smooth, adding extra cold milk or cream.

Flavourings should be suitable for the dish the sauce is to be served with. A vanilla pod is the obvious choice, but consider lemon or orange peel, a bay leaf, some peach leaves, caramelized sugar, powder almond or hazelnut brittle, medium roast coffee beans bruised in a mortar or ground. Strong-tasting liquids can be substituted for some of the milk or cream – fruit pulp or juices, strong coffee (liqueurs and fortified wines are best added at the end, so that they are not overheated).

300 ml (½ pt/1¼ cups) Channel Island milk, or half milk/half cream
flavouring
1 large egg and 1 large egg yolk
sugar

Bring milk, or milk and cream, slowly to boiling point with the flavouring. Cover and leave to infuse for 20 minutes on the lowest possible heat. Taste every so often to make sure the flavouring is not becoming too strong. When it is right, begin the next stage.

Beat egg and yolk together with a level tablespoon (2 tbs) of sugar (or less if a sweet flavouring has been used). Strain on some of the warm milk, beat well. Repeat until all the milk is used up. Rinse out the pan and pour the custard mixture into it. Stir over a low to moderate heat, until the back of the spoon is coated. If you want a thinner sauce, dilute it with boiled milk or cream. If you want it thicker, beat some of the hot custard on to another egg yolk, then return the mixture to the pan, and continue to stir until you achieve the thickness you require.

Taste and add more sugar if you like, or some form of alcohol.

COEUR À LA CRÈME (SOFT)

Very simple to make, the standard French country dish to serve with soft fruit, with tarts, and with almost any pudding that demands cream. In winter, it sets off the best preserves, apricot or strawberry jams, for instance, with large pieces of fruit, and is accompanied by little *tuiles amande.*

Curd or yoghurt cheese can be used instead of *fromage frais*; if they are well drained, you will only need 375 g (12 oz/1½ cups). The mixture can also be frozen as it is, or with flavourings added. Quantities are variable, according to your taste or diet, and what you have to hand. Some people add only a couple of tablespoons of cream. Others prefer one egg white.

500 g (1 lb/2 cups) pot fromage frais, *Jockey or Quark for instance*
150 ml (5 fl oz/⅔ cup) crème fraîche *or whipping or double cream*
caster sugar from the vanilla pod jar
2 egg whites

Drain the cheese in a muslin-lined sieve overnight, or hang up in a muslin bag. Whip the cream to a thick bulky consistency, but not absolutely stiff. Fold in the *fromage frais.* Sweeten with caster sugar to taste. Whisk the whites to the firm peak stage, and fold into the cream cheese. Taste and add a little more sugar if you like: the mixture should not be too sweet.

Note Use the liquid that drains out of the cheese for bread, scones, pancakes.

COEUR À LA CRÈME (FIRM)

By using gelatine and egg yolks, you can make a richer firmer cream that can be set in a large heart mould, or in little pottery hearts. If your mould is the pierced kind, it is particularly important to line it with double tubular stockinette that you can buy in huge rolls from bicycle shops (a huge roll is essential in any kitchen, I find). This prevents the mixture sticking in the holes, and also gives the finished heart or hearts an attractive surface.

You can quite well use cottage cheese, or well-drained *fromage frais* or yoghurt, but avoid the full fat soft cheeses which are too rich and sticky in texture. They should be dryish before you weigh them.

250 g (8 oz/1 cup) curd cheese
2 egg yolks
60 g (2 oz/¼ cup) caster sugar from the vanilla jar
1 packet (½ oz/2 envelopes) gelatine
2 egg whites
250 g (8 fl oz/1 cup) whipping cream, or half double/half single

Sieve cheese if necessary to get a smooth texture. Mix with yolks and sugar. Dissolve the gelatine in 6 tablespoons (½ cup) of very hot (not boiling) water, in a pudding basin. When smooth and almost cool, add the cream(s) to the gelatine and whip until stiff. Fold in the cheese mixture carefully, so that the mixture is well blended, but as light as possible. Finally, whip the egg whites until they are firm, and add into the curd cheese mixture.

Line a ¾ litre (1½ pt/scant 3½ cup) mould, or little heart moulds, with stockinette. Damp and wring it out first, so that it takes the shape of the mould more easily. Ladle in the mixture, tapping the mould on the table to settle it into the curves. Flip the ends of the cloth over the whole thing and put in the refrigerator to set. Turn out on to a dish to serve, and surround with soft fruit, or with poached fruit.

If you are serving *coèur a la crème* with pears, add chopped preserved ginger in syrup – about 6 knobs – and use ginger syrup instead of sugar to sweeten, 4 tablespoons (¼ cup) to start, a little more if you like.

EMERGENCY CREAM

No need to buy a special machine for turning butter and milk into a form of cream that is acceptable in an emergency. Gelatine and blender or processor do the job perfectly well, producing a mixture that can be whipped after chilling, or – if you increase the quantity of milk – be poured.

Emergencies apart, the cream which works out at two-thirds or less of double cream in price can be used discreetly when making fools, ice cream or mousse of strong-tasting fruits (dried apricots, prunes, the first sharp gooseberries). And it can be stirred into sauces, though you may well find that butter alone gives a fresher lighter result if cream is not available. The following quantities make 500 ml (generous ¾ pt/scant 2 cups).

250 g (½ lb/1 cup) unsalted butter,
or half unsalted and half slightly-salted
300 ml (½ pt/1¼ cups) milk
1 level teaspoon gelatine

Cut the butter in rough dice into a small heavy pan. Pour on the milk and heat gently, stirring, until the butter has melted. It must not boil. Sprinkle on the gelatine and stir again until it has dissolved. If you kept the heat low, you can liquidize or process the mixture immediately for 30 seconds.

Otherwise, allow it to cool first to tepid or cold – whichever is convenient – and then liquidize. Pour into a basin and chill several hours or overnight in the refrigerator. It can now be beaten until light and thick.

VANILLA ICE CREAM

It can only have the right flavour if the vanilla is drawn from the pod, and not splashed in from a bottle. To achieve this, use sugar from the jar in which you store the pods. Slit open a fresh vanilla pod to heat with the milk. Further reinforcement can be provided by grinding a pod to powder with 250 g (8 oz/1 cup) sugar, and adding the result in small quantities – it is remarkably strong, even unpleasant in this concentrated form. Use a liquidizer or processor to break up the pod so that it is reduced to dust with the sugar.

1 vanilla pod
300 ml (10 fl oz/1¼ cups) single cream or rich milk
3–4 egg yolks, beaten
vanilla sugar
300 ml (10fl oz/1¼ cups) whipping or double cream

Bring vanilla pod and single cream or rich milk slowly to the boil. Cover and leave over a very low heat to infuse for about 20 minutes. Whisk liquid into the yolks, and return to the pan. Stir over a low heat until the custard coats the back of the spoon. Do not overheat, or the mixture will curdle: should it look even remotely grainy, extract the vanilla pod, and whizz in a processor or liquidizer until it is quite smooth. Sweeten to taste.

Cool the custard, remove the pod if you have not done so already, and strain into the stiffly-beaten whipping or double cream. Mix in gently and add more sugar to taste.

If you use an ice cream making machine of any kind, follow its instructions. If not, pour the mixture into a loaf tin and put into the freezer, or the ice-making part of the refrigerator set at its coldest. When the mixture is firm round the sides, stir sides to middle. When the mixture is consistently set but not hard, remove and whisk, with an electric beater for preference. Return to the freezer to set hard. To serve vanilla ice cream, remove from the freezer a good hour before it's needed. Leave in the refrigerator for an hour, then on the kitchen table until you get a spoonable texture.

SORBETS

Sometimes the amateur cook feels that she – or he – is in a state of perpetual guerrilla war with the skilled professional. No sooner does some domestic machine appear to make a laborious process easier, than an even better one turns up in the professional kitchen. And we are back two steps behind, our new confidence thrown as we lose yet another skirmish.

This is the case with sorbets. For a few years now, we have thought of them as an ideal stand-by dessert, or a dessert that can be made in advance of a dinner party. One less thing to do on the day itself. Now we read in the cookery books by the new chefs, that sorbets should be served no more than three or four hours after they are made. If you want the best flavour, that is, and we all want the best flavour, don't we? What these gentlemen neglect to say is that their kitchens are equipped with the most magnificent new toy, an Italian *sorbetière* that produces ices of the most beautiful soft consistency and keeps them at the right consistency until you want to serve them. You put in the mixture, switch on and forget about them.

The following instructions are for the poor cook at home, who undoubtedly has a refrigerator and an electric beater, liquidizer or processor, and perhaps a freezer. Once machines come into it, special ice-making machines, you must follow their instructions and not mine.

SOFT FRUIT: sieve, or liquidize and sieve, ½ kg (1 lb), then emphasize the flavour with lemon or orange juice or both, and sweeten to taste with syrup (add the two ingredients alternately). To make a syrup, dissolve 150 g (5 oz/⅔ cup) sugar in 300–500 ml (½–¾ pt/1¼–2 cups) water over a low heat, then boil steadily for 5 minutes until the mixture looks syrupy: the quantity of water depends on the strength of flavour of the fruit, more for black currants, less for wild strawberries.

Freeze this mixture at the lowest possible temperature. If you use an electric beater, take the sorbet when it is firm round the edges, slightly liquid in the centre, and tip it by the spoonful into the bowl with the beater switched on to top speed. When the mixture is smooth, and free of icy granules, return it to fridge or freezer until firm. At this beating stage, add liqueur or alcohol. You can repeat the beating once if there is time. After that, it seems to make no difference.

If you use a processor, wait until the ice is just firm, then put it by the spoonful into the switched on processor. It makes a dreadful noise at first, and thumps about, then it turns into a beautifully smooth mixture. The courageous will be able to leave this processing until the last moment, and serve the ice soft yet coherent straightaway. If it becomes too liquid, return it to the freezer until it becomes just firm again, then have another go with the processor. This is something to try on your family first, until you are used to judging speed and texture.

FRUIT JUICE SORBETS: substitute ½ litre (generous ¾ pt/2 cups) of strained juice – orange, tangerine, lemon, grapefruit – and flavour to taste with the syrup (above). When making the syrup, add thinly cut strips of peel from the fruit to flavour it, but keep tasting and remove the strips before the syrup becomes bitter. Gin or an appropriate liqueur are great improvers of citrus ices Freeze as above.

FIRM FRUIT: peel or skin, quarter and core or remove the stone. Put into a pan with 125 g (4 oz/½ cup) sugar for every ½ kg (1 lb) of prepared fruit. Just cover with cold water. Simmer until tender. Remove the fruit, and reduce the liquid until thick and syrupy. When cold, liquidize or process or sieve the fruit and syrup, adding lemon or orange juice to bring out the flavour of the purée. Freeze as above. You might try pouring alcohol over the ice before serving it, rather than adding it to the mixture. Poire William and other pear brandies for pear ice, Calvados for apple, kirsch for peaches, gin for quinces.

In an effort to make sorbets smoother and softer, people used to add beaten egg white. I suspect the processor accounts for its disappearance in many recipes today. You may like its partnership with lemon juice for lemon sorbets, or to soften the strength of black currant purée: this is what you do – whisk one or two egg whites until stiff with an electric beater, then add spoonfuls of almost-frozen sorbet with the beater still going. Be careful at first, or you will have sorbet flying round the kitchen – best to lower the speed a little. However you should aim to get the maximum bulk as quickly as possible, and return the ice to the freezer before it begins to separate into foam and liquid.

FROZEN FRUIT MOUSSES AND PARFAITS

These two recipes for making fruit ice creams are ideal for people without special freezing apparatus, as the mixtures do not need to be stirred as they chill.

MOUSSE GLACÉE: prepare enough fruit, raw if soft or lightly cooked if firm, to give you a $\frac{1}{2}$ litre (generous $\frac{3}{4}$ pt/2 cups) of smooth purée. Bring out the flavour with lemon juice and sugar. If the purée is on the sloppy side, dissolve a packet ($\frac{1}{2}$ oz/2 envelopes) gelatine in 8 tablespoons ($\frac{1}{2}$ cup) of very hot water, and stir it when tepid into the purée. Whip $\frac{1}{2}$ litre (generous $\frac{3}{4}$ pt/2 cups) cream – double or whipping cream – until stiff, with sugar to taste. Fold into the fruit and gelatine mixture, turn into a suitable box and put into the freezer. Liqueurs and fortified wines may be added to the fruit with the lemon juice. Take care to get the flavouring of the purée and the sweetening of the cream exactly right before you mix them together: this is a rule that applies to all cooking.

PARFAIT: prepare and flavour $\frac{1}{2}$ litre (generous $\frac{3}{4}$ pt/2 cups) of fruit purée as above. No need for gelatine. With an electric beater, whisk 3 egg whites and a pinch of salt until stiff. Dissolve 250 g (8 oz/1 cup) sugar in a saucepan with 175 ml (6 fl oz/$\frac{3}{4}$ cup) water over a low heat. Bring to the boil and boil gently for 5 minutes. Pour boiling on to the egg whites, with the electric beater going at top speed, to make a meringue. Beat until cool. Fold in the fruit purée and $\frac{1}{4}$–$\frac{1}{2}$ litre (8–16 fl oz/1–2 cups) cream whipped until stiff – the quantity of cream will depend on how rich you like your ice to be. Turn into a box and put into the freezer until firm.

FLAVOURED SUGARS

In this book, I assume that everyone keeps a jar of caster sugar with four or more vanilla pods embedded in it. And that as the sugar goes down, it is replenished. That as the pods are used, they are washed, dried and replaced, and renewed from time to time. This is the normal baking sugar.

When *vanilla* is specified directly, it means that you can reinforce the flavour with a more concentrated preparation, made by processing 60 g (2 oz) vanilla pods with 125 g (4 oz/$\frac{1}{2}$ cup) of sugar, and then sifting it into 375 g (12 oz/1$\frac{1}{2}$ cups) of sugar. It is better to strengthen vanilla flavouring with some of this concentrated sugar, than to use vanilla essence which may be inferior or fake.

Cinnamon sugar can be flavoured by the same means, sticks of cinnamon in the jar. Wherever a recipe stipulates a cinnamon flavouring, this sugar can be used, either in the weaker or more concentrated form, and the quantity of the spice slightly diminished, to take account of this. Spicing is so much an individual matter that the quantities given in any recipe should be treated only as a guide.

With other spices such as aniseed, caraway and cloves, warm them in the oven in the proportion of 60 g (2 oz/¼ cup) to 500 g (1 lb/2 cups) sugar, before grinding and mixing them as above.

Dried curls of orange and lemon peel make a most useful flavouring for sugar. Cut it thinly from the fruit with a peeler, and weigh out 60 g (2 oz). Hang it up to dry in a warm place – by the open window if it is sunny. When it is leathery, put it with ½ kg (1 lb) sugar into a jar and leave it: or process it as above.

Note If you keep a large bottling jar full of sugar, which holds a kilo (2 lb), you will need to double the above quantities.

APRICOT, QUINCE OR RED CURRANT GLAZE

A fruit glaze can be used as a filling for cakes. It is also brushed over cakes to hold the crumbs in place before the icing is poured over, acting at the same time as a glue to hold everything together. As far as fruit cookery is concerned, it acts mainly as a shiny sweet finish for tarts – red currant glaze is the classic way of finishing strawberry tarts, apricot glaze is brushed over apple tarts and pear tarts. I would also put forward quince glaze for apple and pear tarts.

The aim is to make jam or jelly into a convenient consistency for brushing or spreading. To do this, turn out a pot of apricot jam or quince or red currant jelly into a heavy pan. Add 4 tablespoons (⅓ cup) water. Stir over a moderate heat until jam or jelly has dissolved smoothly. A whisk may be needed to subdue recalcitrant lumps and apricot glaze needs to be sieved free of skin. Add a little lemon juice to taste.

Store in little pots in the refrigerator or freezer. To use, reheat carefully and only add extra water if you must. Brush the warm glaze over the cold tart or cake.

PRALINE

Pralines and praline powder are an old sweetmeat, an almond brittle. The very simple idea came to the cook of the duc de Choiseul, comte de Plessis-Praslin and marshal of France, who produced the little sweets at a banquet given by the duke when he was trying to suppress a rebellion at Bordeaux. In 1630, the cook retired to Montargis and made his fortune with the pralines which he had named in honour of his distinguished employer. The first recipe to be published was in La Varenne's *Le parfaict Confiturier* of 1667, and it hasn't changed.

Put 125 g (4 oz/1 cup) of unblanched almonds and 125 g (4 oz/½ cup) sugar into a heavy pan. Stir in 4 tablespoons ($\frac{1}{3}$ cup) of water, and heat gently so that the sugar melts. Raise the heat slightly so that the syrup gradually darkens to a caramel – stir from time to time. The nuts will pop when they are ready. Have a bowl of very cold water handy and stand the base of the pan into it to prevent the caramel going any darker.

If you want to make praline powder, spread the mixture out on to a greased metal sheet. Leave it to cool and harden, then break it up and reduce to a powder in a blender or processor. Store in a screw-top glass jar.

If you prefer to make pralines, drop small spoonfuls of the hot mixture on to a greased metal sheet, then raise them carefully with a palette knife when they are cold and set. If you store them, use an air-tight box and layer them between sheets of vegetable parchment.

Note Pralines and praline powder can also be made from hazelnuts. Toast them under a grill or in the oven, so that the fine dark skins can be removed; almonds in the recipe above can be skinned if you prefer, by putting them into boiling water for a few seconds and then peeling away the skins.

INDEX

Also by Jane Grigson in Penguin

'One of the three most influential and popular cookery writers in Britain' – *The Times*

JANE GRIGSON'S VEGETABLE BOOK

Written with all the author's customary warmth and erudition, here is a modern kitchen guide to the cooking of vegetables, from the well-loved cabbage and parsnip to the more exotic chayote and Chinese leaf.

FISH COOKERY

There are over fifty species of edible fish and Jane Grigson feels that most of us do not eat nearly enough of them. This book is certain to make us mend our ways as it shows how fish can be cooked with loving care and eaten with appreciation.

GOOD THINGS

Bouchées à la reine, civet of hare, Mrs Beeton's carrot jam to imitate apricot preserve, baked beans Southern style, wine sherbet . . . These are just a few of the delicious and intriguing dishes in *Good Things*. The book does not preclude the use of convenience foods, but puts in an eloquent plea for the type of food that our grandparents enjoyed, which, unless we are prepared to demand, our grandchildren will not know.

ENGLISH FOOD

'Jane Grigson is perhaps the most serious and discriminating of her generation of cookery writers, and *English Food* is an anthology all who follow her recipes will want to buy for themselves, as well as for friends who may wish to know about *real* English food instead of glossified absurdities that are so often trotted out in technicolour as traditional "fayre" of olden times . . . enticing from page to page' – Pamela Vandyke Price in the *Spectator*

THE MUSHROOM FEAST

'Enjoyable, expert guide through woods, fields and every possible combination of food and fungi' – *Observer*

'In this rich and savoury collection of mushroom recipes . . . Mrs Grigson has made a connoisseur's choice' – *Sunday Telegraph Magazine*